MOUNTAIN ENVIRONMENTS
AND COMMUNITIES

Mountains continue to hold a fascination for residents and visitors alike. Around 10 per cent of the world's population actually live in mountain regions and 25–30 per cent directly rely on mountain resources. They are of great religious significance, as well as a global focus for tourism. The distinctive nature of mountain regions is now formally recognised as part of the UN Agenda 21, and 2002 has been declared the International Year of Mountains.

Mountain Environments and Communities explains the background physical environment and then explores the environmental and social dimensions of mountain regions; three themes emerge from the book:

- Mountain communities are not as 'conservative' as some popular accounts portray. Particularly in developing countries, they adopt highly innovative lifestyles and must be resilient in the face of constant challenges and conflicting pressures from their physical and political environment.
- The altitudinal characteristics of mountains generate specific sets of physical processes, but their significance is highly variable in time and space and depends as much on the social as the physical context. Biodiversity and conservation issues are emphasised over and above hazards.
- Although there has been a concerted effort to develop specific policies of sustainable development for mountain regions, there remain considerable problems to be surmounted before these can be fully achieved.

This critical review of the concepts currently employed in mountain research draws upon a wide range of examples from developed and developing countries. The dynamics of mountain life are described through both historical accounts of village-based systems and examples of the contemporary impact of global capital and sustainable development strategies.

Don Funnell is Lecturer at the School of African and Asian Studies, University of Sussex. **Romola Parish** is Honorary Lecturer, Department of Geography, University of St Andrews.

A VOLUME IN THE ROUTLEDGE PHYSICAL ENVIRONMENT SERIES

Edited by Keith Richards

University of Cambridge

The Routledge Physical Environment series presents authoritative reviews of significant issues in physical geography and the environmental sciences. The series aims to become a complete text library, covering physical themes, specific environments, environmental change, policy and management, as well as developments in methodology, techniques and philosophy.

MOUNTAIN ENVIRONMENTS AND COMMUNITIES

*Don Funnell and
Romola Parish*

London and New York

First published 2001
by Routledge
11 New Fetter Lane, London EC4P 4EE

Simultaneously published in the USA and Canada
by Routledge
29 West 35th Street, New York, NY 10001

Routledge is an imprint of the Taylor & Francis Group

© 2001 Don Funnell and Romola Parish

Typeset in Galliard by
Florence Production Ltd, Stoodleigh, Devon
Printed and bound in Great Britain by
Biddles Ltd, Guildford and King's Lynn

British Library Cataloguing in Publication Data
A catalogue record for this book is available from the British Library

Library of Congress Cataloging in Publication Data
Funnell, D. C.
Mountain environments and communities/Don Funnell
and Romola Parish.
p. cm. — (Routledge physical environment series)
Includes bibliographical references and index.
1. Mountains. 2. Mountain people. I. Parish, Romola.
II. Title. III. Series.
GF57 .F86 2001   307.7—dc21
00–062800

ISBN 0–415–18101–1 (hbk)
ISBN 0–415–18102–X (pbk)

CONTENTS

CONTENTS

PLATES

FIGURES

TABLES

PREFACE

The study of mountains has a long history, and these areas hold a strong attraction for many people. This arises out of the recognition of their importance as the basis for livelihoods, for their economic resources, such as minerals, timber and water, or for the aesthetic aspects of the character of mountains and their growing importance as foci for recreation and tourism. In the twentieth century it was Peattie (1936) who produced the first major text on mountains in English, followed nearly fifty years later by Price (1981). These volumes presented a descriptive account of the geography of mountains, and as such are classics of their time. However, both are heavily biased towards the physical environment, reflecting an element of 'environmental determinism'. The current thinking behind mountain research has changed in the last twenty years or so, to include a much greater recognition of the importance of human activity in shaping, conserving and damaging mountain landscapes.

During the last decade in particular, there has been a consistent and concerted effort to raise the profile of mountains and mountain-related issues in the global policy arena. In particular, The Mountain Agenda, comprising a group of researchers with long-established reputations in mountain-related research, took the opportunity offered by the planning process of the UNCED 1992 Conference to make a specific call for attention to mountain regions. This resulted in the formal recognition of the importance of mountains and was marked by the inclusion of Chapter 13 ('Managing Fragile Ecosystems: Sustainable Mountain Development') in Agenda 21 during the Rio Conference.

The establishment of the Mountain Forum and website (http://www. mtnforum.org/) has been a significant step forward in giving voice to the many individuals, groups, and institutions involved in mountain issues but who are outside the standard academic circuit. All have been encouraged to participate in Mountain Forum's on-line discussions, providing a significant volume of experience together with opinion of and reaction to policy-making and academic research. However, still grossly underrepresented in the discussion arena are those people who live and work in mountain regions. Whilst some of the initiatives under Agenda 21 have addressed this problem, it is evident that much remains to be done in the field of empowerment for many mountain communities. Certainly some political movements in the Andes, for example, have used the World Wide Web to provide a viewpoint on various activities in their region. The Mountain Forum has also drawn together various 'satellite' groups concerned with particular issues, such as forests, or with particular mountain ranges, such as the Andes. This permits close interaction on a number of levels of interested parties, and makes these sources of information and assistance freely accessible to the world.

In writing this book, we have attempted to fill a perceived gap in mountain literature by providing a book which doesn't necessarily present all there is to know about mountains, or gather together all the latest research, but rather tries to balance the knowledge of mountains against the various contexts in which mountain research, development and exploitation occur. This is not only highly politicised in today's global environment, but also culturally constrained. Just now we have a period of time when there is a great opening up of international debates and concern about environmental issues and increasingly a growing recognition of the place of history, culture, perceptions and attitudes within these debates.

Although a great deal has been written on the rights and wrongs of dealing with sensitive and yet immediate concerns over resource use and exploitation, there has been less actually written about how the recommendations are to be implemented, or the translation of the learning from previous mistakes into today's policy-making. This is not to say that these do not occur, but to suggest that these issues are not the most visible in the literature. Many proponents of development advocate participation and flexibility, and yet we see few instances where this has actually been put in place. We may have a reasonable idea of what is going on in many mountain regions in terms of environmental and economic change, but not always a clear sense of the priorities of the different actors. The existence of priorities implies exclusions, and in this current climate of complexity – where the ideal is to 'let a thousand flowers bloom' – exclusion is dangerous.

What we have sought to offer here is a book that attempts to understand the complexity of mountain environments within the context of the different histories of evolution and transformation. We do not pretend to include exhaustively all aspects of mountain issues, but we have tried to present those which, reflected in our own fieldwork and in the literature, appear as the most significant. Examples have been drawn worldwide, but there is a concentration on the Himalayas, the Atlas and European Alps, and predominantly texts written in English. This is a limitation, but to widen the scope further would produce a volume of unmanageable scale.

In the process of writing this book, we noted that the arrangement and bias of the text carries a message, which is a reflection of the evolution of the discipline. Part 1 explores the range of theoretical ideas that have been used in mountain studies. In addition the notion of complexity is discussed, which we suggest forms a useful framework to unite the disparate strands of mountain research. The work of early geographers was descriptive, and heavily biased towards physical environmental processes and conditions. This is reflected in Part 2, which deals with what we present as 'traditional' mountain communities and the physical environmental conditions. However, with time, the bias of research turned towards a greater recognition of the role of human activity within the landscape, particularly as a force for change. In Part 3, where we deal with the transformation of the environment and communities, the bias is much more towards socio-economic and political activities, whilst at the same time not dismissing the importance of natural environmental changes, particularly climatic change. This shift in emphasis is, to some extent, integrated with the concepts of scale and time. In earlier work, mountain communities tended to be studied as examples of unchanging entities. In more recent work, the much more far reaching and rapid impacts of globalisation, capitalist penetration and associated economic and social change have been seen to have brought about more substantial and long-lasting changes to mountain communities and their environments than, in many cases, physical environmental processes alone.

It is in Part 4 that we attempt to draw together much of this earlier work and take it a bit further – into the realms of policy and politics. This is not novel, but we hope that by addressing the issues raised at different scales, and critically analysing the approaches to development by institutions ranging from international to individual farmer, we can begin to extend the understanding of policy-making within a complex environment. In particular the book emphasises the importance of a holistic approach, as also is an understanding of the way that policy 'fits' or more often doesn't onto the mechanisms of a complex system.

ACKNOWLEDGEMENTS

The authors would like to thank the many people, including residents of mountain areas, for their patient responses to our questions. Academic colleagues and agency personnel have helped us in many different ways. We would like to express our particular thanks to Hazel Lintott and Sue Rowlands at the Geography Laboratory, University of Sussex for their cartographic work.

The authors and publishers would like to thank the following for granting permission to reproduce material in this work.

Figure 5.1, adapted from Halpin, P.N. (1994) GIS analysis of the potential impacts of climate change on mountain ecosystems and protected areas, in M.F. Price and D.I. Heywood (eds), *Mountain Environments and GIS*, London: Taylor & Francis, 281–301.

Figure 7.1, reprinted from *Global environment Change*, vol 9, Parish *et al.*, 'Climate change in mountain regions: some possible consequences in the Moroccan High Atlas', pp. 45–58, 1999, with permission from Elsevier Science.

Every effort has been made to contact copyright holders for their permission to reprint material in this book. The publishers would be grateful to hear from any copyright holder who is not here acknowledged and will undertake to rectify any errors or omissions in future editions of this book.

Part 1

THE STUDY OF
MOUNTAINS

INTRODUCTION

In every continent of the world we can find mountains. In some cases, the
term is applied to prominent hills of only a few hundred metres. Elsewhere,
ranges towering above 6,000 metres dominate the landscape. Whatever the
case, mankind has always had an interest in these areas. In tropical zones,
mountains have often been the only places where water and land suitable
for cultivation exist, whereas elsewhere they have been the preserve of
nomadic pastoralists. For some civilisations they have been politically
marginal or the place where 'vagabonds' dwell; in other cases, mountains
have been the locus of powerful civilisations that have dominated surrounding lowlands. Many millions pay homage to mountains as part of
religious beliefs, whilst increasingly large numbers seek solace in the environment of mountains as a respite from the stress of lowland living.

Functionally, mountains play a critical role in the environmental and
economic processes of the planet. They influence climate, they provide critical sources of water on which lowland economies depend and they harbour
a wide range of significant resources including minerals as well as plants and
animals. Possibly about 10 per cent of the global population live in mountain regions and, of course, a significantly higher proportion depend on the
resources flowing from these regions. These issues have become particularly
important in the last twenty-five years with the increasing concern about
global ecosystems. Surprisingly it has taken much longer to identify 'mountain problems' than it has those concerning other ecosystems such as tropical
forests, deserts or coastlines. However, today there is a rapidly expanding
movement which embraces mountain populations, global agencies and academic institutions which has as its principal objective the provision of suitable 'mountain policies'. Whilst it is appropriate to debate the wisdom of
many of these activities, they are a key feature of almost all mountain areas.

In setting out to write a book on mountains we were very conscious that recently there have been a number of publications that have focused directly on the question of 'sustainable mountain development'. In the first instance, *Mountain Environments* (Gerrard 1990) provides an important examination of the physical geography of mountains, culminating in a discussion of hazards and the relevance of 'uncertainty' in any model of changing mountain landscapes. This was followed by *The State of the World's Mountains* (Stone 1992) and *Mountains of the World* (Messerli and Ives 1997), which represent major inputs into the debate about the long-term future of mountain areas. These books arose primarily within the context of increasing politicisation of mountain issues associated with the Rio Conference of 1992. In their different ways they have collected together some of the most contemporary work in mountain research and development. However, as is fully admitted by their authors, the UN and associated agencies produced these publications under very tight time constraints; the first as an assemblage of information for the Rio Conference itself, the second to pull together further scientific work for subsequent consideration. Finally, the Mountain Agenda group has produced a very forthright document setting out the principal case for a specifically mountain focus (Mountain Agenda 1997).

More recently a number of studies have explored the developments since Rio (Price and Kim 1999; Price 1999a). In particular, this work has examined the direction of institutional strategies, priorities and the extent to which existing capacity is capable of fulfilling the goals of sustainable mountain development. These issues are considered in Chapter 10. Overall, our objective has been to produce material that is not constrained by the policy deadlines, and to open up debates which were implicit but not always fully explored by these other works. In doing this, however, we fully acknowledge the stimulus that these (and other) writings have given us.

This section describes the development of research approaches used in mountain studies. It highlights a selection of mountain debates that are then expanded later in the book. The discussion also introduces the concept of complexity that offers an interesting framework for handling diversity and uncertainty, both key characteristics of mountain regions. We have constructed a simple model – the cog model – to illustrate how we see the various factors involved in mountain development interrelate, and this forms the underlying concept for organising the book.

1

MOUNTAINS IN GEOGRAPHICAL ENQUIRY

WHAT IS A MOUNTAIN?

Perhaps the key question to be answered is deceptively simple. Is it possible to arrive at a globally accepted definition of a mountain? It is evident that in all parts of the world different peoples retain visual images of mountains which usually, but not always, contain the characteristics of altitude, ruggedness, peripherality and sometimes danger. Often mountains are sacred, such as the famous Mount Kailas in Tibet. In Greek civilisation, Parnassus and Olympus were sacred peaks surrounded by mysterious regions wherein dwelt satyrs and monsters. In the last century, Peattie (1936) set out three conditions that determine mountainous landscapes:

- Mountains should be impressive.
- Mountains should enter into the imagination of the people who live in their shadow.
- They should have individuality.

These characteristics are certainly sensitive to the cultural and psychological factors that build up an individual's perception of mountains. Likewise, one dictionary (Websters) gives this definition: 'any part of the land mass which projects above its surroundings'. This is evident in many parts of the world, where the term 'mountain' is applied to landscape features ranging from huge Andean peaks above 5,000 metres to small local features of only perhaps 1–200 metres. As Peattie writes:

> To a large extent then, a mountain is a mountain because of the part it plays in popular imagination. It may be hardly more than a hill but if it has distinct individuality, or plays a more or less symbolic role to the people, it is likely to be rated a mountain by those who live about its base.
>
> (Peattie 1936: 4)

Other writers have attempted to explore this question further, including Veyret and Veyret (1962) and Bandyopadhyay (1992). In a recent paper, Debarbieux (1999) argues that most definitions, however scientific, arise from *a posteriori* reasoning and rely heavily on intuition in the first instance. This highlights the significance of local perceptions of mountain characteristics and also of the fact that the public image of mountains is, in a sense, manufactured by the very process of mountain research. The result, of course, is that the question 'what is a mountain' is a socially defined concept as much as one arising from carefully established and globally accepted criteria. This becomes apparent when recent efforts at precise definitions are examined. In an attempt to provide a global perspective on mountains, Denniston (1995) states that mountains comprise 20 per cent of the world's landscapes.

A more detailed cartographic attempt to determine the global distribution of mountains and highlands is presented by Messerli and Ives (1997). Using modern electronic (altitudinal mapping) techniques, their book includes a map which shows that 48 per cent of the global surface lies above 500 metres, 11 per cent exceeds 2,000 metres and 2 per cent lies above 4,000 metres. However, not only is this data rather suspect but it fails to help the discussion because of the fact that large areas are plateaux. For example, much of the African continent lies above 1,000 metres but only specific locations above this height (such as Kilimanjaro) are regarded as mountains.

Currently, the most comprehensive cartographic analysis of the distribution of mountain areas is that produced by the World Conservation and Monitoring Service (WCMC 2000). This map uses elevation data from the GTOP030 global digital elevation model and groups the resultant slope and altitude data into six classes. At the time of writing further refinements are in progress, but globally the analysis suggests that the total mountain area (300 metres and above) totals 34 million km^2, about 23 per cent of the global land surface.

Figure 1.1 is a highly simplified map of the global distribution of mountain regions, which serves to indicate the fact that mountain ranges occur in all the major continents of the world. However, at this scale, many small yet important areas do not appear. In most instances, it is much better to consult maps of specific mountain regions. The following statement encapsulates the issue: 'To demonstrate that nearly half of the world's land surface lies above 500 metres tends to support our conclusion that the search for a unitary definition of "mountain" is to chase a chimera' (Messerli and Ives 1997: 8).

However, whilst this approach may be, in some senses, realistic it does pose problems. It is interesting to note, as will be discussed later, that the increasing politicisation of mountains draws on a number of generalisations about their importance as part of the new mountain rhetoric. What is, or

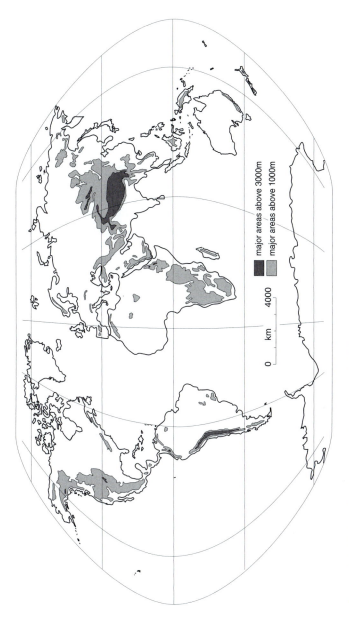

Figure 1.1 Global distribution of land above 1,000 metres

is not, a mountain or highland can be of considerable economic and political significance when resource allocation decisions are under debate. Many European countries have legal definitions of 'mountain regions'. For instance, in Italy, *comunita montane* are defined as areas above 600 metres (Romano 1995), and these have been eligible for special financial treatment at various times. Why is the boundary set at 600 metres? The answer, of course, lies in both the character of the local relief and the perceptions of social reality.

Consequently, most writers have argued that several characteristics, of which altitude is but one, can define mountains. Often considerable emphasis is given to the existence of 'steep slopes' or 'rugged terrain' that suggests the significance of geological and geomorphological factors. However, in plateau landscapes this characteristic may be lacking. Another important factor may be rapid vertical changes in climate and vegetation which are highlighted in the work of the early writers such as Humboldt and Bonpland (1807) and developed into the geo-ecological framework discussed in later sections. This has generated the current use of the term 'verticality' as a key descriptor of mountains, which refers to the fact that many of the characteristics of mountain environments are related to the influence of relative altitude and slope. From a socio-economic viewpoint, mountain areas are often marginal, posing accessibility problems, and have often been considered culturally 'backward'. Until relatively recently, many lowlanders held mountains in awe tinged with fear of the unknown. In all historical periods they have been regions in which outcast groups have nurtured resistance to centralised state power. However, mountains have also been the principal home of some of the major world civilisations – most notably the Incas of the Andes.

Thus we are back to the definition of Peattie quoted on p. 3. In many ways this is the most relevant definition precisely because it is one that encompasses a geographical imagination and creates space for the recognition of a wide diversity of circumstances. However, unlike Peattie, we must face up to the reality that a particular, local definition of mountains may be crystallised into specific, usually altitudinally defined criteria in order to guide state policy. A good example of these problems concerns the management of mountain areas in the EU. Since the 1970s most initiatives for mountain areas in the EU have been established under the provision for Less Favoured Areas where mountains have been considered alongside other 'marginal regions', although there was no specific agenda for the preservation of the natural environment (Danz and Henz 1979). Under the Agenda 2000 proposals for reform, revised structural policies will continue to deal with mountain regions but again within a broader context of general regional backwardness, despite recommendations for more specific treatment (European Parliament: Committee on Agriculture and Rural Development 1998). This will be explored in greater detail in Chapter 10. Thus whilst mountains may be clearly discernible in a physical sense, this does not guarantee specific recognition within the domain of policy-making.

CLASSIFICATION OF MOUNTAINS

Some of the ambiguity surrounding the meaning of 'mountain' lies in the way different languages have developed various specific categories of upland relief. In English, the word 'mountain' is used in contrast to 'hill' to designate something more substantial (although local sites may still be termed mountains). The Cairngorms in Scotland are mountainous for the UK but really no more than minor hills set alongside the Karakorum! In the US, Price (1981) notes that the term 'High' prefixes many of the more substantial mountain ranges to differentiate them from lower ranges. Of greatest influence in the academic literature has been the distinction in German between *hochgebirge* (high mountains) and *mittelgebirge* (middle mountains), the best examples being the Hartz as *mittelgebirge* and the Alps as *hochgebirge* (Troll 1972). The significance of this distinction is the fact that much of the early work on mountain landscapes tended to concentrate only on the *hochgebirge*, with its climatic extremes, limited vegetation cover and rugged landscape. In particular, this emphasised the physical stresses of the mountains, which, in many instances, were thinly populated. Many of the discussions of these mountains concentrate on the definition of zones associated with the timberline. In the tropical areas, however, this distinction does not sit comfortably with greater human settlement, and elsewhere, on the lower slopes of mountains, human influence has generally played a much greater role in the process of landscape evolution. Whilst there have

Table 1.1 Classification of mountains

Long history of settlement and usually high density of population	Recent settlement and thinly populated
1 Traditional subsistence economy prevails: mainly cultivators (Andes/ Himalayas), cultivators and herders (High Atlas, Hindu Kush, Himalayas)	Commercial agriculture, forestry and other market-oriented activities, usually with relatively large-scale units
2 Declining traditional activities but rapid development of new initiatives, e.g. tourism (Alps, Norway, Pyrenees)	Modern and ancient mining Tourism very important
3 Collectivised agriculture (former USSR, e.g. Tajikistan, China and eastern Europe)	(New Zealand, Australia)

Source: Modified from Grötzbach (1988)

been many attempts to classify mountains according to their physical characteristics few exist in which the nature of human populations constitutes the principal focus. One of the best known is that suggested by Grötzbach (1988) where he argues that the principal dimension should be that of density of population and the length of human settlement (see Table 1.1).

This classification, whilst by no means perfect and itself subject to constant change (e.g. the Soviet Union), emphasises the human use of mountains. It is worth noting that the underlying physical characteristics of each of the ranges included do not form a direct match with the population groups.

MOUNTAINS AND SOCIETIES

To some extent the problems over definitions of mountains and their classification are associated with the question of 'voice'. It depends exactly who is laying out the definition and for what purpose. If we return to the statement by Peattie (p. 3), it is clear that this becomes very significant because he allows for a very broad spectrum of voices in his definition. Those who often have little personal affiliation with mountains, or if they do, construct their views to reflect a wider perspective, prepare most of the literature available in academic or government publications. This is particularly true where mountains form only part of a state and the stronger 'voice' is that of the lowland rather than the highlands. This may not be the case where the majority of citizens of a territory actually live in mountain zones, but even here 'official wisdom' often prevails to the long-term marginalisation of mountain communities. Recent discussions in Europe (ARPE/CIAPP 1996) suggest that most mountain people consider that their long-term development is closely associated with 'empowerment' and that long-term security is not best left to outside agencies. However, this populist position has to be evaluated against two important facts:

1   Mountains perform key physical functions globally, within the environment of individual states, and at the local scale. Of particular significance are the supply of water, the provision of minerals and the special role of plant resources in the maintenance of biological diversity.
2   Today most mountain populations are integrated, to varying degrees, economically, socially and politically with lowland communities and the wider world. Market relations and modern communications are often available, if not always predominant features of many mountain communities.

These points highlight the fact that today one of the most important issues facing mountain areas concerns the nature of their relationship with the lowlands. As nation-states, and now international bodies, seek to determine the economic and environmental status of highlands, so the autonomy of mountain people is reduced. Interestingly, much of the recent emphasis on sustainable mountain agriculture includes strong pleas for local initiative and control, whereas for most mountain zones state management of the principal resources such as water has become the norm. Consequently, one of the biggest challenges in the current debates on mountain policies lies in the need to provide management institutions operating at different geographical scales.

THE PROBLEM OF SCALE: HOUSEHOLD, COMMUNITY, VALLEY AND THE WIDER WORLD

In many circumstances there is a close link between spatial scale and the relative significance of physical and social processes. Moreover, an understanding of scale effects helps us focus on manageable options in the process of developing new social or physical activities. For example, it is often possible to respond to highly local temperature differences due to aspect and plant crops accordingly, but it is the mountain range itself that may influence global circulation. At an abstract level the scale question remains central to how we understand social and ecological systems. Almost all our theorising is grounded in some model of scale relations, and Haila and Levins (1992) suggest that the question of appropriate scaling remains one of the most important issues in our understanding of nature–society relationships. This has an important bearing on later discussions about modelling mountain environments and communities because an increasing number of writers are arguing that there is no one way in which we can define either temporal or spatial scales. However, most studies end up working with a hierarchy of scales which leaves open a major political question as to how we 'arbitrate and negotiate' between them (Smith 1992).

In a mountain environment it is suggested that the four important scale components are household, community, valley and the wider world. These somewhat arbitrary units of analysis represent different levels of political control and also have recognisable physical counterparts. For example, hydrological systems can be treated at the level of the local well, at the level of a collection of sources providing water for a particular community, and, for the valley, the watershed. The interplay of scale effects is particularly significant in those studies which have emphasised highland–lowland relationships; for example, the relationship between the Himalayas and the Indo-Gangetic plain (Hofer 1998). More important perhaps is the fact that particular scale relationships change over time, a factor which is today of considerable importance with the growing impact of globalisation.

Historically, households have been the principal production and consumption units in most mountain communities as well as being the source of labour. Different patterns of land ownership have existed, but in almost all cases the utilisation of the land depends upon decisions taken at the household level. The concept of the community is less sociologically and geographically adequate and may be best associated with the village, with its framework of associated dwellings and functions. Nonetheless, a combination of household units and kinship links, often dating back many generations, produced an identifiable physical presence in the form of a village community. Membership of this community and residence rights were closely linked with access to land and other resources as well as rights of participation in the decision-making institutions that existed at this level. It is here that local customs and practices become very significant in ordering the life of a community, and they also provided the benchmark against which individual behaviour and attitudes to 'outside' activities may be assessed. The 'valley' represents a wider geographic scale, which has both physical and social dimensions. Village communities often had to link with others in the management of natural resources, particularly water but also grazing and trees. Marriage patterns regularly depended upon good relations with other villages in the neighbourhood and, in turn, helped to maintain relatively stable links between villages. The most significant element of many traditional mountain communities is the complex web of political and kin affiliations associated with tribal links that are often determined by valley scale linkages. Examples can be found in the Yemen, in the High Atlas and medieval alpine Europe. In the past it was through such affiliations that links with the wider world were mediated, although one of the influences of 'development' has been to foster more direct communications between the village and the outside.

The emphasis on household, community and valley linkages is at best only partial and today increasingly irrelevant. First, it is quite clear from studies in historical demography (see Chapter 6) that few mountain communities were completely isolated. In some cases documented for the European Alps extensive networks of migration stretched into the lowlands from a very early date. Today, in Europe and the USA, but also in some parts of the Himalaya, with the increasing urbanisation of mountain valleys and the integration of local economies, the critical influences on individual and social behaviour derive more from the wider-scale relationships with the world than with local influences. Exactly how, and in what form this transformation takes place is highly variable but can be seen with the penetration of formal, state-sponsored systems of education, and the increasingly ubiquitous mobile phone! In each of the subsequent chapters examples are drawn at different scales to try to explore the interplay of this important factor.

APPROACHES TO MOUNTAIN RESEARCH

How therefore have geographers and others set about accounting for the specific patterns of environment and settlement in mountain areas? In the ancient world, mountains were key elements of Greek cosmology and were treated both as sacred and remote places. Varro, an essayist in the Roman period, described the Scythian mountain range by the term 'inhospitable solitude', whilst Lucretius described such regions as the home of primitive peoples (but much more preferable to the degraded Romans!) (Tobias 1986). Later, other 'western' writers tended to picture mountain communities as remote, 'backward', and living in conditions of extreme hardship (Peattie 1936). However, in the fourteenth century, the Arab writer Ibn Khaldun (1852) wrote not simply of the physical hardships of the mountains of North Africa but of the powerful Berber peoples whose base was in those mountain areas. This work suggests that in warmer climates mountain areas take on a more positive role for settlement. In the US nineteenth-century exploration towards the Rockies not only included pioneer work by the US Geological Survey but also the images of the mountain ranges by painters, for example Seymour's work 'Distant View of the Rocky Mountains (1820). These overwhelmingly depicted an image of 'wilderness' which was to influence US conceptions of the mountains for many generations. In the scientific world, the nineteenth and early twentieth century saw writers such as von Humboldt (1769–1859) and Semple (1923) develop a concept of 'natural regions' in which the physical characteristics of a region provided the key determinants of any settlement or social behaviour. It was, of course, precisely this approach which has been largely rejected by many geographers since the 1950s because of its associations with determinism considered by then to be inappropriate. Interestingly, other disciplines, including various anthropologists, have flirted with these ideas, which have once more become fashionable under the guise of the post-1970s concern for environmental problems. So too have a number of international agencies such as the FAO who increasingly adopt a 'natural regions' approach to their interventionist agenda. This is particularly relevant today with the upsurge of interest in mountain policy following the Rio Conference, where the very success of the 'mountain case' depended upon identifying mountains as distinct regions. This issue will be elaborated later.

It is usually Alexander von Humboldt (1769–1859) who is credited with the first substantially scientific work on mountains in a two-volume account of the Andes. The essence of his work was to identify the impact of altitude and latitude on other physical characteristics of mountains such as climate and vegetation. In the Andes he recognised the existence of altitudinal belts, each the result of changing environmental conditions with an increase in altitude. Carl Troll (1959, 1968, 1972, 1988), who introduced the term 'geoecology' to the geographical literature in 1938, continued this work.

His studies covered a considerable range of mountain environments and later incorporated a social dimension by emphasising the adaptation of mountain peoples to their environment. Troll's lifelong work emphasising 'comparative high mountain geography' and the 'three dimensional classification of mountain regions' (Kreutzmann 1998) inspired many other geographers and others in this tradition (Uhlig 1995). In 1998 the journal *Erdkunde* produced a special volume of papers in recognition of the centenary of Troll's birth, and a recent collection of essays (Ehlers and Kreutzmann 2000) has been dedicated to his work.

THE CONCEPT OF VERTICALITY

The concept of verticality has played an important role in the development of a specifically 'mountain perspective' on the evolution of landscapes and social organisation. In the early nineteenth century, von Humboldt and Bonpland (1807) developed a description of the physical landscape of Ecuador. They identified a series of zones, from the low hot *tierra caliente*, stretching from the coast to about 1,000 metres, and, in order of ascending altitude, the *tierra templada*, the *tierra fria* and the highest, coolest *tierra helada*. A similar altitudinal pattern of vegetation can be seen on the Pic de Teyde in Tenerife (Figure 1.2). This diagram is adapted from the many produced by von Humboldt and his co-workers that stylistically represent the altitudinal changes in vegetation and land use. It shows particularly clearly the fact that the lower zones are cultivated but that even at latitude 27° N occasional snow cover can be expected at relatively low levels.

Carl Troll (1968), one of the most distinguished contributors to mountain geography, extended this concept. He argued that the 'three dimensional space' of mountains was characterised by the fact that the specific vegetation in any given altitudinal belt varied according to latitude, broadening the original von Humboldt work from its tropical origins. This geo-ecological school has been influential in many subsequent studies of mountains that examined the relationship between the altitudinal patterns of soils and vegetation and the organisation of agricultural production. In fact, Stevens (1993) has argued that the ideas of verticality and production zones are often intertwined in cultural ecology but that they have different emphases. Whereas the discussions of verticality tend to focus on broad scale factors that influence whole regions, the literature on production zones concerns itself with the local detail of farming systems. Nevertheless, the links between these two have been well developed in the ideas of *Alpwirtschaft* in the Alps (Peattie 1936) and by Murra (1972), Brush (1976b) and Guillet (1983) in the Andes, and Uhlig (1978), Stevens (1993) and others in the Himalayas.

However, this *staffelsystem*, in which production zones are organised into altitudinal layers, has been criticised by Allan (1986). He argues that whilst

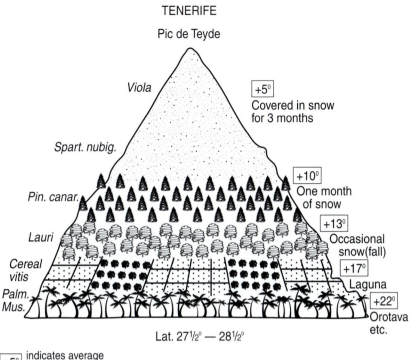

TENERIFE

Pic de Teyde

Viola

+5⁰
Covered in snow
for 3 months

Spart. nubig.

Pin. canar.

+10⁰
One month
of snow

+13⁰
Occasional
snow(fall)

Lauri

Cereal
vitis

+17⁰
Laguna

Palm.
Mus.

+22⁰
Orotava
etc.

Lat. 27½° — 28½°

+5⁰ indicates average
monthly temperature (°c)

Figure 1.2 An illustration of A. von Humboldt's recognition of vertical zonation.
This is a highly simplified adaptation of a drawing of the Pic de Teyde,
Tenerife, taken from *Kosmos*.

Source: Hein (1986)

the vertical model might well have applied to mountain areas which were
largely inaccessible, the advent of improved communications makes its
usefulness strictly limited. This comment sparked an unusually lively debate,
especially within the pages of *Mountain Research and Development*. This still
remains somewhat unresolved, the more so since Forman (1988) has argued
that 'verticality' should form an important element of any state intervention in mountain areas. Despite the wide-ranging criticism of some writers
(Hewitt 1988), Ives and Messerli (1990) remark,

> This concept served us well during the first three-quarters of the
> present century when most mountain regions had only tenuous con-
> nections with the main centres of population and political power,
> and when mountains were truly peripheral in a physical sense.
>
> (Ives and Messerli 1990: 105–6)

Thus, through the work of Troll and his followers, the concept of verticality has been introduced into the vocabulary and has been used as a defining element justifying mountain-centred research. However, in this book we take an eclectic view of this concept in so far as its use is really related to the scale of investigation. We prefer the idea that the geographical pattern of mountain vegetation and land use is better represented by the notion of a mosaic in which some elements of the pattern may reflect altitude and others the prevailing cultural norms or socio-economic pressures. In the steep-sided valleys that penetrate the northern face of the High Atlas, for example, altitude can be seen to play a role as vegetation changes both up the valley sides and as the height of the valley floor increases. However, land use patterns are not so clearly structured: barley can be found throughout the valley, often to relatively high locations, and tree crops such as apples can be found both in low and high zones, reflecting localised environmental factors (such as aspect), entrepreneurial flair and accessibility. Furthermore, the relative importance of these factors may depend upon the scale factors already noted. Spatially detailed investigations may reveal a complex pattern determined more by socio-economic behaviour. On the other hand, a more synoptic view of a mountain region may reveal broad patterns of activity in which altitudinally related factors provide a useful explanation.

MODELS OF MOUNTAIN SYSTEMS

The International Biological Programme of the 1960s attempted to provide greater scientific rigour to the understanding of how ecosystems evolved, especially in harsh environments such as mountains. Unfortunately, although computer simulation techniques had by then been introduced, further research was limited by the conceptual complexity of appropriate ecosystem models and also by the lack of time-series data which is critical in any attempt to develop models of system change. Nonetheless, Ives and Messerli (1990) identify the work of Troll and associates as being particularly critical in the process by which the 1970s and 1980s saw the development of formal models of the evolution of mountain environments. In 1968, Troll established the IGU Commission on High Altitude Geoecology which led to the influential Man and Biosphere (MAB) programme. The theoretical advances introduced as a result of research under the MAB-6 programme were not necessarily deep but consisted of attempts to formalise some of the key relationships embedded in earlier notions. However, in principle the main models developed under MAB-6 placed considerable emphasis on trying to incorporate both natural and social dimensions of mountains and, in some early formulations, contain the elements of what a later generation would call reflexive modelling.

One of the greatest problems in modelling a complete mountain ecosystem is to adopt a metric that can express both social and ecological functions. Ives and Stites (1975) describe a model based upon energy flows throughout a system of animal herding and associated land use, which has been developed elsewhere – for example, the study by Dougherty (1994) in the High Atlas. However, perhaps most influential in subsequent thinking, particularly with respect to the economic development of mountain regions, was the work associated with the Obergurgl region and slightly later with a Swiss MAB project (Ives and Messerli 1990).

Situated in the Austrian Tyrol, Obergurgl is a village at about 2,000 metres that has experienced rapid growth of tourist activity from the late 1950s. The MAB-6 project in the 1970s developed a simulation model of the possible long-term impacts of further expansion of tourism on the natural ecosystem. A simplified version of the model is shown in Figure 1.3, indicating the fact that land use is treated as the dependent variable and is the result of interaction between natural and socio-economic systems. It is based upon the analysis and simulation of four interlinked processes:

1   The level of recreational demand.
2   The impact on farming and associated ecological change.
3   The dynamics of population growth and associated changes in economic activity.
4   The impacts on land use and the role of formal mechanisms of control over new physical developments (construction).

However, in order to understand the contribution of tourism to the local economy and environment, it was necessary to model the impact of external pressures, especially those of an economic nature which would generate and specify the level and nature of demand both for tourism and also for other economic activities. However, the model still tended to work in simple physical quantities – for example, the rate and scale of population growth, and the linkage between land availability, herd size and carrying capacity. Unfortunately, the concept of carrying capacity remains doggedly undynamic in this model as it is assumed that a fixed capacity could be justified from available evidence. Thus one of the crucial elements of the model remained insensitive to the changes in the other parts of the system. In addition, the attempt to simulate the effect of tourism over the following forty years was severely limited by the fact that no financial parameters appear to have been included. Nonetheless, the Obergurgl model (Figure 1.3) constitutes one of the most important contributions of the 1970s in another dimension: the recognition of the importance of involving the local population in the process of determining the main variables of the model and considering alternative scenarios. Although since that time there has been considerable fanfare about 'participation', the Obergurgl

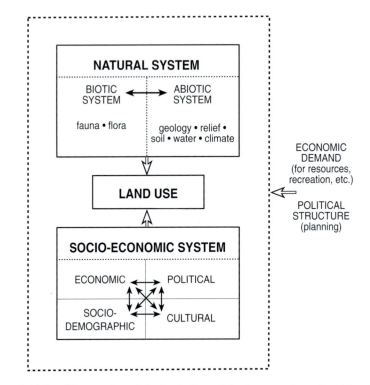

Figure 1.3 The Obergurgl model: simulation of the impact of tourism (simplified from Ives and Messerli 1990)

experiment was certainly conceptually advanced for its era (Moser and Moser 1986).

The Swiss approach to MAB-6 began in the 1970s with strong support from the federal authorities. The overall objective focused on the relationship between socio-economic development (particularly tourism) and the ecological carrying capacity in the mountains, and the Swiss programme placed strong emphasis on the development of interdisciplinary teams of researchers, despite the fact that it was predominantly physical scientists who showed most interest. Initial discussions concerned the selection of research sites. Ultimately four were chosen: Aletsch, Davos, Grindelwald and Pays d'Enhaut. Although each investigation followed broadly similar objectives their research methodologies were somewhat different (Price 1995). For example, the Davos work utilised GIS whereas that at Grindelwald used extensive computer simulation of the impacts of man on environmental variables. The Aletsch study, which involved a relatively large number of communes, did not utilise a particular modelling approach. In addition, although the work at Grindelwald utilised sophisticated modelling there was relatively little consultation with the local populace, whereas in the Pays d'Enhaut this aspect was a major component and the detailed research initiatives reflected local concerns. The impact of linkages established through tourism was seen to push communities into a new dimension of development initiatives, especially with the diminishing contribution of Alpine farming to household incomes. Overall, however, the research work suggested that different methodologies were needed to encompass the relatively quantitative analysis of population change on the one hand and, equally, the important issues of cultural quality and diversity.

A German MAB-6 project commenced in 1983 for eleven years. It was located in the Berchtesgaden region and involved four test areas. It utilised theories of landscape ecology along with the systems approach developed in the Swiss studies. GIS has formed a key element of this work and the key issues concern linking the explicit spatial dimension of the GIS data model with data from environmental psychological investigations into perceptions of landscape quality and change. As yet, however, an explicitly dynamic model of this process has not been developed (Price 1995).

In all of these examples the principal approach has been to examine the relationship between the onset of new economic activities, especially tourism and changes in the ecology of mountains. For current planning purposes in the countries examined, the time-span has been confined to the post-Second World War period. However, some of the work from these MAB-6 studies has been extended to Kenya where local research pointed to problems with the Swiss model, particularly in the methods of assessing pluri-activity. The German experience has been extended to mountain reserves in China (Price 1995). It is likely that other areas will also see various components of these models adapted to local conditions as many more mountain areas are seeing the influence of tourism expand. However, it is worth noting that beyond very general model frameworks, it is not yet clear that the systems approach or indeed GIS can offer real gains in situations where very little quantitative data are available and where the dynamics of the situation are as yet unclear.

An alternative approach to the analysis of change in the agricultural sector has been explored by McCrae (1982) in the Andes. Here the emphasis was on the capacity of agriculturalists, working at high altitude (4,000 metres), to adapt their agro-ecosystem. This approached the problem through modelling the extent to which this high-altitude economy could evolve from one based largely on herding, to one which included much more labour time allocated to employment on haciendas. Once again a transformation model was involved based upon a recognised historical experience. In addition the model tried to predict the outcome of replacing the predominantly sheep herd with one based upon the much more valuable alpaca. This shift served as a useful test of the model in its attempt to examine the resilience of the farming system in the face of technical change. In fact, the basic dynamics of this model were drawn from other well-known herding models, adapted for use in the High Andes. As had been found elsewhere, the model indicates that a significant fall in the animal population (greater than 10 per cent) would have important repercussions on the supply of organic fertiliser, which in turn was the basis of fertility maintenance in these areas for both crops and grass.

Looking at all these attempts of the period before 1985 to explore, by modelling, the nature of mountain ecosystems, it is possible to draw several conclusions:

- First, the design of the models assumed 'perfect knowledge' and closure as defined by the model system itself. Hence, refining the model meant redefining the variables themselves and linkages between them. Once 'set going', the final state could only arise within the terms established by these basic variables.
- Second, calibration has been a serious problem. Mountain areas tend to have poor data sets, both of physical and human variables, and this is particularly apparent where, for example, time-series plays a critical role in simulations. In special circumstances, for instance in parts of the European Alps and perhaps in areas of predominantly Tibetan culture where monasteries may have relatively ancient records, some long-term social data can be found (Viazzo 1989; Crook and Shakya 1994). Here environmental reconstruction techniques are beginning to provide enlightening information on the natural landscape.
- Third, the process of 'transformation' was usually modelled in a strictly 'unilinear' fashion, the results suggesting a simple division between a downward path for the environment (degradation), and upwards (development) for those fortunate to take part in new ventures (e.g. tourism).
- Finally, the models tended to pay particular attention to the 'fragility' of the environmental variables. Some disquiet about this can be found in the Swiss MAB model where it was discovered that ecosystems had much greater resilience than was previously thought. The implications of this for the process of long-term transformation are, of course, considerable.

The theoretical framework for all these models relies upon the assumption that human activities should adapt to the physical environment in a manner reminiscent of the arguments of environmental determinism. It should be remembered that the 1970s also witnessed the increasing political acceptance of the findings of global scale modelling exercises by Meadows *et al.* (1972) and the influence of the work of Eckholm (1976). Therefore, policy directives for mountain regions built upon the largely negative image of environmental crises resulting from the inappropriate management of natural resources by communities living in these areas. The result has been interventions that remain highly controversial.

THE HIMALAYAN DILEMMA

It was in the 1980s, however, that several alternative perspectives were beginning to form as a result of what was to become known as the Theory of Himalayan Environmental Degradation (Ives and Messerli 1989). One of the main features of this work has been to recognise the multifaceted nature of environmental change rather than focus almost exclusively on human agency. Moreover, ideas of resilience, absent in much former work, are introduced. This theory is discussed in greater detail in Chapter 8.

However, for this introductory section it is important to recognise that this phase in theoretical analysis shifted the emphasis further away from sole concern with the physical environment. Not only does the human dimension of the mountain regions become central but also – tentatively at first, but later with considerable force – the whole question of mountain environment and development is re-addressed as a political issue. Whilst Ives and Messerli systematically examined the specific evidence concerning the Himalayas, perhaps the most important theoretical contribution emerged from the work of Thompson and Warburton (1985a, 1985b). This team, working also within the broad confines of systems analysis at IISA, stated two things:

- that previous models and theories had failed to handle the diversity of mountain regions as an explicit characteristic of their development;
- that the key to understanding the evolution of mountain regions was to acknowledge the uncertainty of future prediction. This element was in stark contrast to the highly deterministic models of the earlier period.

In addition to providing a trenchant critique of the advocates of an environmental crisis within the Himalayan region, these writers question the fundamental formulation of mountain problems. In a deceptively simple section of their 1985 paper 'Knowing Where to Hit It', Thompson and Warburton write:

> The starting point for our analysis has been the recognition that man does not interact directly with his environment. It is not just perception but cognition – seeing and knowing – that brings man and his environment together and, since there is more than one way of knowing, there is more than one way for him to come to grips with his environment.
>
> (Thompson and Warburton 1985b: 204)

Moreover, their work highlights the significance of scale, the way in which macro and micro institutional structures interact – and sometimes clash – to produce the framework in which particular environmental management decisions emerge. Most of the existing approaches assume that these institutional relationships, and the attitudes associated with them, are immutable. In fact, they argue that a key to understanding these relationships is to accept that they do indeed change. However, these changes occur not so much directly as a result of straightforward alarm about environmental or social distress but usually as a result of deep-seated shifts in cultural and political thinking. Consequently, typical policy initiatives generated by contemporary, often external agencies, fail to work because they do not (cannot?) address the underlying issues.

PERCEPTIONS OF MOUNTAIN ENVIRONMENTS

It is appropriate at this point to note that the preceding, largely historical account of the development of theory about mountain regions has concentrated upon writing which has fallen into what might be called a conventional social or natural science framework. With the focus of attention beginning to look more closely at decision-making, institutional frameworks and a deeper understanding of the transformation of mountains as a social phenomenon, another literature can be introduced. At the same time as accounts by scientists of the theoretical development of mountain geography pay particular homage to the contribution of Alexander von Humboldt, it is equally important to recognise the role of writers and artists in developing a cultural imagery of these regions. Particularly from the eighteenth century onwards, anthropological writing – and also the work of historians and other more general commentators – has influenced perceptions of those in lowlands towards the mountain environment and habitat. In western Europe, the influence of writers and painters on the perception of the Alps has been well documented. Tobias (1986) cites the work of Gavin de Beer who claims that around the turn of the nineteenth century there were more than seven hundred individuals who wrote of their experiences of the Alps and were on the lecture circuit. Attitudes formed by these writers have had a profound effect. Byron and Shelley wrote of the 'goitered idiots' who populated the lower slopes of the mountains, thus focusing most attention on the upper slopes which continued to resonate through later scientific studies of mountains. As Tobias remarks:

> By the industrial era that very sentiment had entered the working vocabulary of social psychologists who stated resolvedly that alpine peoples were short, brown, broad headed, with limited mentality.
> (Tobias 1986: 184).

Without doubt it was the scenic charms of the Alps, later to be transferred to the Himalayas, the Atlas and other mountain ranges, that attracted comment, visits and approval. Early visitors to Tibet in particular returned with glowing accounts of strong governance and ordered people, and admiration can be found for the 'hardy, fiercely independent' local inhabitants, often soon to be subjugated in wars associated with colonial expansion. Isaiah Bowman, who in 1916 was Director of the American Geographical Society, wrote graphically of the 'character' of the Peruvian Indians and this style of writing continued to reflect what became known as the 'Heidi complex'. Thus an idealised form of strong, but independent person living in a relatively untouched environment typified a widely held view of mountain peoples. However, it was also true that this same image was compounded with one of 'simplistic technology and unsophisticated society', an image that others have shown dominated much western writing on the question of indigenous knowledge (Richards 1985).

In North America at least two traditions can be discerned. On the one hand the extensive studies of pre-European Indian communities in mountain areas (Dozier 1970) and, on the other, the exploration, settlement and exploitation of the main mountain ranges which was all part of the 'frontier tradition' in popular mythology. A good example of such work is discussed in Dick (1964).

The significance of these contributions to an understanding of mountain communities lies in the fact that they attempt to refocus attention on the deeper cultural attributes of populations, both of those who live in mountains themselves and, equally importantly, of lowlanders who have generated influential images of highland society. These cultural images, far from being merely decoration, tie in powerfully with the points made by Thompson *et al.* (1985a, 1985b) who, as noted above, were beginning to reformulate a new approach to work in mountains.

THE NATURE OF SOCIAL AND ENVIRONMENTAL CHANGE: CULTURAL THEORY AND COMPLEXITY

In his review of the literature on mountain development, Ken Hewitt (1988) pointed to the problems posed by the different epistemologies used by the physical and social sciences. He noted that much of the writing had, up to that date, either blatantly ignored the specific social or cultural characteristics of mountain communities or followed the path of explicit or tacit acceptance of the dominant control of the physical environment over social organisation. This, in essence, was seen to be the key characteristic of mountain areas because of the 'harshness' of the environment. These writers failed to note that, in many tropical areas, the mountains represent the most favoured sites for human activity! Equally, it is also noticeable that many anthropologists and other social scientists working on mountains have paid scant regard to the physical environment, choosing instead to interpret the social and cultural structure and behaviour through models that privilege interpersonal, kinship or community relations. Few of these works enter the canon of accepted 'mountain specialists'!

Recent work of Thompson and others has begun to reformulate our approach to problems in mountains and he has been a strong advocate of the role of 'cultural theory'. At the same time, given the problems of epistemologies noted by Hewitt, other writers from both an ecological and sociological background have been exploring the relevance of 'complexity' to environmental problems. This work is relatively new but would seem to address some of the problems raised by Thompson.

Cultural theory and mountain communities

Thompson *et al.* (1990) develop the idea of behavioural myths, arguing that both society and individuals hold a number of generalised ideas or 'myths' about the way nature behaves. He proceeds to use these myths to generate several categories of social or organisational behaviour:

1   Nature can be 'benign', and we believe that the physical environment will be robust in the face of all the methods society uses to exploit it. This results in a *laissez-faire* attitude both culturally and economically so that a demand-driven 'market' determines the form of use of the environment.

2   An alternative perspective is based upon a belief in a 'constrained utility' in so far as use of nature can be tolerated within certain limits. This means that users (society) will need to respect those limits in order to prevent environmental damage. The market will not function satisfactorily to keep exploitation within 'manageable' limits without some regulatory framework, usually provided through a hierarchy of institutional controls.

3   A third 'myth' views the environment as highly sensitive to any exploitation. It is essential to have carefully planned use, and decision-making is a communal responsibility. Such an egalitarian viewpoint rarely sits well in modern capitalist societies, but elements of this approach can be found in many cultures.

4   Finally, there are those who believe that nature is capricious, and whatever action is implemented will be ultimately subject to the constraints of the physical environment. This fatalist approach, whilst not immediately recognisable, is contained in some of the behaviour of communities living at the margins of industrial society and, conversely, by some whose ideology may be identified in today's environmental movements.

This group of myths leans heavily on a social typology developed originally by the anthropologist Mary Douglas (1985) which has been criticised
because it suggests a deterministic linking of culture and social organisation
(Milton 1996). However, Thompson *et al.* (1990) maintain that their development of this approach is not deterministic because individuals combine
these myths and behaviours in different ways according to circumstances
and specific outcomes remain highly uncertain. A crucial element of cultural
theory is the notion of surprise. This occurs when the prevailing myths of
nature increasingly fail to match up to the behaviour that people see around
them. For instance, if after some time it becomes apparent that local institutional rules (perhaps with regard to a particular land use practice) do not
deliver the necessary results, then some members of the community will
suggest changes. Inevitably there will be resistance, not just for personal
reasons but also because new initiatives may well undermine the underlying
myth of nature on which the current practice is founded – for example, the
view that the use of resources needs controlling to avoid environmental
damage.

In a series of studies, Price and Thompson (1997) and Thompson (1997)
show how these ideas throw a rather different light on the process of social
transformation in mountain areas. For instance, in a study of Davos in the
Alps, they reveal how, over several centuries, families have shown remarkable swings in attitudes towards their environment. They argue that we can
trace the evolutionary dynamics of communities through the ever changing
balance of 'myths' within the community, and that the consequent evolution of institutional structures (social rules, political behaviour, etc.) changes
in a way that is uncertain. They note that much of our understanding of
the social transformation in mountain regions still assumes a unidirectional
and largely predictable trajectory, which is often summarised in the term
'modernisation', and focuses heavily on the specification of successive stages
which are presumed to occur. However, cultural theorists would argue that
there is no 'organisational climax' to which society is evolving, in the same
way that many ecologists, after Holling (1985, 1994), reject the static
notions of climax community in vegetation studies. Referring to conventional models of social transformation (e.g. modernisation) that often
underlie work on mountain communities, Price and Thompson (1997: 78)
argue that: 'these models are beginning to be seen as less than satisfactory
as they explain change by getting rid of it and are increasingly incapable of
making sense of what is going on'.

The contribution of cultural theory to understanding social change has
been examined by Milton (1996), and the version proposed by Thompson
et al. (1990) is one particular strand of this idea. According to Milton, one
of the key tenets of cultural theory is the need to maintain cultural diversity in the face of the process of globalisation. In this context cultural
diversity relates to the many and varied sets of beliefs and behaviours that

inform our interactions with nature. She argues that by recognising, and if necessary protecting, this diversity, we are more likely to be able to handle unfamiliar circumstances which are bound up with long-term environmental change. The genesis of this approach lies largely within the discourse of anthropology and, as such, tends to focus on the 'small scale' social frameworks of individual, family and community. It fits neatly into a conception of mountain societies which emphasises their particularistic, highly local social and economic character. It also emphasises the diversity of these societies. However, in other ways it has close affinities with much broader models of change which have become the focus of interest in both physical and social sciences and which may offer new insights into the way we examine mountain communities.

The consideration of cultural theory has raised many issues in the study of mountain development, which are considered in recent special edition of *Mountain Research and Development* (Volume 18, No. 2, 1998). In particular, this approach raises the question of the meaning of 'myth' and its interpretation within different branches of human endeavour. For instance, the Canadian literary critic Northrop Frye explored multivalent interpretations of key words such as 'myth' and argued that its meaning derived from situations where the external world is represented by personalised forces such as Gods. More significantly, he argued that myth is often used in an ideological context and employed as part of rhetoric. In this instance, myth represents the creation of images that only make sense once particular contexts or social norms of behaviour are accepted (Frye 1964, 1990). Recently Forsyth (1998) highlights the use of the term 'myth' in discussions surrounding the integration of social and natural science. For natural sciences a 'myth' is some statement that can be subjected to formal scientific examination and, as the word implies, would likely be debunked. However, as Thompson uses the idea it is a substantive statement of people's beliefs, which is valuable in its own right and not subject to scientific demolition. Forsyth emphasises that an improved understanding of mountain environments and their development must be based on the recognition of plural explanations and it is therefore appropriate that we now turn to a consideration of the ideas of complexity where this issue is at the forefront of the debate.

Complexity and social theory

The complexity concept originated in the natural sciences, in particular physics (Nicolis and Prigogine 1989) and also biology (Holling 1985, 1994). It has been used to explore problems that have proved difficult to handle by conventional reductionist models. For some years social scientists have been exploring the application of this 'complexity thinking' to the problems of social transformation (Streufert 1997). The ideas examine the commonalities that exist between the behaviour of all non-linear dynamic systems where it is not sufficient to understand just the functioning of an individual component but also the operation of cross-system linkages. It is these joint effects that are relevant to many (if not all) social science analyses. Kauffman (1993) and others emphasise that complexity is not the same thing as 'complication'. A clock might be complicated, with many linkages, but we understand the mechanism and once set going it will work in a predictable manner. Lo Presti (1996) cautions against the confusion of complexity and complication and notes that complexity is concerned both with models of knowledge and also the nature of the objects under investigation. A very useful approach is provided by Fioretti (1996), who suggests that if a phenomenon can be described by more than one logical model then it is complex, emphasising that it is our difficulties of understanding that are at issue. This is a crucial consideration in any attempt to use these ideas in social science. Some writers, such as Amin and Hausner (1997) in their discussion of the application of complexity to economics, suggest that more recent economic structures and behaviour are more complex than those of the past. However, this view is not shared by all, as it can quickly deteriorate into an analysis of modernism rather than an attempt to offer an alternative way of looking at problems.

Ideas of complexity are particularly relevant for problems of evolution or transformation over time. It has been developed to handle non-linear dynamic systems in which change over time is subject to 'sensitive dependence on past events' but not completely determined by this. Random or novel events arise which can move the system to a new behavioural trajectory. Multiple systems co-evolve and it is their interactive behaviour that is important to our understanding of the feedback process across system boundaries. In this context, scale (both temporal and spatial) becomes important, as the linkages vary according to their place within the system and their duration, and systems often establish a hierarchical organisation. Within this, different processes may be important at different levels but interaction between levels may itself be part of the feedback process. Living systems possess 'adaptive self-organisation' (Allen 1994), but behavioural complexity in social systems operates within a self-reflexive framework (Streufert 1997). Delormé (1997) maintains that complexity thinking not only allows for interaction between social actors but also applies at the level of constructing or reforming systems and institutions. Thus Clark et al. (1995) argue that it is the capacity to make errors, to learn and reconsider that is the key process in dynamic change. Persons, groups and institutions possess self-reflexive properties and it is this use of recursive knowledge that provides the indeterminacy that challenges our linear conceptualisation of the evolution of social systems.

A number of writers have begun to employ these ideas to examine problems in social sciences, for instance in social theory (Allen and Lessor 1991), in economics (Clark *et al.* 1995; O'Connor 1995; Amin and Hausner 1997), and in social psychology (Streuffert 1997). In geography Zimmerer (1994) has indicated that the trend towards dynamic ecological models, which use ideas from the 'complexity stable', has forced a re-evaluation of some key concepts such as 'carrying capacity', and the notion of 'climax community' in biogeographical work (see Leach *et al.* 1997). The implications for an understanding of long-term landscape change are important, and it is worth while exploring the extent to which our understanding of mountain environments and communities might benefit from using these ideas.

Although we have chosen to separate the notions of complexity and cultural theory, both theoretical constructs approach the problems of dynamic social systems with somewhat similar basic presumptions. The language of cultural theory is perhaps much more comfortable for many social scientists, but the ideas behind complexity also provide a behavioural mechanism for understating the evolution of communities. The key to the approach is 'always learning, never getting it right'. These ideas place great weight behind the fact that a precise future path of development cannot be predicted and so strategies that assume a linear path cannot provide an appropriate framework for understanding the development of either social or ecological relationships.

A THEORETICAL PERSPECTIVE ON MOUNTAIN ISSUES: THE 'COG MODEL'

Some of the models developed under the MAB-6 programme have already been discussed briefly. These all attempt to capture the dynamics of the transformation process through the interaction of man and environment using a variety of approaches, including GIS and simulation. The work of Krippendorf (1986) and Messerli (1989) in particular incorporates both structural change and uncertainty into models of alpine transformation. In a somewhat simpler form this book uses what we call the cog model as a framework to present the concepts of complexity, uncertainty and diversity in environment, perception and understanding. This model is introduced here in order to explain how these concepts are critical for an understanding of how mountain environments and societies operate, and to place into some context the realities of our knowledge and understanding of the variables or components and the linkages between them.

The model comprises a mechanical set of cogs whose teeth interlink, and the driving mechanisms and the power supplies that induce movement in the components. Obviously, this is a mechanistic approach that is flawed in the context of a dynamic, social and biological system. The boundaries and units within the system are far more fluid than can be represented diagrammatically. The idea of 'fuzziness', taken from mathematics, might be more

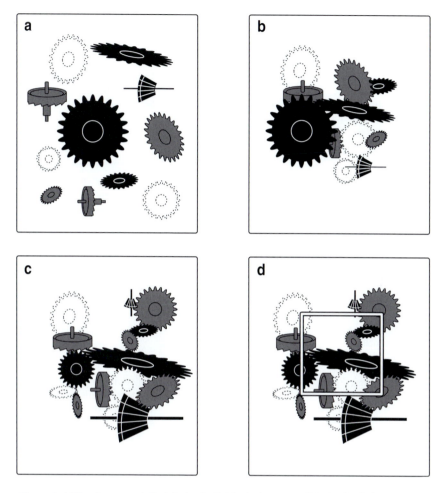

Figure 1.4 The 'cog model': (a) the individual elements; (b) the basic interaction; (c) the dynamics of growth and change; (d) the impact of administrative closure. For detailed explanation see text and Table 1.2

appropriate here than the imagery of clockwork cogs, which are well defined and in a closed system as can be seen in Figure 1.4. However, we feel that it represents a starting point by introducing the collection of ideas and different approaches discussed in the book and applied in mountain research. These flaws are discussed later.

Table 1.2 presents the outlines of four stages of the model, which correspond to the four parts of this book, and thereby serves as a 'map' of the material presented. First, however, it is necessary to explain the way that

Table 1.2 The 'cog model' and the organisation of the book

The cog model	Approaches	Book section
'Frozen cogs'	Descriptive:	Part 1
Components/variables: • black = known • white = unknown • levels of grey = levels of knowledge	Verticality Geo-ecology Systems	Definitions and framework Classification Scale
'Cogs interact and run'	Static and descriptive 1950s-70s:	Part 2
Mechanisms and interactions	Mountain environments The traditional mountain society	Elements of climate, geomorphology, ecology, society, population and economy
Dynamics of change Cogs interact, which, involves dynamic evolution Cogs change in size, in 'visibility' and linkages New cogs appear New perceptions	1980s-90s: Principal forces for change Dynamic of growth and decline Recognition of uncertainty Holistic models	Part 3 Transformation 1 Demographic 2 Climatic 3 Economic
Managing dynamics Cogs framework bounded by bureaucratic pluralism	1990s: Adaptive environmental management Complexity Cultural theory and myths of nature Postmodern concepts Self-reflexivity	Part 4 Policy development Himalayan dilemma Conservation Agenda 21 Political structure, pluralism

mountain environments (including both physical and socio-economic aspects) are represented by the model. The basic idea is to consider a machine of cogs where each cog represents a component of the mountain system (for example climate, geomorphology, economy, culture, etc.). These cogs are linked by a series of levers and pistons that represent the processes and interactions between the variables (for example, the influences of climate on geomorphological processes, or the impact of cultural change on land use practices). In Figure 1.4 some of these cogs are coloured black, representing a relatively significant level of understanding. Others, which are white, represent variables of which we are unaware, or of which we have only a vague idea of their existence and significance. Those coloured grey represent variables that have been 'discovered' or more recently acknowledged in mountain research. Whilst this is a slightly different way of showing components to that employed in much systems modelling it has the advantage of emphasising the knowledge levels particularly clearly. Examples might be a black cog for climate and a grey cog for erosion processes. The importance of including the white cogs is borne out by the fact that during the last couple of centuries there have been 'new' variables continuously 'discovered' in the process of research which have proved significant in the understanding of the total structure. An example might be the 'discovery' of indigenous institutions that regulate resource use such as forest and grazing management. These are now grey cogs, partly understood in some contexts.

There are two issues that immediately arise. Obviously, each cog represents in itself a collection of components, such as temperature, precipitation and winds in climate. This model therefore applies on different scales, in the understanding of the details of climate or socio-cultural variables, as well as on a broader cross-disciplinary scale. This in itself demonstrates the hierarchical nature of mountain studies. Each cog is in itself complex. The second issue is that the degree of understanding of each cog, linkage and driving mechanism is relative, not absolute. That is to say that the black cogs do not represent absolute, total knowledge, but relatively more complete knowledge than the grey or white cogs. Also, for simplicity we have only shown the cogs in the three tones, but in fact they are a continuum from an invisible white through the grey to black. Strictly speaking, none can be truly black, as this implies complete knowledge, but cogs or linkages can be completely white representing as yet unknown or unacknowledged variables and processes. With respect to this concept of relative knowledge, two further points are relevant. First, that the general shade of grey of the whole model becomes darker or lighter as the overall level of understanding of mountain systems changes with further research. The relative darkness of the different cogs is therefore set upon a fluid background. This has also been omitted from the diagrams for simplicity. Second, the shades of the cogs will also change relative to each other. For example, the 'discovery' of indigenous resource management institutions required a significant reappraisal of the knowledge already accumulated of cultural integrity and socio-economic rationality. This is reflected in our changing understanding of observed elements of mountain landscapes. A good example of this, in the case of forests, concerns the western perception of deforestation in Nepal as an entirely negative process that needed to be reconsidered in the light of later work. Observations by Gilmour (1995) and others indicate the 'rearrangement' of trees in the landscape and the fact that the indigenous population place greater value on forests as convertible rather than renewable resources. This not only requires that we need to adjust our understanding of farmers' perceptions of forests, trees and potential productivity of land under growing population pressures, but also our views of indigenous rationality.

In the first diagram (Figure 1.4a) we have 'stopped' the model in order to examine the various components and linkage mechanisms. In mountain terms this represents stopping the clock on traditional societies at different times. Where we are dealing with essentially traditional societies in the Alps this may be 200 years ago, in North America perhaps 250 years ago, whilst in the Himalayas only 75 years ago. In the Andes, it may represent the pre-conquest times 400 years ago. We are therefore looking at the society that existed before the impact of rapid and far-reaching economic, demographic and political changes in relatively recent times. This 'snapshot in time' does not deny that mountain societies and environments are continuously evolving systems, and therefore at the time of our snapshot the situation is the product of long periods of time during which different environmental, economic and cultural influences shaped the traditions we examine.

Figure 1.4b represents a situation of the machine running and the various basic processes in operation. Part 2 of the book covers this element. Obviously we have chosen to emphasise particular components of the physical and social environments as they reflect the currently held wisdom about important processes in mountain areas. Although different components are examined they are treated in a relatively static framework, primarily to help exposition. Particular attention is paid to the interactions between the components because the resultant environmental and social landscapes in mountains reflect the specific way the elements have combined in a particular locality.

In Part 3 we move to examine the process of transformation which occurs as the rate and direction of change alters. This is suggested in Figure 1.4c where factors such as capitalist penetration, population growth, etc. become very significant. The dynamic effect of these factors is uncertain. They may cause individual cogs to run unevenly, to be decoupled from their linkage mechanisms, or to be unbalanced and change size as different variables become more important than before. For example, the change from subsistence to cash-based economies can shift the relative values of crops grown, which in turn alters the pattern of land use and also the balance of advantages surrounding tenure issues as privatisation becomes more favoured than communal mechanisms. Of course, in attempting to describe the dynamics of mountain environments we can only record what we know! This is not so trivial as it sounds because, if we follow the underlying ethos of the cog model, we must proceed with an open mind, and accept our current knowledge level as a 'working hypothesis'.

The final part of the book, Part 4, corresponds with Fig 1.4d and represents the point at which the intervention of policy-makers, along with conservation and development agencies, impinges upon the functioning of the system. Here we have the placement of a rigid square over part of the mechanism representing the development 'target', the state policy objective or the conservation issue at the heart of the intervention. The placement of the rigid square does not rest easily within the complex model, and while the central elements are clearly identified, around the edges only parts of components are included. The figure illustrates that there are dangers in applying inappropriate perceptions, priorities and views, based upon incomplete understanding of the entire model. As our understanding grows, we are able to incorporate more complete coverage of the 'relevant' cogs. However, the key issue is how to tackle the target effectively, incorporating sufficient adjacent cogs, and allowing for the fact that there will be unknown entities and processes that are not yet visible. At this point, the problem of scale becomes evident and all such models have tried to grasp the need to construct mechanisms of governance and management that operate at different spatial scales yet are able to overcome the inherent inertia of typical institutions in the face of rapidly changing situations. For example, we cannot effectively tackle the problem of snow leopard preservation without taking into consideration the habitat, ecology, cultural niche, etc. that it holds in the life-world of the local inhabitants. On the other hand, a rigid model of institutional protection may not cope with the dynamics of public interest and economic pressure. Part 4 therefore addresses the imbalances and impacts of external and internal changes in the mountain system, the effective, ineffective and potential ways of coping with these impacts, and the differences in appropriate and inappropriate policy interventions.

It was mentioned above that the model was essentially flawed, a result of the mechanistic nature of cogs and levers, etc. This imagery suggests a view of finite systems, rigid interactions and variables, and well-defined limits and boundaries to the components, the linking mechanisms and the shape of the whole model itself. These issues of finality and rigidity are crucial. A real mountain system comprises variables that change their nature through time. An example could be the distribution of trees in the landscape, or a feature may become increasingly dominating – for instance, the increase in frequency of large-scale landslides, earthquakes and similar geomorphic processes. It may be that the landscape is radically altered through human activity – for example, the construction of dams. Also, the mountain system is not closed and complete in itself, but intricately associated with wider continental and global systems. The importance of latitude in modifying climate and the changing influence of the monsoon in the Himalayas are two such examples in physical terms. In the socio-economic realm, the intervention of external pressures to conserve aspects of the environment, such as forests and wildlife, and the political and economic penetration of mountain valleys and their increasing connectivity with lowland political power centres also demonstrate the openness of mountain systems.

This openness and connectivity leads to the issue of the specificity of mountain environments and the viability of considering them as unique and special entities, requiring specific approaches, understandings and knowledge, with clearly defined borders and problems. This will be discussed in Part 4.

This chapter has explored some of the basic elements of the study of mountain areas, noting the fact that mountains have always assumed a special place in mythology, literature, and later in natural and physical sciences. It has emphasised diversity, in that some mountains have, as far as we know, never experienced high population densities – partly because of their environment but also due to their inaccessibility. Other areas have been thickly populated for generations and constitute the core areas of some contemporary nation-states. Scientific enquiry has long noted the importance of the relationship between altitude and vegetation and land use. Various models of man–environment relationships have been used to understand how the landscape has evolved. In more recent times, human intervention, now in the form of international and national agencies, has concerned itself with the long-term sustainability of these areas and the attempt to establish management regimes to maintain environmental stability. However, it is doubtful that such an objective is actually appropriate in so far as environmental stability is itself a chimera, and the increasingly dominant role of human agency subject to quite remarkable behavioural shifts when viewed over a long timescale. The development of the cog model serves to focus attention on this continual dynamic of social and environmental systems and highlights the problem for agencies attempting to construct viable institutional frameworks to deal with threats to mountain environments and communities.

Part 2

COMPONENTS OF MOUNTAIN ENVIRONMENTS AND LIVELIHOODS

INTRODUCTION

In order to appreciate the diverse environments and social histories of mountain regions this section introduces some basic characteristics of these areas. This forms the background of much of the discussion on transformation that occupies the remainder of the book. In terms of our basic model, Part 2 somewhat artificially focuses more or less on what might be called traditional or pre-modern mountain societies. It is divided into three chapters. First, there is systematic examination of the physical factors influencing mountain environments. Then we examine the principal cultural and social dimensions that have provided a framework for mountain communities. Finally, there is an extensive examination of the production environment, farming systems and other economic activities that once dominated mountain society, and in some areas still do so. The reader is reminded again that we have 'stopped the clock' somewhat arbitrarily in our choice of examples, primarily for pedagogic reasons.

2

THE PHYSICAL ENVIRONMENT
OF MOUNTAINS

CLIMATE AND MOUNTAINS

Climate is a key influence on the nature and rates of geomorphological processes and on the potential of the agricultural environment. The extreme variability in time and space of temperature and precipitation are not unique to mountain environments, but the combination of these with the topography presents particular difficulties in sustaining a livelihood.

Climate operates at different scales from global to local, with each element interacting with others on all these scales. Mountains can modify global atmospheric processes significantly, and generate their own climatic conditions, thus affecting the climates of adjacent regions. For example, the Himalayas are of fundamental importance to the occurrence of the monsoon in northern India, and of the continental arid conditions in Central Asia.

At the valley scale, the effects of the interaction of conditions at different scales are to produce a diverse mosaic of microenvironments, which have been exploited by traditional agricultural populations. This mosaic arises from variations in soil moisture, temperature, frost susceptibility and solar radiation intensity, all of which affect the growing season. Many plants and animals exist in mountain areas at the limits of their environmental tolerances, and relatively small shifts in climatic conditions can cause extensions or shifts in species distributions or even their disappearance from the area. Small changes in conditions tend to have disproportionately large effects. Conditions tend to change rapidly as well – for example, the fall in temperatures when clouds block the sun. Indigenous endemic species are abundant in mountains, and have evolved in order to adapt to fluctuations, extremes and the instability of mountain environments. Such diversity is critical in the maintenance of floral and faunal viability through time.

Climate data for mountain regions is sparse, and records do not usually extend over long periods of time. The Alps has the longest records, extending for around a century, with the earliest appearing to be at *Hohenpeissenberg* (Germany) starting in 1781. Relatively dense networks exist for the Alps and parts of North America. Elsewhere, problems of access, finance and political disputes have limited the efficacy of weather stations in places such as the Himalayas. Barry (1992) tabulates the principal observatories in mountain regions, and of thirty listed, nineteen are in Europe and none in the Himalayan arc. Even where such stations exist, they cannot record re-alistically the whole diversity of conditions within the typical valley.

Influences on climatic conditions include latitude, altitude, continentality and topography. These may be expressed at a number of scales, but are particularly significant at a continental scale. These will be considered first, then the variables of climate: solar radiation, temperature, precipitation, humidity and evaporation. Finally, the significance of these conditions for livelihoods is considered. Climate change and its impact is reviewed in Part 3. Comprehensive reviews of mountain climatology can be found in Price (1981) and Barry (1992).

Global and continental scale factors

Latitude

Latitude affects solar radiation receipts, temperature, seasonality and modi-fies the influence of altitude, causing treeline and snowline altitudes, and the occurrence of permanent snow and ice to descend polewards. Latitudinal effects produce temperature and precipitation regimes which result in high-altitude permanent snow on the equator but seasonal snow cover at lower levels in higher latitudes. The interaction between altitude and latitude can also give rise to more favourable conditions for agriculture in highlands within subtropical desert regions, such as the mountains in the Sahara region.

Global atmospheric circulation gives rise to four broad regions (see Figure 2.1): Equatorial low pressure (0–20° N and S), Subtropical high pressure (20–40°), Subpolar low pressure (40–70°) and Polar high pressure (70–90°). High-pressure zones tend to be drier, and include the Himalayas, Tibet and the Atacaman Andes. Low-pressure zones tend to be wetter, including the lush mountains of Borneo with almost daily precipitation.

Seasonality and day length varies from equator to pole. On the equator day and night are of similar length and seasons of little account, giving rise to more or less constant conditions throughout the year. Mount Wilhelm, Papua New Guinea (5° S), has a 7–8°C diurnal range, and 0.8°C seasonal range at 3,480 metres. Niwot Ridge (40° N) has a seasonal range of 21°C, although the diurnal range is similar (Barry 1992). The High Andes can experience snowfall or frost at any time of the year, but in mid-latitudes this is concentrated in the winter months – commonly from October–November to March–April – when either snow cover prevails or low temperatures limit vegetation growth. The main implication of this for livelihoods is the length of the growing season – near the equator the growing season is continuous, but is increasingly restricted further north and south.

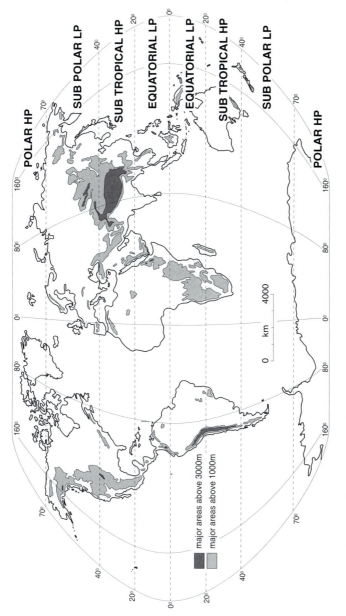

POLAR HP

SUB POLAR LP

SUB TROPICAL HP

EQUATORIAL LP

EQUATORIAL LP

SUB TROPICAL HP

SUB POLAR LP

POLAR HP

major areas above 3000m

major areas above 1000m

0   km   4000

Figure 2.1 Regions of global circulation

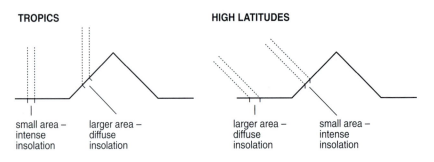

TROPICS   **HIGH LATITUDES**

small area –   larger area –   larger area –   small area –
intense   diffuse   diffuse   intense
insolation   insolation   insolation   insolation

Figure 2.2 The effect of latitude on solar radiation (midday sun) in mountain
areas

The impact of latitude on solar radiation concerns the height of the sun and the angle at which its rays hit the surface of the earth (see Figure 2.2). In tropical regions, the sun is overhead at midday, and thus will be perpendicular to horizontal surfaces and at an angle to slopes. The greatest intensity of solar radiation occurs when the rays are perpendicular to the surface, so tropical mountains receive less at the surface than adjacent lowlands. There is the additional propensity for cloudiness in the tropics, which will cause further reductions to radiation receipts. However, at higher latitudes, where the sun is much lower in the sky, the rays will be more perpendicular to slopes facing the sun than to horizontal surfaces. Thus mountain slopes facing the sun will receive more solar radiation than those in the tropics. The intensity is accentuated by the thinner atmosphere and reduced cloudiness, and as the daylight hours are longer, net daily receipts will be greater than for the tropics, but only during the summer months. At the scale of the individual valley, daylight hours – or at least hours of sunshine at any one location – will vary considerably as a function of aspect (see below). On a global scale, Price (1981) gives figures for daylight hours for the summer solstice (21 June) of 12 hours 7 minutes for Mt Kenya on the equator, 13 hours 53 minutes for Mt Everest (28° N), 15 hours 45 minutes for the Matterhorn in the Swiss Alps (41° N) and 20 hours 19 minutes for Mt McKinley in Alaska (63° N).

The influence of global circulation also contributes to precipitation; the trade winds, for example, are an important source of precipitation in the Andes (see the case study on pp. 70–73). This can also affect the distribution throughout the year. Some regions have a clear annual wet season, such as the Himalayas where the monsoon dominates. Elsewhere there may be a bimodal distribution arising from more than one source of wet winds. In the tropics, precipitation may be daily throughout the year, as in Papua New Guinea.

Altitude

The general trend with increasing altitude is a reduction in temperature, air density and pressure, proportions of carbon dioxide, water vapour and concentrations of impurities such as dust. The intensity of solar radiation, and especially the UV component, increases with altitude. The effect of these changes is to increase the stresses of high-altitude environments – for plants in terms of growing conditions, and for animals and humans who suffer from physiological stresses related to oxygen deficiency. Plants, animals and native or acclimatised human populations have adapted to compensate for these factors. A number of studies, particularly of Sherpas and high-altitude Andean Indians, have concluded that such people tend to have greater lung capacities and higher red blood cell concentrations than lowlanders. Plants also adapt to the high wind speeds and exposure factors by their growth habits (see pp. 61–66).

The lapse rates (change in conditions with altitude) will vary as a result of wind mixing and turbulence. Wind chill and exposure creates additional stress to living organisms and can increase temperature lapse rates. Water vapour is an essential component in the air contributing to heat absorption and retention. In mid-latitudes water vapour comprises about 13 per cent by volume of air at sea level, but 0.08 per cent at 8,000 metres (Price 1981). Particles such as dust are also agents of heat absorption. Below 500 metres there are approximately 20,000 particles per cm^3; at 5–6,000 metres there are about 80 particles per cm^3. However, this quantity will vary according to the source of particles – for example, in mountain regions adjacent to heavy pollution sources the average values may be much higher. Local winds, vegetation cover, clouds and relief all play significant roles in modifying lapse rates.

Continentality

The distribution of land and sea in relation to the location of mountains is important. Oceans have the effect of moderating climate. Coastal mountain ranges tend to be wetter and cloudier due to the effect of humid air blowing onshore and being forced up, thus giving rise to precipitation. This creates the differences observed in the Andes between the west-facing coastal slopes and the east-facing inland slopes. On the former, the occurrence of cloud forest demonstrates the importance of oceanic moisture sources. The presence of the oceans also creates a lag time in response to seasonal changes, as water takes longer to heat up and cool down. Temperature ranges therefore tend to be less than for continental interiors. This moderating influence is particularly significant for west coast mountain ranges in mid-latitudes, for example the Andes.

Conversely, continental interiors tend to be more extreme, with greater temperature fluctuations occurring more rapidly, drier conditions, less cloud and consequently higher solar radiation receipts. The upper limits of tree growth (timberlines) tend to be higher on continental interiors, partly as a result of the greater distance from significant sources of moisture. In many arid region mountains such as those of the Middle East, therefore, a lower limit of tree growth occurs as a result of the increase in moisture upslope compared with the more arid lowlands. Examples of this are discussed later. Continentality can be enhanced by the mountain mass effect – where large mountain ranges and high plateaux such as that of Tibet exist, they act as heat islands, generating their own climate. The greater surface area absorbs more radiation and creates convectional storms. Barry (1992), however, points out that the complexity of continental effects is such that generalisations are of little value, and that other factors become more important on a regional or local scale.

Topographic and barrier effects

There are three main aspects for consideration here – the mountain mass effect mentioned above, barrier effects and the importance of relief. First, the mountain mass effect alludes to the significance of the dimensions of the mountain range. Isolated peaks tend to be less intrusive into atmospheric circulation and lack sufficient surface area to absorb enough heat energy to modify the surrounding climate significantly. Temperatures on isolated summits tend to be closer to free air temperatures. Summits are also subject to high wind speeds as frictional drag is reduced.

The case is rather different with large mountain ranges such as the Himalayas and the Tibetan Plateau. Here the large surface area of the extensive range absorbs large amounts of heat energy and acts as a heat island. Maximum heating occurs in June, where a net heating of +1.7°C per day is experienced (Yeh 1982). These higher temperatures counteract some of the radiative cooling, and thus permit agriculture at higher altitudes than might otherwise be possible. This heating creates anticyclonic conditions that cause air to be displaced towards the adjacent plains, and compressional heating which can give rise to thunderstorms. Upper air anticyclonic conditions forming over the Tibetan Plateau also have a much wider impact on regional climate – they are sufficiently strong in Tibet to weaken the westerly jet stream, which is replaced by tropical monsoons (Flohn 1968; Price 1981). The effect of mountain mass also includes an impact on the agricultural potential – for example, rice can be grown at higher elevations in the subtropics than in the tropics (2,500 metres as opposed to 1,500 metres). In the latter case, lower cloud base levels, and the effect of obscuring direct solar radiation, result in lower temperatures in the tropics (Uhlig 1978; Brookfield and Allen 1989).

Large mountain ranges are also important barriers to air circulation. The magnitude of this effect is dependent upon the dimensions of both mountain range and air mass. Isolated peaks are much less effective barriers, as air masses will divide and rejoin around them. The effect of the range is also dependent upon its orientation to the air mass movement – if perpendicular, the air mass is blocked, either in its lower parts, or in its entirety. The blocked air is then usually subject to compressional heating, forced ascent, and results in precipitation. If the range is at an angle to the direction of air movement, the air mass will be deflected rather than blocked.

A classic example of the effect of blocking and deflecting air masses is that of the Himalayan range. Here the cold, continental air to the north is blocked, allowing the subtropical monsoon air masses to penetrate much further north than would otherwise be the case. The monsoon air is deflected to the westwards and its influence and associated precipitation weakens westwards, giving a pattern of decreasing precipitation. Similar blocking of air occurs in Europe where the Alps block cold air from the north-west, giving rise to the cold mistral wind of the Rhône Valley. Compressional heating of air on the windward side and associated precipitation causes accelerated descent of air on the lee side, thus the wind.

Climatic elements

Temperature

This is perhaps the most important environmental variable, and is closely related to solar radiation. Variation in temperature regimes results from aspect, altitude and seasonality. There is a distinctive reduction in temperature with altitude: the lapse rate. A generally accepted figure is 1–2°C/300 metres (6.5°C/1,000 metres; Vincent 1995). Lapse rates are steeper under daytime, summer, clear sky, sunny and continental interior conditions (Price 1981; Barry 1992). All these conditions affect the ability of the air to absorb and retain heat. Lapse rates vary with latitude and season, reaching their highest in tropical deserts during summer. Turbulence, differential heating and global atmospheric circulation also influence lapse rates. Eddy, albedo, topographic and radiation effects are significant on a local scale (see following sections). On regional scales, mountain mass effects are important in modifying temperature regimes.

Inversions are common features of mountain valleys at night. Valley scale circulation patterns of warm and cold air exchanges are complex but rhythmical, following seasonal and diurnal variations. Cooling of air at higher altitudes causes it to sink and flow below the level of the warmer valley air that is displaced upwards. This causes the temperature of the valley floor to be colder than the slopes above, resulting in greater risk of frost on the valley floor than on the adjacent slopes. Farmers exploit this in decisions concerning which crops to plant where – for example, frost susceptible horticultural crops will be planted on slopes rather than on valley floors.

In Obergurgl in the Austrian Alps, inversions gave rise to differences in temperature of 3°C in winter, and 1.5°C in summer between slopes and valley floor (Aulitsky 1967; Baumgartner 1960, cited in Barry 1992). In Bavaria the inversion made a difference of 1–2 weeks in the growing season length. Inversions are best developed in clear, calm conditions. The flow of air may be channelled in gullies, or impeded by features such as walls. Inversions tend to be in the region of 300–600 metres deep. Temperature ranges decline with elevation, as at high altitude the density of air is much less, and surface temperatures remain much closer to free air. Heating is primarily a direct response to sunshine. Intense heating and rapid cooling are characteristic of mountain environments. Lower altitudinal limits of frost depend on latitude, precipitation, continentality and local topography. Sharply defined limits occur in tropical mountains, and frosts are found daily above 4,700 metres in Peru. In the central Andes, night frost is rare below 2,500 metres but above 4,000 metres 80 per cent of days experience frost (Vincent 1995). The number of frost days increases with latitude, but also, of course, so does the increasing seasonality of their occurrence.

Clouds have an important impact on temperature, by blocking the sun's rays. The greater cloud cover of mountains compared with lowlands as a result of uplift and condensation enhances the effect of thinner air and reduced heat retention. This contributes to the rapid fluctuations in temperature experienced in mountain regions. Cloud formation may have a daily pattern; early morning stratus is common, but usually dissipates within an hour of sunrise. Afternoon cumulus tends to affect primarily the lower slopes and may be a daily feature of humid tropical mountains. West coast mid-latitude mountains often experience hill fog, which is discussed later (see p. 44).

Precipitation

There are two main sources of precipitation – convectional and synoptic. The former is a local scale feature involving vertical movement of air as a result of surface heating. Synoptic rainfall is associated with blocked air masses and forced ascent. The type and amount of precipitation depends on the moisture content of the air, the rate of ascent, wind speed, and degree of uplift. Atmospheric water vapour content declines with altitude, but precipitation increases. This arises from the reduced density of air and lower temperatures, thus the air is less able to retain moisture. The actual process of precipitation formation involves the uplift of air, which cools adiabatically with elevation to the dew point where, given sufficient nuclei, the drops form and fall. The air then continues to rise over the mountain crest and descends on the lee side, warming as it goes (see Figure 2.3). This warm, dry air on the lee slope can form strong winds (e.g. the föhn winds) and, because it is dry, gives rise to the characteristic rainshadow effect.

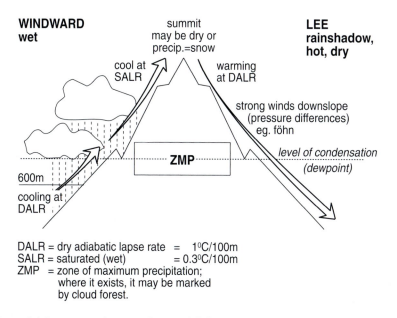

WINDWARD
wet

summit
may be dry or
precip.=snow

LEE
rainshadow,
hot, dry

cool at
SALR

warming
at DALR

strong winds downslope
(pressure differences)
eg. föhn

ZMP

level of condensation
(dewpoint)

600m

cooling at
DALR

DALR = dry adiabatic lapse rate = 1°C/100m
SALR = saturated (wet)   = 0.3°C/100m
ZMP = zone of maximum precipitation;
            where it exists, it may be marked
            by cloud forest.

Figure 2.3 Processes of orographic rainfall formation

The difference in precipitation between windward and lee slopes can be dramatic – the Western Ghats receives more than 5,000 millimetres, but the Deccan Plateau of India on the lee side only 380 millimetres.

Convectional rainfall is generally greatest at the level of the cloud base, especially in the tropics. Topographic controls on precipitation include the orientation of range to an air mass, elevation, relative relief and slope angle. Large deep valleys develop their own localised circulation and precipitation patterns from local convection currents. Not all regions have a clearly identifiable zone of maximum precipitation, as local effects can be highly significant. Precipitation is generally higher near sea level in polar and equatorial regions. In mid-latitudes there is a tendency for increased precipitation with altitude, but in the subtropics there is no clear relationship due to local effects (Lauscher 1976a). However, local and regional conditions are significant in modifying these general trends. For example, on Mount Cameroon, West Africa, the southern slopes have a maximum at low altitude because of the monsoon source, but the north-eastern slopes have a maximum at 1,500 metres where trade wind sources predominate.

The type of precipitation – rain or snow – is determined by a threshold temperature. This varies considerably – in lowland Britain it may be 250 metres while in Central Asia it increases from 1°C at 500 metres to 4°C at 3,500–4,000 metres. The proportion of precipitation falling as snow increases with altitude. The altitudinal limits of snowfall may remain static, as in the aseasonal tropics, or may change seasonally as in the mid- and high latitudes. In the eastern Alps 65 per cent of precipitation falls as snow above 3,000 metres even in the summer, but this falls to about 15 per cent at 2,000 metres (Lauscher 1976b). Snow depth and duration of cover may be highly influenced by topography as well as snowfall patterns, and reworking by drifting or avalanche also occurs. However, variability in snow depth and coverage decreases with elevation, reflecting longer accumulation periods, for example in the San Juan Mountains, Colorado (Caine 1975).

Other forms of precipitation include condensation – dew, fog, frost and rime. Fog may be either a low or high hill fog caused by condensation at the base of cloud level. Fog drip is the condensation of water as moisture-laden air passes through obstacles, such as vegetation. In some areas, fog drip forms a significant proportion of the total precipitation – in eastern Nepal it adds 22 per cent to annual totals at 3,100 metres (Kraus 1967), and in Austria 100–150 per cent of winter precipitation values have been recorded (Grunow and Tollner 1969). A particular example of the importance of fog drip is that it is closely associated with montane cloud forests. For instance, these can be found on the western coastal ranges of the Colombian Andes between 600 and 750 metres, and depend for all their moisture upon fog drip. Removal of the forests also removes the mechanism by which the moisture condenses, resulting in increased aridity of the area. Rime is supercooled fog, and forms as ice crystals, needles or glaze over surfaces. Significant accumulations may develop, particularly on the lee side of obstacles. This may cause damage to structures such as weather stations, trees, cable car apparatus and the like.

Orographic precipitation is also associated with the deposition of pollutants. Where industrial activity in adjacent lowlands produces airborne particles, these drift up into the valleys, and act as condensation nuclei that are washed out of the air and deposited into the valley. This is evident from studies in the mountains of central Europe, such as the Tatra, which are adjacent to the industrial complexes of Poland. These demonstrate strong correlations between orographic processes and wet deposition of elements such as lead (Konèek et al. 1973), giving rise to increases in the lead contents of soils and consequently in herbivorous birds and chamois (Janiga 1999). The formation of summit or cap clouds and resultant precipitation has also been shown to increase the deposition of pollutants in high-altitude environments of the Sudety Mountains, between Poland and the Czech Republic (Dore et al. 1999). Defoliation of trees was concentrated in spatially and altitudinally defined areas corresponding to the dominant wind and precipitation patterns bringing airborne pollutants from adjacent industrial centres and their deposition in zones of maximum rainfall and cloud formation.

Solar radiation

The thinner air at higher elevations causes rapid fluctuations in the response of temperature to changes in solar radiation receipts. Two issues are important with respect to mountain environments – the quantity, and the quality of radiation receipts. These are affected by cloudiness, aspect and topography. Snow cover increases the surface albedo, and therefore reduces the absorption of energy, and so snow cover duration affects total radiation receipts. These increase with altitude possibly at about 7–10 per cent per 1,000 metres under clear skies in the Alps according to Müller (1985). Moreover, radiation receipts are highly variable diurnally, especially in middle latitudes. It is also a function of aspect, cloudiness and surface albedo so that the presence of cloud usually indicates a fall in surface radiation receipts (Barry 1992).

The quality of radiation also changes with altitude. There is an increase particularly in the UV component. This retards plant growth, and of course is recognised as a cause of cancer in humans. Plants tend to be specially adapted to survive the harsh environments of alpine zones, and thick leaf cuticles and bark help to reduce UV penetration.

Differential heating gives rise to complex local circulation patterns, which are discussed below. In the mid-latitudes, aspect (orientation), slope angle and season combine to create high diversity in soil moisture conditions and temperatures. In the northern hemisphere, south-facing slopes receive more direct solar radiation, and this is reflected in the distribution of settlements in many alpine valleys. Garnett (1935) mapped the distribution of settlements in relation to the shadow at noon in midwinter. The majority of these settlements are located outside the shadow zone. Differences in the location of agricultural activities may also reflect aspect. In the Hunza valley, Pakistan, the south-facing Hunzakuts enjoy much more favourable agricultural conditions than north-facing Nagar. In Nagar, however, higher soil moisture conditions and reduced sunshine permit more forest growth than the more arid Hunza slopes. East- and west-facing slopes are also subject to differential heating. East-facing slopes receive the morning sun, but this must dry out any dew or clear stratus cloud and fog before heating of the soil can take place. West-facing slopes will have had time to dry out before direct sunshine reaches them, but in some areas local afternoon convectional activity causes cumulus clouds to form, obscuring the sun's rays. Differences in radiation receipts can lengthen or shorten the effective growing season by several days a year (Plate 2.1). In marginal conditions, every day counts, and thus locations where topographic shading occurs may be avoided, or used for less critical crops.

Plate 2.1 The impact of aspect on snow melt, Reraya valley, Morocco

Evaporation and water balance

There is a complex relationship between evaporation, transpiration and precipitation. Evaporation data is lacking for high mountain regions but is generally quite low, despite the high wind speeds and the aridity of many high summits. Although precipitation might increase with altitude, relative humidity decreases. This is in part a reflection of the lower density of air, but the lower water vapour content has been demonstrated for many mountains. Rapid cooling of air and surfaces causes an increase in relative humidity, and therefore less evaporation but more condensation, and hence fog. Local wind effects cause turbulence and mixing and give rise to aridity on many summits.

Winds

Winds occur at different scales, ranging from the disturbance of upper air circulation due to peaks penetrating these levels, through regional scale effects of winds such as bora, föhn and mistral, to local valley circulation. Upper air circulation may not be dammed or blocked as synoptic air masses often are, but wave effects similar to those of a wake in water can arise due to frictional drag over mountain regions. Additionally, local thermals may produce clear air turbulence, well known to pilots and air travellers.

Winds are either synoptic or thermally induced. The former operate on global and continental scales, whilst the latter are more significant at the local, valley scale. Synoptic winds related to pressure gradients arise due to the blocking or deflecting effect of mountain ranges. When air is piled up on one side of the range, compressional heating and uplift occur, together with precipitation. The air warms rapidly down the lee slope. This mountain barrier effect gives rise to temperature and pressure differences on different sides of mountain ranges (Barry 1992) – for example, the föhn wind. In the European Alps, a southern föhn affecting the north side of the Alps is induced by high pressure in the Mediterranean. In addition, there is a northern föhn blowing off the southern slopes of the Alps which arises from high pressure in northern Europe. The former case is much stronger and more significant, as the air is considerably warmer than the air of the region through which it passes. The air is hot, dry, dusty and associated with much folklore. The northern föhn is much cooler and less meteorologically significant. The same conditions occur in the Rockies, causing a hot dry wind, known as the Chinook, to pass over the Great Plains. The cold mistral arises as a result of pressure differences between highs in north and west France and lows in the Mediterranean. The resultant wind is channelled between the Massif Central and western Alps, and has reached speeds of 145 km/hour in the Rhône Valley, where some trees have the flagged or one-sided appearance characteristic of vegetation subjected frequently to such winds.

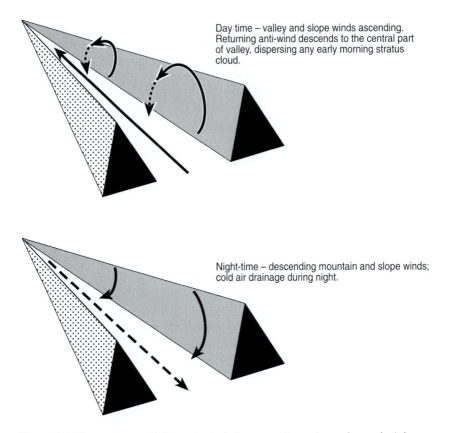

Day time – valley and slope winds ascending. Returning anti-wind descends to the central part of valley, dispersing any early morning stratus cloud.

Night-time – descending mountain and slope winds; cold air drainage during night.

Figure 2.4 The structure of thermal winds in mountain regions: day and night-time effects

Thermally induced winds comprise slope winds, mountain and valley winds, and other more locally modified winds such as glacier winds. These tend to operate simultaneously, giving a complex pattern of circulation in three dimensions. Most thermally induced winds operate on a diurnal basis, tending to reverse their daytime direction at night. Slope winds move up and down the valley sides as a result of heating of the valley floor (see Figure 2.4). They are strongest in the daytime; at night they tend to take the form of a downward drainage of cold air rather than a more energetic uplift. They tend to be strongest where heating of the surface is greatest. During the day the location of the strongest winds may shift as a result of the effective heating of the surface and change in direction of incident rays. South-facing slopes in the northern hemisphere might be expected to have the strongest winds, but this pattern may be modified by cloud patterns, moisture content of the ground (and therefore effective heating), and other local wind

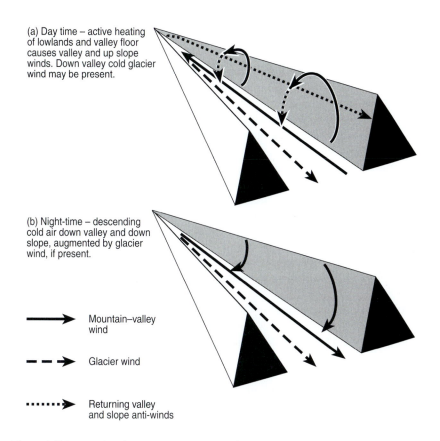

(a) Day time – active heating of lowlands and valley floor causes valley and up slope winds. Down valley cold glacier wind may be present.

(b) Night-time – descending cold air down valley and down slope, augmented by glacier wind, if present.

→ Mountain–valley wind

– – ➤ Glacier wind

⋯⋯➤ Returning valley and slope anti-winds

Figure 2.5 Interaction between mountain–valley, slope and glacier winds

patterns. All circulation patterns have a compensatory anti-wind. For slope winds this takes the form of descending air along the axis of the valley floor (see Figure 2.4). This descending air dissipates early morning stratus or fog from the centre outwards.

A larger-scale thermal wind is the mountain–valley wind circulation. During the day, differential thermal heating of lowland and upland causes a wind to blow up valley, beginning shortly after sunrise (see Figure 2.5a). During the night, cooler air from the mountain flows down valley towards the plains (see Figure 2.5b). The wind is strongest in narrow valleys during summer when temperature differences are more significant. The geometry of the valley affects not only wind speed but also heating – V-shaped valleys are heated less than more open U-shaped valleys. As with slope winds, there is a compensating anti-wind above which blows along the axis of the valley.

The combination of slope and mountain–valley winds creates a general daytime movement up slopes and valley floor, with anti-winds blowing down valley and towards the centre. Strengths of each component vary with intensity of solar radiation, valley topography and the influence of other larger-scale wind patterns. This pattern is reversed during the night, but the air tends to drain down slope and down valley by gravity, rather than by more active processes. Detailed relief and structural features such as walls, terraces and water channels may direct the flow. Local heating effects and cloud cover will influence the speed of uplift on different valley slopes, and thus compensating subsidence.

The combined effects of aspect and slope angle therefore significantly modify climate on a local scale, giving rise to the mosaic of microenvironments mentioned above. Freytag (1987) identified the pattern of circulation in the Inn Valley in the Swiss Alps which comprises the following:

0500–1100   UTC upslope begins, continuing mountain wind down valley
1100–1500   UTC upslope and valley winds, with compensating subsidence (Figure 2.5a)
1500–2100   UTC continuing valley and down slope winds; less subsidence needed
2100–0500   UTC down slope and mountain wind (Figure 2.5b)

These complex patterns of diurnal circulation are important with respect to agricultural activity – cold air drainage increases the risk of chilling and frost on valley slopes. The problems of air pollution and the dissipation of plumes of smoke and particles are also highly dependent on local winds. The pattern of valley circulation can cause recycling of pollutants for long periods of time in enclosed valleys, as well as concentration of heavier particles in sheltered pockets and under inversions (Hanna and Strimaitis 1990).

Turner (1980), in a study of alpine valleys, identified four main types of irradiation and wind conditions in relation to relief, as shown in Table 2.1. Superimposed on these thermally induced winds are other winds such as the glacier wind. This is a constant drainage of cold air from ice. It is strongest during the day when temperature differences are greater. It flows down valley, and will tend to flow underneath the rising, warmer valley wind. Glacier winds cause severe ecological stresses, as they effectively reduce heating of the ground surface and create a constantly cold air environment. Vegetation may be stunted and growth severely retarded in such cases. Seasonal cool winds blowing off snow patches may also occur, but these have often cleared from seasonal snow as it melts and is therefore less ecologically devastating during the growing season (Hoinkes 1954).

Eddies and deflection by obstacles such as forest and other vegetation, walls, rock outcrops and surface features cause local variations in temperature at the ground surface. This is important in relation to duration of snow

Table 2.1 Aspect and radiation in mountains

Slope	Irradiation	Wind speed
Sunny windward	High	High
Sunny lee	High	High
Shaded windward	Low	High
Shaded lee	Low	Low

Source: After Turner (1981)

cover in sheltered, shady hollows, soil moisture, etc. Local air flow and drainage are diverted and channelled by small topographic features, causing ponding of stagnant air in small hollows, affecting plant growth. Surface features may also cause backscattering of incident radiation and the release of absorbed heat, which can have significant effects on growing conditions.

The influence of relief is particularly significant at the valley scale. Valleys can develop their own microclimates as a result of sheltering from larger-scale atmospheric conditions, as well as by internal relief effects. Turbulence arises as a result of frictional drag – both from surface details and from ridge–valley configuration. Abrupt breaks of slope give rise to greater turbulence than gentle ones. Slope angle and aspect play an important part in differential solar radiation receipts, for given altitudinal and latitudinal locations. Valleys perpendicular to prevailing winds, protected by a ridge, may be more fully sheltered than those perpendicular to the prevailing winds.

GEOMORPHOLOGY

On geological timescales mountains could be considered ephemeral features (Powell 1876), constructed by tectonic processes and gradually destroyed by subaerial denudation and erosion. Most of the major mountain ranges are associated with plate boundaries, and are uplifted as a result of plate collision. Volcanic and seismic activity is also closely connected with mountain ranges. The process of mountain building arises from the movement of the earth's crustal plates. The convergence of two or more plates causes buckling up of the intervening sediments and upthrusting to form up-standing relief. The process of formation provides the basis of some classifications of mountain systems (Fookes *et al.* 1985), and the geological structure and lithological influence on landscape form is another basis of classification (Fairbridge 1968).

A good example is the Himalayan arc which stretches from Indonesia through south-east, south and central Asia into central and western Europe (Gerrard 1990). This was formed primarily around 30 million years ago as a result of the subduction of the northern-bound Indian plate below the

Eurasian plate. The intervening marine sediments were buckled up in a series of waves, with the highest peaks in the centre, successively lower ranges to the south forming the lesser and sub-Himalayan foothills, whilst to the north the great Tibetan Plateau slopes gently down northwards. The formation of the Himalayas is part of wider orogenic activity encompassing the Zagros mountains of Iran, the Taurus of Turkey, the European Alps and North African Atlas range. A similar process formed the Andes, which is part of the Pacific 'Ring of Fire', and formed from the subduction of the Nazca oceanic plate beneath the South American plate, beginning about 190 million years ago. The European Alpine system was formed from a series of complex movements of the European and African plates over 200 million years, with uplift at rates of 1–2mm/year (Selby 1985).

The accompanying volcanic activity has formed, for example in the Himalayas, hard resistant cores beneath the softer marine sediments, and once the latter are stripped off by erosion it is these hard igneous masses which form the highest peaks. The eroded sediments are deposited on adjacent lowlands – in the case of the Himalayas some 5,000 metres or more of sediments have been deposited to form the Ganges and Indus plains. Uplift is a continuous process, accounting for the frequency of earthquakes in mountainous areas. Block faulting, rather than plate collision, forms other mountain ranges, such as the Vosges and Black Forest Ranges of the Rhine valley and the Sierra Nevada.

The landscape of mountains, consisting of dissected and differentially eroded surfaces with abundant steep slopes and high absolute and relative relief, creates an environment of high energy that gives rise to high rates of erosion. Most mountain systems are in disequilibrium with their current environmental conditions and many regions are still adjusting to the legacy of the Quaternary climate change. This means that there is an abundance of loose sediment in the system that is gradually reworked, and an inherent disequilibrium between current climate and inherited landforms.

Mountain environments are characterised by a preponderance of high-magnitude, low-frequency events that punctuate periods of apparent stability. Such events, including major landslides, earthquakes and floods, have critical roles in the production of loose sediment, and in transporting this material, and reshaping the geometry of mountain valleys. Valley geomorphological activity therefore can take the form of 'punctuated equilibrium', and due to elevation could be said to remain out of equilibrium with environmental controls.

There have been a number of classifications of the geomorphological activity in mountain environments. Barsch and Caine (1984) discuss the classic geomorphological division between form and process, whilst Brunsden and Allison (1986) describe the different processes which predominate at different altitudes. Other workers have classified activity on the basis of formations (Fookes et al. 1985), or more fundamentally on the basis of geological structure and lithological influence (Fairbridge 1968). In the literature a dichotomy exists between processes and rates, but in reality they interact in the process of landscape formation and alteration. Geological structure and rock type, together with climate, are the principal controls over process and rate, which both vary considerably in time and space in their magnitude.

Sediment is produced within the valley system, and follows a pattern of 'temporary' storage whilst being removed from the system and deposited onto the adjacent lowlands. Movement of sediment, for example, may be continual in aseasonal climates, but highly temporally constrained in climates where fluvial activity is limited seasonally, especially when augmented by meltwaters. Major floods can move more sediment than normal annual flows. In many valleys where active downcutting occurs as a result of uplift, as is the case in the Himalayan range, rivers have cut deep into sediment accumulations, the upper surfaces of which are occupied by settlement and cultivation. These cliffs can be subject to undercutting during times of peak flow.

The main processes operating consist of frost action, glaciers, fluvial and mass movement. Weathering is important, particularly in humid tropical environments with high temperatures and abundant water, and aeolian processes tend to be restricted to summits and arid zones where vegetation cover is restricted.

Processes

There is no single classification of geomorphological processes in mountains: Brunsden and Allison (1986) distinguish processes spatially (valley floor, slopes and foothills, intermediate slopes and high summits), whereas Barsch and Caine (1984) distinguish various sediment systems (glacial, coarse, fine and solute systems). Coarse debris is estimated to account for 10–60 per cent of the 'work' of the systems, whilst solutes comprise a relatively minor component (10–12 per cent). This reflects the importance of physical transport of abundant material compared with the more minor role of chemical weathering. According to Brunsden and Allison, high summits are dominated by freeze–thaw, periglacial and glacial activity; intermediate slopes by erosion and deposition processes of glaciers and rivers as well as mass movement processes. By contrast, the foothills of many mountains are substantially altered by human activity such as terrace agriculture, and are often affected by gullying and sheet erosion. These processes, however, are not exclusive to any one zone. The four main agents of landscape change – glaciers, rivers, mass movement and weathering – arise from complex interactions between geology, climate and land use, and are highly variable in time and space.

Glaciers

Glaciers are abundant on many mountain ranges, and have fluctuated over time, reaching their last maxima during the Quaternary period. The last glacial maximum ended around 14,000 years ago, but since that time there have been fluctuations, such as a general expansion during the Little Ice Age. During the last century, or since more detailed records have begun (over the last 50 years or so), glaciers have generally been shown to be in recession.

Glacial activity is one of the most powerful forces on the landscape. Erosion produces characteristic features such as U-shaped valleys, sharp ridges, hanging valleys and cirques. Deposition of unsorted glacial till as terminal and lateral moraines, and the deposition of stratified fluvio-glacial material, gives glaciated valleys a somewhat 'untidy' appearance. Many mountain regions are still adjusting to the effects of deglaciation, by isostatic rebound and in the reworking of the vast quantities of material deposited in the lower reaches of their valleys. In the European Alps, glaciers tend to be of the cirque and valley types, being topographically constrained. On the isolated volcanic peaks of central Africa the glaciers are small, but are sensitive indicators of climate change (Rostom and Hastenrath 1994). In the Karakorum, however, despite the aridity of the lower slopes, the glacial coverage is very extensive and provides important resources for agriculture.

The activity of glaciers depends not only on topography but also on the accumulation of snow over ablation. In the mid- and high latitudes, accumulation and ablation have one season each year, but in the tropics there may be a bimodal distribution for each. Rapid accumulation facilitates movement and hence erosive activity. Erosion occurs by plucking and grinding or scouring. The production of material is supplemented by freeze–thaw and solifluction activity on the slopes above.

Fluvial activity

Fluvial activity is closely related to precipitation and to snow and ice melt, and is thus variable in time and space. Erosion and transport of sediments are therefore punctuated by periods of relative stability on a diurnal and/or seasonal basis. For example in Hunza, Pakistan, flow rates vary seasonally with a peak in spring from snow melt and a larger peak in late summer from glacial melt. Flow varies diurnally also, peaking in later afternoon where the source of water is primarily meltwater. In addition, mountain streams respond quickly to rainfall events, and flash floods arising from heavy rainfall (unusual or seasonal such as the monsoon) are common.

Movement of sediment in river systems, and erosive activity, is therefore highly variable (Griffiths 1981). Braided channels and reworking of glacial, alluvial and mass movement debris are common. Continued uplift of mountain systems means that the energy for downcutting of valleys is sustained – in the upper Indus, the river is currently downcutting through sediments deposited previously, producing terraces that have become foci of human occupation. The interaction of land use, vegetation cover and fluvial activity is subject to much debate and research within the context of mountain degradation (see Chapter 8). In tropical regions, seasonal variations are negligible, and flow much more constant, but still susceptible to unusual precipitation events.

Mass movement processes

Mass movement processes are perhaps the most important in terms of amount of sediment produced, and their effects on landscape and on human activity. The hazards associated with landslides and avalanches are discussed below. Mass movement may take the form of creep, evidenced by bent tree trunks and fence lines, or the more devastating, larger-scale slides, slumps and mudflows. Mass movements are commonly triggered by earthquake activity or by the cutting of slopes for road construction. The relationship between increased landslide activity and deforestation is complex, and is discussed in the context of Himalayan Degradation Theory in Chapter 8.

Key influences on landslide activity include the dip of strata, rock type and tectonic activity. Climatic variables, particularly precipitation, are important as further rain on saturated soils, as in storms or monsoon rainfall, is a common cause of slope failure. Land use, types of terracing and vegetation cover are also important. Landslides of various forms are a fact of life in many mountain regions. Many are relatively small, but unusually heavy rain can cause severe disruption in communications – one period of such rain in Pakistan in September 1997 caused eleven major slides to block the Karakorum highway between Karimabad and Gilgit. Although it was unusual to have rain in September, such blockages are a common feature during the winter. Other mass movements are more devastating – the Huascaran landslide in Peru following the 1970 earthquake ran out for 16 kilometres and buried two towns and an estimated 18,000 people (Patzelt 1983).

Weathering processes

Weathering processes in mountain regions, whilst not the most significant in terms of volume of sediment produced and transported, are locally significant. In the high summit areas, physical breakdown by freeze–thaw action predominates, and deflation which removes fine material. Insolation weathering may be important, particularly with granitic rocks, which may suffer granular disintegration or spalling. In the High Atlas of Morocco, high temperature variations are thought to contribute to the formation of desert varnish on sandstone. The varnish, which comprises accumulations of manganese and ferric oxides, is thought to take centuries to form but provides a protective surface layer on the weak sandstone (Robinson and Williams 1992). Nivation hollows are sometimes foci of chemical weathering, as the snow insulates from extreme temperature variations and also allows water to reside in contact with the rock surface for longer. Rates of weathering are relatively under studied in mountains, but the process is more significant in limestone areas than igneous areas, although in the tropics deep weathering of granite is also a result of chemical processes. In the tropics, however, warmer temperatures in the lower regions and lack of significant seasonal variation permit more constant rates of chemical breakdown.

Rates of change

Actual rates of different processes are difficult to obtain. Poor accessibility, the high spatial and temporal variability, and the difficulty in measuring certain variables contribute to this. General rates of uplift of mountain ranges have been estimated by dating rock surfaces. The European Alps are estimated to have been uplifted at rates of 300–600 millimetres per thousand years, and the Bolivian Andes and Himalayas at about 700 millimetres per thousand years. Areas experiencing active plate collision have the fastest rates (Goudie 1995).

Surface erosion rates of the Indus–Kosi valley systems are estimated as 1 mm/yr (Hewitt 1967). However, these mean rates mask the high variability in time and space; for example, storms in Darjeeling in north India give rise to unusually high spatially and temporally constrained rates of 10–20 mm/yr, compared with a mean rate of 0.5–5 mm/yr (Starkel 1972). The Alps are estimated to erode at about 0.6 mm/yr (Clarke and Jager 1969).

The key controls on process rate are rates of uplift and degree of deformation. Deformation, including folding and faulting, introduces zones or planes of weakness which are exploited by erosion processes. The nature of the rock surface and rock type will dictate how the rock will respond to weathering processes – granites are mechanically strong but susceptible to chemical weathering, fine-grained igneous rocks such as basalt are very resistant. Sedimentary rocks such as sandstones tend to disintegrate rapidly.

The other main control on weathering and erosion is climate. Many processes require the presence of water, and temperature fluctuations are the main agent of weathering on high summits. Climate not only influences the rate but also the nature of the processes themselves. In addition, climate will influence soil formation and the nature of the vegetation cover that may enhance or retard weathering and erosion processes.

Finally, the influence of land use on erosion rates has been much studied, particularly in the context of mountain degradation (see Part 4). Disturbance causing changes to the natural vegetation cover, such as through the use of forests and grasslands, or by cultivation generally, increases the vulnerability of surfaces to erosion. The intensity and type of land use and its timing in relation to seasonal wet or dry periods are key factors. Mountain farmers have developed a range of strategies to cope with the vulnerability of the surface to erosion and disturbance, and these are discussed in Chapter 4.

MOUNTAIN SOILS

The nature of mountain soils is highly variable, reflecting microclimatic variability. Soils on summits develop from sediment collected in pockets and on slopes, anchored by vegetation and protected from high wind erosion. They tend to be thin, stony and often low in nutrients and in need of improvement. On slopes, soils are a mantle of weathered scree and debris washed down from above and incorporate an aeolian component which produces a fine texture. Finer, deeper material accumulates in alluvial fans and in valley bottoms with a better developed A-horizon. Glacial tills provide the basis for many soils. Soils on slopes containing high scree elements tend to be free draining, increasing their stability, but in some areas, such as valley floors and where clayey tills occur and where fines accumulate, waterlogging may be a problem. The presence of rock flour and silt ensures that some soils are fertile, but they generally require considerable improvement to increase their productivity.

In high alpine meadows, the development of dense meadow turf protects the surface as long as it remains intact. Excessive trampling, or other surface disturbance, can quickly lead to deep gullies once the turf mantle is broken. In addition, the covering of snow during part of the year, and its slow melting, often helps to protect it. In the valleys, steep slopes can retain their soils by vegetation cover, or by artificial means. The underlying scree on many slopes can make the mantle highly susceptible to erosion and relatively minor disturbance can upset the precarious balance of the mantle on the slope.

Mountain soils are generally poorly developed compared with lowland soils, but under certain conditions, such as arid zone mountains, they may actually be more fertile and productive. Soils follow a 'climosequence' in mountains, responding to falling temperature, increasing precipitation and wind speeds with altitude. As with other aspects of mountain environments, great changes can occur over short distances. The increase in precipitation causes greater leaching, and falling temperatures restrict the biological decomposition of organic matter. Soils, therefore, have a tendency to become increasingly acid with altitude, and litter becomes more important than organic matter incorporated within the soil. Bedrock and surface deposits significantly influence the nature of soils, however; in the Alps there is a distinct difference between soils developed on limestone in the south and those on igneous rocks in the north. In the same way, soils developed on glacial deposits will vary depending on whether they are developed on moraines, with larger clasts and mixed fabric, or on tills with a higher fine (clay) content. Soils on alluvial fans tend to be freer draining, whereas valley bottom soils may be subject to regular waterlogging. Soils formed on lava and volcanic ash tend to be very rich and fertile, and often free draining.

Aspect can also affect soil development – soils tend to be deeper on south-facing slopes in the northern hemisphere where greater insolation results in deeper weathering and more active micro-organism activity. However in hotter, arid areas, such as the subtropical mountains, soils may in fact be more productive on shaded slopes where they are less desiccated and subject to smaller temperature fluctuations. In the tropics, mountain soils may improve as a result of better drainage at higher altitudes, up to a point where low temperatures become a limiting factor for soil development. Windward and lee slopes may also differ. This is partly a result of the close feedback mechanism between vegetation and soils. This is also apparent between coniferous forests, which promote more acid soils, and broadleaf forests where more nutrient rich soils predominate.

Seasonality, particularly the occurrence of nightly frosts at high altitudes in ranges such as the northern Andes, can have significant effects on the activity of soil organisms and nutrient cycling. Persistent soil moisture also promotes leaching of many elements, including iron, aluminium and silica under different conditions.

The classification of mountain soils can be broadly aligned to vegetation zones, although great variation occurs locally. The formation of soils is dependent on climatic conditions and substrate. At the highest altitudes the slow rate of weathering – primarily by physical breakdown – and the low temperatures restrict chemical and biological processes, and the high wind speeds contribute to deflation and desiccation. Soils therefore tend to be poorly developed, and concentrated in tiny pockets and crevices. Studies of deglaciated surfaces and moraines demonstrate the succession from lichens to mosses and alpines as soils form alongside vegetation colonies (Miehe et al. 1996). High-altitude soils reflect more closely the parent material, but deeper soils, especially on tropical mountains, may be more influenced by subaerial processes and deposition than by substrate.

On upper mountain slopes, and coinciding with grasslands, fertile turfs with good organic A-horizons and waterlogged, mottled and gleyed B- and C-horizons dominate, with pockets of boggy, peaty soils in damper areas such as around spring seepages. These soils are protected by a thick turf mat, which enhances stability unless disturbed. This helps maintain a soil mantle on steep slopes, and retains soil moisture more effectively by preventing evaporation from the soil surface.

Forest soils tend to reflect the characteristics of the trees growing on them and the lithology of the substrate. On igneous or metamorphic rocks, soils tend to be more acid podsols and are associated with coniferous species that tolerate and sustain acidic conditions, whereas broad-leaved trees prefer, and sustain, more base-rich brown forest soils. Under coniferous forests soils tend to be nutrient poor, with a thick litter layer and subject to considerable leaching. Most of the nutrients are retained within the trees. Broad-leaved deciduous trees return much higher nutrient levels to the soil, and after clearance these soils are therefore more productive. They tend to be deeper, less leached, and much richer in organic matter.

FAUNA AND FLORA

Geoecology

This section seeks to outline the main features of changes in vegetation with climate-related factors and to consider the adaptations of forests and meadows to their environment. The emphasis here is primarily on flora rather than fauna, although the latter is a critical component of conservation strategies discussed later in this book.

A key element for consideration is the zonation of mountain flora. We have already seen the broad altitudinal zones that can be applied to climate and geomorphological environments. It was noted in Chapter 1 that the study of altitudinal zonation, originally by von Humboldt and Bonpland (1807) and continued by Troll (1971), established the concept of geoecology. The broad transition from forest to grassland to snow is in fact highly variable and diverse in form, composition and continuity. There is some debate in the literature over the degree to which such zones are in fact realistic. Huggett (1995) describes the identification of zones as the coincidence of the limits of range tolerances, and argues that in fact in many areas, the vegetation composition changes much more gradually, forming a continuum, rather than distinct zones. Where zones occur, they may be the function of recent disturbances and human activity, or as a result of the nature of underlying soils and drainage patterns. Microclimate and habitat niches play an important part in the survival of many species. The interdependence of vegetation and soils may sustain patches of distinct vegetation. The complex superimposition of different effects and influences from climate, relief, soils and vegetation and fauna operate within a dynamic timeframe (Troll 1971; O'Connor 1984). The application of strict zones may only represent the current status, and suggests a more rigid nature of mountain geoecology than is really the case. That said, however, some ecosystems, such as the cloud forest and forests associated with fog-drip water sources, are highly spatially constrained and closely allied to cloud and precipitation conditions at specific altitudes, producing very clear zones. Some examples are given at the end of the chapter.

The distribution of mountains, either in contiguous chains running gener-ally north–south or east–west, or as isolated peaks (such as the East African volcanic peaks) will determine the source of the flora, and the potential for it to migrate during times of environmental change. The biogeographical concept of island ecosystems has frequently been applied to mountains, as their vegetation tends to differ substantially in both affinity and form from the adjacent lowlands (Cox and Moore 1985). Specialisation and adapta-tion to the environmental conditions of high mountains has produced distinctive floras and associations with high rates of endemism. Affinities between adjacent lowlands are greater in the subtropics where other sources of flora were too distant to reach isolated peaks (Jeník 1997). In addition, the fewer glaciers and smaller distances permitted more widespread move-ment of floral and faunal species. The orientation of a mountain chain, which follows the directions of critical environmental factors, can allow free move-ment of species in response to climatic changes. By contrast, mountain chains that are aligned across environmental conditions form distinctive barriers to migration. Such differences are manifest in the different evolutionary events between east–west oriented European ranges and north–south aligned American ones (Jeník 1997).

Most of the major mountain ranges (Andes, European Alps, and Himalayas) are the result of tertiary uplift during the last 65 million years. They have been subject to considerable climatic change during the Quaternary, when the Alps were largely overrun with ice. The Himalayas, due to much greater aridity, escaped comprehensive ice cover and thus served as a refugium for many floral elements, whereas a complete flora had to recolonise the initially barren landscapes of the Alps. In each mountain range, therefore, the flora is made up of a combination of local speciation, migrants from near and far and more recently introduced cultivars. This produces a mixture of old endemic relicts, new endemics and more widely occurring species. Therefore mountains are an important source of biodi-versity and thus a focus of conservation (see Chapter 9).

Colonisation of mountain surfaces occurs by succession. Studies of the colonisation of volcanic surfaces such as Krakatoa, and of recently de-glaciated surfaces and moraines which can be dated, have given indications of the pattern of colonisation – for example, the Langtang Himal (Miehe, 1990). Initially lichens appear, and then mosses and grasses, alongside edaphic development. Rates of soil formation, snow cover, and seasonal or diurnal climatic conditions can slow the rate of succession, and in addition the sources of plants may be distant and slow to migrate. Eventually grass-lands will be succeeded by forests if conditions permit. The concept of a 'climax' community is disputed, and human activity in many mountain regions, which often controls the maintenance of grasslands and affects the regeneration of forests, or soil development, may result in an 'anthropogenic' of edaphic climax rather than a climatic controlled one. This is an important factor to consider in the planning of conservation measures which may change the patterns of human activity, and thus the nature of the ecological communities which are the subject of conservation.

The main components of mountain vegetation are forests, alpine meadows and tundra. Forests and meadows have significant economic and domestic value, including, more recently, their importance for tourism. The transition from forest to grassland (timberline) may be either diffuse or highly distinct. Above the meadows is the lower limit of permanent or seasonal snow, which is usually closely controlled by climatic and relief factors. The presence of steep slopes and frequent surface disturbance by volcanic, tectonic or mass movement activity, including landslides and avalanches, creates a relatively unstable condition for ecosystem development, and patches of recently disturbed surfaces will carry a different flora and associated fauna than adjacent undisturbed areas.

Forest types and forms

There are three basic types of mountain forest: evergreen conifer (*Pinus*, *Picea abies*, *Larix*, Juniper), evergreen broadleaf (Magnolia and Banyan) and broadleaf deciduous (*Quercus*, *Acer*, *Fagus*, *Castarea*, *Fraxinus*, *Carpinus*, *Betula*), although great heterogeneity occurs within many regions, particularly in the tropics (Hamilton *et al.* 1997). Many mountain ranges have their own distinctive species and communities which vary with altitude and local conditions – aspect and topography for example. As altitude increases, ecological complexity reduces, as does the density of forest stands. In the Himalayas, rhododendron is an important component, which also occurs in the Middle East and Hengduan mountains, China. In the southern hemisphere, the species are different – *Nothofagus*, the southern beech, and *Araucaria*, the monkey-puzzle tree, predominate. Endemic faunas are often closely associated with tree species – for example, mountain gorillas in Zaïre, Uganda and Rwanda, and the red panda in the eastern Himalayas.

In middle and high latitudes, and temperate and high altitudes of tropical mountains, coniferous forest comprising predominantly pine, spruce and fir occurs throughout the northern hemisphere. These needle-leaved conifers are well adapted to temperature fluctuations, short growing seasons, deep snow, poor nutrients and high winds (Hamilton *et al.* 1997). The tapering shape sheds snow effectively and retains flexibility to withstand high winds. The sloping branches also promote reproduction by layering, where tender young root and shoot stock can be protected under the insulating snow during the winter. This makes up for the generally poor viability of seeds and somewhat haphazard dispersal potential. The evergreen needles have thick cuticles that not only reduce desiccation and protect from cold but also are ready to photosynthesise as soon as temperatures and solar radiation conditions permit. This is important in regions where the growing season is too short to permit maturation of buds before food reserves need to be accumulated. In addition the problem of late frosts or cold spring winds kills many early buds. Conifer trees are also less nutrient 'hungry', and survive on poor, acid soils. Dense conifer forest effectively shades out any competitors, preventing the establishment of deciduous forests even if conditions permitted.

Broadleaf deciduous trees tend to occur in lower altitudinal belts than the conifers in the same regions as above. However, species tend to be more diverse. Birch tends to be an early colonising species to newly cleared ground, and occurs in the highest forest zones in the Himalayas. At the timberline, dwarf birch woodland may occur. Oak, chestnut, maple, hornbeam and beech may all occur, either as single species stands at timberlines or more commonly as mixed stands. Some species have evergreen varieties that occur at higher altitudes – for example *Quercus ilex*, the evergreen oak in Morocco and the Mediterranean.

These trees occur in lower altitudes where growing seasons are long enough to allow the development of buds and growth of leaves. These trees also reproduce by insect and wind pollination as warmer conditions favour insect populations, and the more abundant soils ensure that wind pollination is less of a 'hit and miss' affair than it is in the patchy soils of higher altitudes. Warmer conditions also ensure more effective breakdown of organic matter by micro-organisms, maintaining higher soil fertility and better developed horizons.

The form of forests in the tropical mountain areas comprises rounded crowns, which are not significantly modified by prevailing winds. In the lower regions forests may be three-storeyed, but become less complex with altitude, ending in simple single-storeyed forests. Foliage is relatively luxuriant due to abundant moisture, but leaves tend to lose their drip tips at higher altitudes, particularly in the transition from tropical rain forest to montane forests. In moister areas, noticeably in cloud forest ecosystems, epiphytes and mosses are abundant. The full extent of the diversity of tropical forests has yet to be established. Above the tropical forest species, forests tend to comprise temperate species similar to those of higher latitudes. For example, Mexico has some 200 species of *Quercus* above the 1,000 metre altitude, and some forty species of pine (Price 1981).

Cloud forests

A specific type of forest in mountain regions is that of montane cloud forests. These are characterised by very high rates of diversity and endemism, and have a specific and important hydrological function, gathering their water supplies directly from the rising air from, for example, oceanic coastlines.

Cloud forests are particularly significant in the Andes (at various altitudes – in Ecuador between 1,500–3,500 metres), the Ruwenzoris in East Africa (2,000–3,500 metres), and as low as 100 metres in Hawaii (Hamilton *et al.* 1997). They also occur in south-east Asia, in Indonesia, Malaysia, Papua New Guinea, the Philippines and Sri Lanka.

Cloud forests have a distinctive flora and structure, and occur in a locally narrow altitudinal zone. Clouds forming at dew point from moist rising air, usually off subjacent oceans, not only provide the moisture but also reduce evaporation by shading from the sun. Between 500–10,000 millimetres moisture per year is captured by such forests (Hamilton *et al.* 1997). They are usually less dense than lowland forests but rich in epiphytes and climbers (Aldrich *et al.* 1997). The forests possess a rich avian and insect fauna, and other fauna – for example, the mantled howler monkey and jaguar in the northern part of the Ecuadorian cloud forest (Parker and Carr 1992). Long (1995) estimates that about 10 per cent of all identified restricted range species occur in cloud forests. Once they are removed, moisture capture is lost and the associated environment, so conservation is a current issue of great concern for reasons of biodiversity, hydrology and landscape stability.

The timberline

The most dramatic ecotone (transition from one community to another) is that of the timberline. In some locations the cessation of trees is very abrupt, but elsewhere it may be a gradual thinning of tree density, stunting of shape and appearance of dwarf, twisted and flagged forms obviously damaged by the severe environmental conditions. Beyond the timberline lie shrub or meadow floras. The timberline has been the subject of considerable investigation in terms of its cause (Daubenmire 1954; Morriset and Payette 1983). It is variously defined as the forest line (upper limit of contiguous forest), tree line (upper limit of erect, 2–3 metres high, arborescent growth), the *krummholz* (German term for stunted tree forms), and the species limit (upper limit of a particular species). Many tree species at the timberline indicate by the tiny annual incremental rings that they grow extremely slowly, but they can also survive for long periods. The bristlecone pine (*Pinus aristata* var. *longaeva*) in the Sierra Nevada, so important for dendrochronology and radiocarbon calibration, has individual trees of some 4,000 years old. The slow growth and twisted forms imply environmental stress as the main cause of cessation of growth at a given altitude.

The timberline varies from tropics to temperate latitudes. In the tropics it tends to be highest in the valleys. Here soils are deeper, more fertile and collect more moisture. The risk of frost and wind exposure may also be reduced. In temperate higher latitudes, however, timberlines are higher on the ridges, despite the greater exposure. This is attributed to the greater accumulations of snow, its longer lying in the valleys, and the propensity for cold air, which is ecologically devastating, to flow down valleys and gullies throughout the year (Wardle 1973). The altitudes of timberlines vary with latitude, from about 3,500–4,500 metres in the tropics to near sea level in polar regions. Examples include the timberline altitude of 750 metres in the northern Urals; in the European Alps it rises from 1,500–1,600 metres in the northern section to 2,000 metres in the central section, and can be observed at 3,250 metres in the High Atlas Mountains. In the continental interiors of Iran the timberline is found at 4,000 metres due to the aridity of the plains. The highest timberlines occur at 4,700 metres in the *puna de Atacama* in the High Andes and at 5,000 metres on the high Tibetan Plateau (Troll 1973a, 1973b). Timberlines tend to be lower on windward slopes, and those influenced by maritime conditions and on the margins of mountain ranges. Within major ranges, such as Tibet, the altitude is higher.

Our understanding of timberlines remains imperfect. As a general rule, timberlines globally coincide with the 10°C July isotherm (Daubenmire 1954), but it is clear that temperature alone is not responsible. In arid regions where precipitation increases with height, there is usually a lower timberline defined by the cloud base or altitude at which sufficient precipitation occurs to support forest. This may be very clearly delimited, for example, in cloud forests. In this case, it is precipitation not temperature that is the limiting factor for forest growth. Elsewhere, where tree species are less specific in their ecological requirements or environmental conditions less extreme, timberlines may be hard to define. In some areas, poor soils and abundant bare rock, or the recent passage of glaciers, may have prevented subsequent forest development. Breaks in the timberline may occur at sites of recent disturbance such as mass movement or avalanche tracks, where different species may have colonised. In most areas, it is the activity of humans clearing or selectively cutting forest species that has been the most significant agent of change in timberline levels in recent years. For example, the timberline is some 400 metres below its climate optimum in the European Alps (Holtmeier 1994).

The timberline is probably the result of a combination of factors, including temperature and precipitation, length of growing season (in temperate latitudes), negative carbon balance and poor resistance to disease and environmental stress (Jeník 1997). Many of these variables have both positive and negative effects on growth. Snow, for example, insulates and provides water, but also damages by weight; also, slow melting may retard the development of lower growth. Wind causes obvious damage to trees – flagging, accumulation of rime and wind-chill effects, desiccation – but also contributes to the arrival of viable seed from lower altitudes where more viable seeds are produced. It is estimated that whereas a lowland conifer may produce viable cones annually in the lowland, in mountains this may occur only one in four or five years, and at the timberline, perhaps only one in ten years. Solar radiation, particularly the UV component, is known to retard plant growth, but high intensities can also speed up photosynthesis during the short summers, and leaves with thick cuticles provide adequate protection. This permits survival where insolation is very strong, such as on the Tibetan Plateau where timberlines are highest. Cloudiness may obscure sunlight, but the presence of clouds is the sole means of water for cloud forests. For trees at their upper limits, most of their energy goes into survival rather than reproduction. The survival of very old trees, such as the bristle-cone pine, testifies to their ability to adapt. It certainly accounts for their stunted forms in some areas, but not always explains fully why they stop growing where they do.

Meadows and tundra

Above the tree line the open meadows and tundra comprise a treeless plain, dominated by herbs and grasses, and in some cases shrubs. In the lower altitudes the grass mat may be highly diverse and continuous, but at higher altitudes it is replaced by more rocky terrain, with patches of herbs and alpine flowers as far as the snowline where lichens dominate. The flora is varied according to soil type, disturbance, drainage and grazing by wildlife or by domestic stock. Environmental gradients can be very sharp in this zone, and are often reflected in the vegetation. The importance of micro-relief in providing shelter, pockets of soil, insulating snow patches and protection from the wind is critical at these altitudes. The highest vascular plants occur in the Himalayas and Andes at 5,800–6,100 metres, although whether this is primarily a function of the occurrence of land at this altitude is not certain. Meadows often have a very rich flora – the Kashmir Himalaya has some 1,610 species identified in the alpine and subalpine zones (Dhar and Kachroo 1983). Many other mountain zones still lack any proper botanical investigation.

Similar species occur in widely separated mountains – *Ranunculus* occurs in the Alps at 4,270 metres and in the Himalayas at 6,100 metres. Around half of the alpine flora of the middle latitudes is arctic in origin, although this falls off considerably at lower latitudes. In Scandinavia 63 per cent of a total flora of 180 species are circumpolar in affinity; in the Swiss Alps this is 35 per cent of 420 species, and in the Altai of Central Asia this is 40 per cent of 300 species. The isolated East African peaks have some 80 per cent endemics (Price 1981).

The harsh nature of high-altitude environments requires a specialised plant structure and life in order to survive. Within about one metre of the surface the environment is modified from the free air, being more sheltered from the wind, and having reduced temperature fluctuations due to heating and heat retention of the ground surface. Snow cover (depth and duration), however, can further restrict the already short growing season in higher latitudes, whereas the alternation between hot day and cold night in the tropics also raises environmental stresses. The thin soils and aridity, together with the low nutrient status of the soil, provide poor resources.

Alpine plants show a spectacular range of adaptations to these environmental conditions. Almost all are perennials. The growing season in middle to high latitudes is too short to permit the full life cycle of annual plants to be completed, resulting in the possibility of loss by late frost or snow, or by the early onset of winter. Perennial plants, like the coniferous trees, can respond rapidly to warmer conditions in spring, and thereby make the most of the short growing season. They tend to have a low, compact growth form, often as tussocks or cushions near to the ground (Troll 1959; Rougerie 1990). This has a number of functions – making the most of warmer conditions near the ground surface, avoiding the full force of the wind, and thereby reducing evaporation and desiccation by wind or sun and physical damage. Leaves have thick, waxy cuticles that provide further protection from moisture losses. Many are thorny to repel grazing animals. In the tropics, diurnal rather than seasonal fluctuations from frosty nights to very hot days can be accommodated by woolly or hairy leaves trapping warm air at night, and also by the leaves of rosette plants folding over the tender central buds at night (Rundell *et al.* 1994).

Below the surface, the biomass of the alpine flowers may be between two and six times the surface biomass. The extensive root system has a number of functions. It provides anchorage, and draws water and nutrients from a wide area, which is critical in a low nutrient and often arid environment. In addition, the roots provide a significant store of nutrients that can be mobilised early in the growing season when conditions are particularly harsh. Finally, the roots provide an important means of propagation, by rhizomes. The uncertainty of reliable insect faunas and the problems of lack of suitable unoccupied sites for colonisation by seeds mean that rhizomes may be the most reliable means of propagation. The colourful and often large flowers indicate that insect, or in the tropics, bird, pollination remains a possibility, though a risky one. The new plants remain attached to the parent until they are established and can thus share in the stored nutrients whilst they develop. In some areas, one single plant may cover tens of metres in area.

The productivity of the biomass has been estimated at 1–3 gr./m²/day over a growing season of 30–70 days (Price 1981). This compares favourably with temperate lowland productivity but is remarkable for the fact that this only accounts for surface productivity. When the rootstock is included, this figure may be three times as much and highly concentrated within the short growing season. This high productivity of meadows and tundra provides important grazing resources for mountain farmers who, in turn, have a significant role in their long-term maintenance (Chapin and Körner 1995)

Wildlife

The mobility of mountain fauna makes it more difficult to establish the diversity of species than for flora. Certain birds and mountain gorillas are highly specific to their habitats, and so their distributions coincide with a particular ecosystem, as is the case for cloud forests mentioned earlier. The origins of endemic fauna are discussed by Huggett (1995), who summarised the debate surrounding the small mammal populations of montane 'islands', and their extinction and speciation, with reference to the mountains of the USA. During the Quaternary these ranges were largely joined, but were subsequently broken up following warming conditions and the retreat of boreal floras to higher altitudes. Faunas therefore suffered a degree of extinction due to increased competition within a smaller area, and due to changing environmental conditions, but the disjunct distributions prevented immigration of new species. It has been found that species richness correlates with area rather than isolation. However, immigration is not prevented in all cases, and woodland was shown to be less of a barrier to immigration of small mammals than grassland and shrubland, leading to greater variation in the affinities and evolution of mammal communities.

In poorly developed ecosystems it is advantageous for species to be more generalist in their habitat preferences in order to maximise options for survival, just like the human livelihood strategies in mountains. In the Swiss Alps, for example, ninety-six species of birds are associated with the coniferous forests, twenty-seven with the shrub zone and eight in the high meadows. This also reflects the increasing marginality of the zones with increasing altitude, with respect to productivity, year-round opportunities for occupation, later onset of summer and earlier winter, as well as habitat diversity. Huggett (1995) also refers to this pattern of decreasing richness with altitude for birds and rats on Mauna Loa, Hawaii. Here distributions of birds are closely related to particular ecosystems in which individual species are specially adapted to live and feed. Second, other environmental factors operate, including vegetation and its structure, which controls the nature of microenvironment and availability of suitable food. The rats, on the other hand, are less spatially constrained, but all were introduced by humans and have survived by adapting to a range of habitats rather than specific niches. They overlap in distribution, but occupy different ecological optima.

At higher altitudes in the tundra and meadows, mountain faunas tend to be dominated by rodents, scavengers and insects (Franz 1979). Predator birds and large carnivores may range far over these regions. In shrub and forest habitats, the fauna becomes more diverse and the food web more complex. Some species are resident only during summer, and migrate downslope to adjacent lowlands during winter; in the case of birds, migration may span continents. Other species are permanent inhabitants and thus require adaptations to cope with winter conditions in mid- and high latitudes (Jeník 1997). Migration is more commonly associated with seasonal climates and represents one of these adaptive strategies. In the tropics the populations tend to remain more static and diverse adaptations have developed to cope with the diurnal fluctuations in temperature. Diversity in fauna often reflects diverse floras, such as in the Andes and Himalayas, although there is no causal link between them.

Life at high altitude in seasonal latitudes revolves around a regime of alternating feast and famine. During the winter mammals and some insects that are permanently resident either hibernate or are physiologically capable of withstanding extreme cold and exposure (Franz 1979). Hibernating species include insects and rodents that remain in burrows from the first snows until the spring thaw. During hibernation, their metabolic rate is slowed by up to 60 per cent, conserving energy resources. Mammals and birds that remain active during the winter often change plumage or coat to match the snow cover – partridge, hares and arctic foxes for example. During the change from summer to winter plumage and vice versa, the mottled appearance is equally effective camouflage. Fur can be shed each spring and regrows in autumn in response to changing daylight hours. Other mammals have the ability to constrict the blood vessels to their extremities, permitting sufficient blood flow to prevent frostbite but preventing excessive heat losses and thus maintaining the body core temperature. Some foxes can rest in snow to temperatures of –40°C until shivering occurs to raise metabolic rates. In the tropics high diurnal fluctuations mean that a heavy coat might be needed at night. However, this would be too hot during the day, and so strategic areas of the body, such as under the legs, remain relatively bare. This permits the animal to keep cool during the day, but these areas are covered up when the animal curls up at night.

Some larger mammals are apparently well able to cope with hypoxia. Llama and alpaca (*Lama* spp.) are native to the high Andes and survive at altitudes over 5,000 metres, and the yak (*Bos grunniens*) can be found up to 6,000 metres in the Himalayas (Franz 1979). Goats and some native sheep including the ibex (*Capra ibex*) and snow leopard (*Panthera uncia*) also survive at high altitudes all year. However, it is also clear that domestic species do not survive at some altitudes; lowland cattle, for example, only survive below 3,000 metres, which has implications for pastoralism. This is in part overcome by cross breeding of yak and cattle in the Himalayas to produce animals well adapted to the zone above the limits for cattle and below the limits for yak (Bishop 1998).

There is a seasonal migration pattern of fauna. Birds arrive first – carnivorous species after the herbivorous ones to allow time for insect populations to be built up. Predators, such as birds of prey, arrive later, coinciding with the season of young birds and rodents. The sense of timing of activities is also demonstrated by the fact that for many species the period of migration does not overlap with reproduction, thus spreading the energy demands of individuals more evenly, which is significant in landscapes where food may not be abundant. Reptiles are cold blooded and unable to moderate their own body heat, needing sunshine directly on their bodies to keep warm. Thus even shadow from cloud cover can moderate their activity. This relative inability to cope with temperature fluctuations makes them less well suited to mountain living, particularly at higher altitudes and in more exposed environments. Consequently, reptiles are not well represented in high-latitude mountain ecosystems but can be found in warmer Mediterranean mountains.

HIGH ALTITUDE AND HUMAN PHYSIOLOGY

One of the most important issues for any understanding of human occupation of the mountains concerns the extent to which any biological population is physiologically adapted to living at high altitude. Numerous classic studies exist for both plants and humans (Baker and Little 1976) and it is an area of very active medical research (West 1998). One of the most important parameters for determining stress to human biology is the fact that atmospheric pressure reduces with altitude and limits the oxygen-absorbing capacity of the blood. With the decline in pressure the process by which oxygen is bound into the haemoglobin in the bloodstream does not work so efficiently. This is known as hypoxia. Some writers argue that 2,500 metres represents the altitude at which oxygen stress can be clearly identified, and there has been considerable work in this field associated with mountaineering and aviation. The British Mountaineering Council (http://www.thebmc.org.uk/) suggests that the effects are likely to be evident at 3,500 metres. There is no doubt that beyond 4,650 metres severe effects are manifest in the form of 'mountain sickness' – loss of breath, nausea, dizziness, and insomnia. Many of these studies have been carried out on native Andean populations. A number of physiological changes are identified as assisting the transport of oxygen to the tissues and cells. These include lung capacity, which is frequently greater; naturally faster breathing rates, even at rest; higher concentrations of haemoglobin in the blood; and more efficient transfer to tissues (changes in pH of blood) and utilisation by cells. Visitors to mountain regions tend towards hyperventilation until they have acclimatised by producing more haemoglobin (Frisancho 1993; Lahiri 1974).

Evidence suggests that oxygen absorption is also linked to other bodily systems, so that for a new resident in high-altitude areas the impact will be quite complex. For example, stresses associated with cold are linked to hypoxia and even in tropical mountains low temperatures at night can cause considerable physical discomfort. Another issue concerns radiation, which at high altitudes can be 50 per cent greater than at sea level. Little is known about the long-term effects of exposure, although it is increasingly evident, from mountaineers and others, that protection from ultra-violet radiation is necessary. For long-term residents of high mountain regions, physiological and psychological adaptation to these stresses has occurred, though the precise form this takes is often in dispute. Whereas the visitor to mountain regions over about 2,500 metres is likely to experience respiration problems with any exertion, the native born would not find difficulties. Interestingly, the literature suggests that this difference remains, even if the outsider lives for lengthy periods in the mountains, suggesting some form of physical adaptation. Bishop (1998) refers to differences in physiology between mountain and lowland populations as resulting from an apparent stunting effect on growth. This may arise not only from the physiological stresses directly but is combined with poverty, poor nutrition and hard physical work. The combined effects of all these factors contribute to low birth rates, higher mortality and slow growth of individuals. However, others (Pawson 1976) did not find significant growth differences between sea level and high-altitude populations – although this was in Ethiopia where the mountain environment is more agriculturally productive than the lowlands, which may even out the effects of malnutrition associated with other mountain regions.

The physical stress of a low technology and hard labour life, together with poor health and low nutritional status is also reflected in earlier studies such as that of Groser (1974). This is thought to affect the size and mortality of mountain populations, the age of menarche and the reproductive capacity of women. This topic remains controversial but is often cited in demographic studies (Viazzo 1989). Whilst some early studies in the Andes did suggest that fertility was negatively related to altitude, others have argued that there is no clear proof (Stone 1992).

CASE STUDIES

Two examples are examined in more detail: the Andes, which provides an example of a north–south range where latitude has a marked effect on environmental conditions, and the Himalayas that stretch roughly east–west and form a significant barrier to air masses moving north–south. The examples demonstrate the interactions and interdependence between climate, geomorphology and ecology.

The Andes

The Andes range provides an excellent example of the interaction between latitude, altitude, proximity to coast and local effects such as aspect (Figure

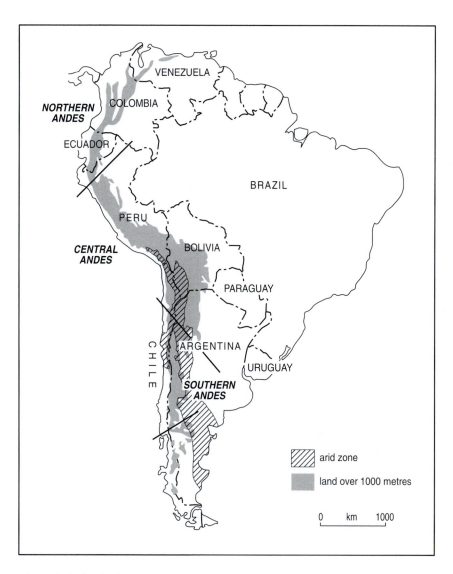

Figure 2.6 The Andes: principal climatic regions (adapted from Stone 1992)

2.6). They have also been widely studied, most notably by Troll (1968). Both N–S and E–W gradients are superimposed on altitudinal gradients. The climate is affected by high pressure in the surrounding oceans and the southern extension of the ITCZ (Intertropical Convergence Zone) during December–January (see Figure 2.7). The Andes can be briefly considered in three main parts.

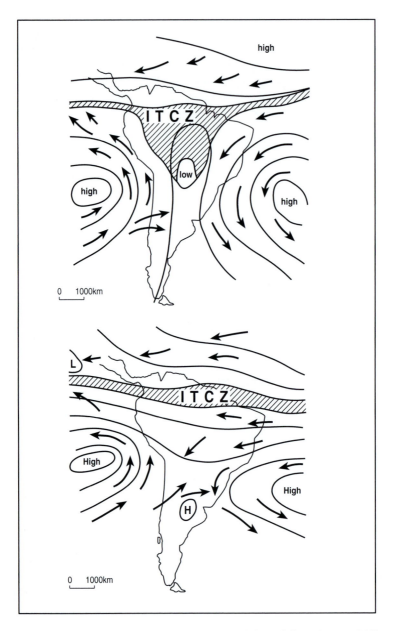

Figure 2.7 South America: pressure systems (adapted from Stone 1992)

The northern Andes of Venezuela and Ecuador are wet all year, falling in the tropical belt. Zones tend to be symmetrical on both sides of the range, and seasons are much less significant than diurnal variations. The transition is from tropical to montane rainforest, and then upper cloud forest that occurs at the level of the cloud base between 2,000 and 2,800 metres, with a mean temperature of 16°C. Species diversity is high, according to Klötzli (1997), with between fifty and sixty species per 500 m² plus ten epiphytes in some tropical mountain rainforests. Soils may be more fertile than the deeply weathered lowland soils. An evergreen sub-alpine ericaceous forest occurs between cloud forest and *paramo*. The *paramo* (wet grassland) occurs between 3,500 metres and 4,500 metres, particularly in Venezuela, Colombia and Ecuador, with 30–40 species per 500 m² dominated by cushion and rosette forms (Chauverri and Cleef 1997). Frost can occur all year round in the *paramo* belt, and precipitation is daily in the lower regions. Convectional rainfall is a dominant component of precipitation.

The central Andes is a zone of transition between the tropical, through subtropical to the more Mediterranean climate of the southern Andes. On the west side, the climate gets drier southwards as far as the Atacama desert, the driest place on earth. The east gets progressively wetter due to the influence of the NE and SE trade winds. These influences create a NW–SE ecological gradient from humid grassland to dry grassland (*puna*), to desert conditions at higher elevations. On the east the cloud forest, the '*ceja de la montana*', exists at the point of condensation of ascending air. Seasonality becomes more pronounced in this region, and increases southward through Peru and Bolivia into Chile. Seasons alternate between wet and dry, and frost risk at high altitude becomes more temporally constrained.

The southern Andean zone is more temperate and seasonal. The climate gets progressively cooler with increasing latitude. The east becomes drier whilst south of the Atacama the climate becomes wetter. Vegetation changes from desert scrub to Mediterranean forest vegetation. At higher altitudes, the *puna* gives way again to humid *paramo*. The Patagonian region has a polar climate and environment with abundant glaciers.

These transitions between environmental zones with altitude and latitude are reflected in the adaptations of the local populations. Agricultural activity in the equatorial region continues all year with patterns of fallowing to allow soil nutrient regeneration. The nightly frosts at high altitude are used for freeze-drying potatoes, and the indigenous camelids are adapted to large diurnal fluctuations in temperatures. Further south, agriculture becomes more seasonal, although still combining pastoral with cultivation activities. These implications are discussed in more detail later.

The Himalayas

The Himalayas stretch for some 3,000 kilometres NW–SE and encompass an enormous diversity of ecological conditions (Figure 2.8). The orientation and altitude of the mountain range separates the monsoon climate of southern Asia from the continental climate of northern Asia. The mountains thus permit a more northerly extension of the monsoon, but deflect it westwards. Its influence decreases westwards, as reflected in the change in vegetation. The northern ranges are dominated by the continental conditions and are much more arid and affected by cold polar air masses. Schweinfurth (1984) distinguishes several zones. The southern outer Himalaya is the wet lower foothills, especially of the eastern end, including lower Bhutan and Sikkim. The inner Himalaya includes Upper Bhutan, Sikkim and Kashmir, which comprise both moist and dry temperate forests, and where local variations in soil and climate conditions are important. Finally, the Tibetan Himalaya consists of the northern ranges, which are more arid.

The pattern of vegetation in the Himalayas has been studied both along the parallel ranges, and in transects (Schweinfurth 1954, 1984; Troll 1972; Bhatta 1992; Miehe *et al.* 1996; Richter *et al.* 1999). These studies reveal a complex mosaic of vegetation types, reflecting regional and local edaphic and climatic conditions, differentially modified by centuries of human activity. It is difficult to summarise this complexity, but in the same way that the Andes can be divided into roughly three sections, the Himalayas can be divided into east, west and a transitional central zone.

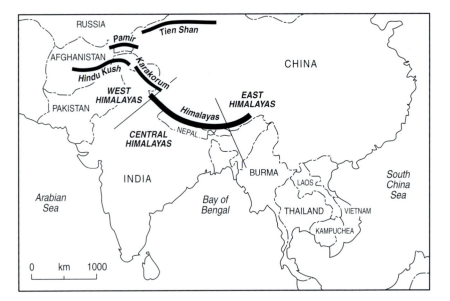

Figure 2.8 The Himalaya–Karakorum–Hindu Kush mountains

The eastern Himalayas, incorporating the hills of Myanmar, northern Thailand and southern Bhutan, are the most affected by the monsoon and by subtropical conditions. The lower slopes are clothed with tropical rain-forest up to 900 metres in Myanmar, 800 metres in Bhutan (Troll 1972; Bhatta 1992), and succeeded by tropical deciduous forest up to 1,000 metres. These forest groups are succeeded by subtropical montane forests – broadleaved up to 2,000 metres in Bhutan and 1,500 metres in Myanmar, and above that pine or evergreen forest, dry temperate forest (up to 2,800–3,000 metres) and finally alpine scrub and grassland up to 3,000 metres in Myanmar and above 4,000 metres in Bhutan.

The central Himalayas form a transitional belt between the wetter east and drier west, and comprise areas reflecting both these climatic conditions. In Nepal, some tropical deciduous forest in the east grades into subtropical forests up to 2,000 metres. The subtropical forests predominate at lower altitudes further west as the influence of the monsoon decreases. It is succeeded by temperate forests – moist and dry – at about 2,000 metres, and by alpine forest and scrub between 3,400 and 3,900 metres.

The western Himalayas are much more arid. The effect of local conditions in modifying regional trends is important (Miehe *et al.* 1996). Both the moist and dry forms of subtropical forests are present on the lower slopes, but not the tropical forests. In the subhumid areas, closed coniferous forest occurs in shady north-facing slopes (such as the Nagar region of the Hunza valley), with closed juniper forest above (some 600–700 metres higher). Above this are grasslands dominated by Cyperaceae mats. In the more sub-arid areas, open forest of similar species dominates, with a dwarf scrub of Artemisia at higher elevations. In Afghanistan on the western end of the range, temperate forest occurs at lower altitudes (up to 1,800–3,600 metres) and is succeeded by sub-alpine forest (2,200–3,600 metres) and then scrub. Richter *et al.* (1999) examined a transect running from the western Himalaya into the Tienshan. The transect across the mountains reflects the change between the oceanic and continental influences, with a clear border between the Karakorum and western Kunlun, differentiating the Tibetan Himalaya from the Tienshan. Some altitude belts were missing from the transitions from lower-altitude moist to higher-altitude drier vegetation zones. This was attributed to local air and humidity patterns. The Karakorum and eastern Kunlun were driest in the foothills and characterised by high precipitation gradients, accounting for the occurrence of substantial glaciers at high altitude in the Karakorum.

Conclusion

The physical environment of mountains is highly diverse. The mountains themselves arise from large-scale plate tectonic movements operating on very long timescales. The uplift of mountains modifies their climate and that of the surrounding area, and contributes to the continuation of very active geomorphological activity. This results in landscapes which are highly diverse and dynamic.

The main factors affecting climate on a large scale include latitude, altitude and continentality. Altitude and aspect are important in determining differences across mountain ranges, and within valley systems. The balance between rain and snow and the variations in temperature and cloudiness, together with the geological structure of a mountain, determine the nature of its environment with respect to the geomorphological processes in operation and in terms of its fauna and flora. Local and regional scale processes, such as thermal winds and orographic precipitation, interact to produce a complex, diverse and dynamic physical environment.

In the two examples of the Andes and Himalayas, climate, geomorphology and ecology combine to provide the diverse physical environment of mountains. However, this diversity and dynamism is not just a function of the physical factors involved but also much depends upon the pattern of human occupation. For the human population, the physical environment is a resource or habitat to be moulded to suit the panoply of human requirements and provide appropriate livelihoods. This requires us to explore the cultural characteristics of mountain societies that provide the framework of beliefs, ideas and practices that govern the way the physical resources are exploited.

3

THE CULTURAL
FRAMEWORK

This chapter explores the basic social and cultural dimensions of mountain areas. As a broad framework for the discussion, the chapter begins by examining the process and pattern of settlement in mountain areas. Simple models of settlement type and various economic functions are described. This is followed by an examination of patterns of land use and wealth creation. We then move to review the discussions about a 'mountain culture' which is closely associated with claims concerning the distinctive qualities of mountain life and therefore with the politically significant pressures for specific mountain policies. This leads to an analysis of some of the political factors which today are a major influence in many mountain regions.

SETTLEMENT IN MOUNTAIN AREAS:
EVOLUTION, PATTERN AND INTERNAL
STRUCTURE

Since the arrival of man on earth, the mountainous regions of the globe have provided a habitat for human communities. In some instances these communities were, and remain, relatively small, often nomadic or semi-nomadic as in parts of the Tibetan Plateau. In other cases, highland populations grew to become the heartland of particular civilisations, for instance the Inca and Aztec societies of the Andean range. Archaeological research in African mountains, particularly the existence of rock carvings, has confirmed the existence of dwellings in prehistoric time, attracted by the relative abundance of water and lush vegetation. At that time the climate of some of those areas was far more propitious for settlement than it is today.

Settlement evolution

The evolution of settlement in mountains is an integral part of the dynamic of these areas, which is related to the economic, political and cultural characteristics of the population. In addition, particular lifestyles, village structures and dwelling designs are associated with the nature of the local physical environment, placing particular emphasis on the needs of protection for both humans and animals during the harsh periods of the year, and for defence. Villages are not usually built on good agricultural land, but can be found on defendable sites such as the Yemeni hill forts or the Moroccan *agadirs*. However, the evolution of settlement patterns owes much to the particular dynamic of population change. The trajectory of demographic growth, migration, the development and adaptation of farming systems and other activities, and the specific social and cultural framework which linked each settlement to the wider world, play an important part in the evolution of settlement patterns. These issues will be examined below.

As an exercise in historical geography the settlement narrative of any mountainous area poses particular problems, as often the small, remote communities are even less well documented than their lowland equivalents. However, there are some exceptions. There is a relative abundance of information on the European Alps, especially Switzerland, and also in North America. To a lesser extent, pre-Conquest archaeological work in the Andes has produced interesting material on the settlement sequences.

In North America an increasing volume of work now exists to document the dynamics of settlement in the Rocky Mountain Range (Wyckoff and Dilsaver 1995). In Colorado, for example, pre-European settlement was mostly temporary, as the Native American Indians used grazing and hunting resources. It was in the nineteenth century that early mining communities established more substantial settlements, such as the aptly named Leadsville. Inevitably, these were located in relatively inaccessible upper valleys, and many were simply deserted as the minerals became uneconomic to exploit. A few however have become recreational resources, which will be discussed in Chapters 6 and 7. Other small settlements became trading posts to support the 'frontier economy', which included hunting (fur trapping) as well as early exploitation of natural forests. More important for subsequent settlement development was the sequence of events that were associated with ranching. The 1872 Homestead Act provided large areas of land at relatively low cost, and whilst this was attractive to many frontier pioneers, much of the land remained in government hands, especially in Colorado state.

Consequently, in many areas of the Rockies subsequent development of settlement has consisted of mineral and energy locations subject to cyclical fortunes. Associated service settlements developed, located along valley bottoms, but it has been mainly in the period since the 1960s that new patterns have emerged. Of particular importance has been the growth of 'ranchette' development, where former pastures have been subdivided, thus leading to the widespread growth of settlement not only in valley bottoms but also across hillsides (Riebsame *et al.* 1996). This, along with recreational expansion, is explored in greater detail in Chapter 7.

The evolution has also been described for some areas within the Himalayas. In the case of the Sherpas of the Khumbu area of Nepal, Stevens (1993) writes that each of the families recites tales of their ancestors coming from either Tibet, in the north, or the lower valleys to the south. Whilst this oral evidence is contradictory, Oppitz (1973) suggests that the area was settled at least by the 1500s with Sherpas originating over 1,200 kilometres to the north-east on the far borders of Tibet. Linguistic evidence also suggests such links, if not the precise dating, and as a consequence the current population of the area still possesses many cultural affinities with that area. Attitudes to land management, the importance of Buddhism and the relatively egalitarian structure of society in which both men and women are property owners provide the essential backdrop to the functioning of the 'traditional society'.

Monastic records can offer some clues to the historical sequence of settlement, in particular documents such as the *Bo.yig* that describes the early history of the Zangskar valley in Ladakh (Crook 1994a). Certainly settlement in this area is ancient, although the particular structure of the population and distribution of habitation is far from certain. Buddhist influence can be traced at least to the seventh century AD and most probably earlier. Many monasteries, often initially using cave sites, were founded throughout the region in the eleventh century and settlement expanded into more remote valleys.

Another example of the way different groups come to occupy mountain valleys is provided by Kreutzmann (1994), who describes the evolution of the settlement pattern in the Hindu Kush–Karakoram. Permanent settlement sites can be found in the arid floors of the main valleys, whilst groups of huts at higher levels are occupied only in the summer season. The permanent sites are associated with the availability of water and cultivable land, usually on the valley floor or on terraces to which water can be brought using gravity fed irrigation systems. These settlements are sited at altitudes between 1,400 metres and 3,500 metres and depend mainly upon the flow of water from glaciers high in the mountains. Seasonal settlements, which usually can only be reached by mule paths, are located near to the summer pastures. In some cases the occupants cultivate hardy grains as well as manage livestock at the higher sites, but more often they consist of a few stone shelters and animal pens around which grazing is organised. Kreutzmann describes the settlement expansion in the Hunza valley from the nineteenth century and also points out that many sites have been subject to natural hazards, including landslides, earthquakes and floods.

In the High Atlas of Morocco it is possible to trace the history of Berber settlement. The work of the fourteenth-century writer Ibn Khaldun (1852), drawing on earlier texts, describes the history of Berber settlement of North Africa and highlights the role of the Almohad dynasty whose power-base lay in the Atlas Mountains. The city of Marrakesh, which lies almost in the centre of the Haouz Plain, still remains a symbol of the ebb and flow of

Plate 3.1 Tinmel mosque, the High Atlas, Morocco

Berber civilisations between the mountains and the plains of Morocco. However, the birthplace of the twelfth-century religious reformation led by Ibn Tumart, and associated with the rise of Almohad power, lies at Tinmel deep in the High Atlas along the Oued Nífis.

The Tinmel mosque (see Plate 3.1) has been recently restored and become a popular tourist site. Sadki (1990) comments that the Berber communities developed a settlement pattern that involved both the occupation of the high mountain valleys and the use of the neighbouring plains, especially for grazing. The subsequent history of the High Atlas found successive tribal confederations establishing power in the lowlands and these all developed links with smaller confederations within the mountains. Central to this process were the differing attitudes to Islam and the nature of interaction with the Arabic population of the lowlands. The increasing arabisation of the lowlands, including the lands adjacent to the High Atlas, led, by the fifteenth century, to increasing hardship for the mountain population who retreated further into the valleys as their livelihoods were attacked through the occupation of their important grazing lands.

The significance of this background to the settlement of the High Atlas south of Marrakesh is that it describes a constantly evolving situation. Often,

in the work of French writers of the 1930s and 1940s, the Berber communities are considered as solely mountain peoples. However, it is evident that
the degree to which this characterisation is appropriate depends upon the
historical period under review. Equally, and in reverse, Arab influence has
penetrated strongly into mountain valleys, raising difficult questions about
the 'correct' designation of the peoples of the Moroccan Atlas. It is in this
context that we can examine in more detail the evolution of settlement in
the Reraya valley, in particular that of the Ait Mizane (Miller 1984). Most
of the evidence comes from oral sources, with the occasional fragment of
written material.

Miller (1984) studied the Imlil area that lies some 50 kilometres south
of Marrakesh. Evidence suggests that the valley was already populated by
the twelfth century when the Reraya confederation (the group of tribes
within the watershed of the same name) were recorded as adopting the
fundamentalist doctrine emanating from Tinmel in the neighbouring N'fis

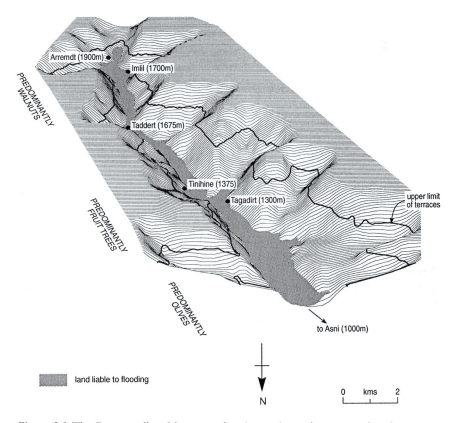

Figure 3.1 The Reraya valley, Morocco, showing main settlements and cash crops
Source: Parish and Funnell (1999)

Plate 3.2 Imlil village, Morocco

valley. By about 1500, three clans occupied the upper valley: the Ait Takhsin, the Arremdt and the Ait Mzig. According to legend, it is at about this time that a major flood occurred which destroyed the village of the Ait Takhsin. Although this event is not confirmed by written evidence the destructive power of the recent flood in 1995 suggests that this story is highly credible. The clan then built several new villages on higher land and today they form part of the settlements that occupy the upper Reraya valley. Later, probably in the nineteenth century, a new clan, the Ait Souka came to the Mizane valley probably from the Sous valley and settled in what was a relatively unpopulated eastern extension of the basin. This clan is now part of the Ait Mizane tribe. The accompanying map, Figure 3.1, shows the settlement pattern of this mountain valley, and Plate 3.2 illustrates the character of the settlements.

Settlement patterns, ecological characteristics and verticality

One of the key factors influencing the locational pattern of settlements in mountains has been the linkage between the physical environment and the pattern of agricultural activity on which the settlement has depended for its survival. Consequently, many writers have argued that the close relationship between the altitudinal organisation of ecological zones, the vertical pattern of agricultural activity, and settlement structure constitutes one of the principal characteristics of mountain geography. In Part 1 the basis of this 'verticality' concept was examined and in this section we detail examples which use this idea to explain settlement organisation.

Murra (1972, 1995) has developed models of vertical organisation for the Andes. He argued that these vertical zones provided the basic framework for political and economic organisation of Andean societies. In the Peruvian Andes pre-Incan settlement indicates that patterns of exploitation utilised the different ecological zones to provide agricultural products. By the sixteenth century, there is considerable evidence that not only was the production system organised vertically but so also was the institutional framework of political power, trade and cultural relations. According to Murra the basic spatial structure within the mountains revolved around the establishment of a 'nucleus settlement' at a given level and then 'colonies' within the other ecological zones. Settlers living in the colonies kept all their rights to land and property in the 'nucleus' and therefore these settlements tended to contain closely related families. By contrast, groups from other nuclei could also occupy the 'colonies'. From the perspective of a given family or community, its property and production relations were a 'vertical archipelago' with islands of production between one and four days away from the nucleus settlement. Brush (1976b) has argued that Andean systems of production zones are either:

- Compact: where different zones exist in close proximity and access to each zone is relatively easy.
- Archipelago: where unused areas (unusable) separate zones and travel is in the order of several days or more. Here colonies are established and trade relations, based upon reciprocity or redistribution, operate through kinship links.
- Extended: each group exploits only one zone so is, in effect, a specialised group. Access to products from other zones is via barter/market.

Although this pattern mirrors the physical environment it was also paralleled by an administrative structure. Economic and political control between the different zones could be exerted through two basic mechanisms:

1   Through the ownership of property at different altitudinal zones. Murra noted that different communities within the Andes had different patterns of property relationships. Some owned property at the different zones and therefore the utilisation of the produce was organised within the extended family or kin-groups.
2   A particular group dominated each zone and exchange relations provided the mechanism for interchange of goods.

Drawing upon an example from Fioravanti-Molanié (1982), an interesting pattern of exchange is illustrated in Figure 3.2. The flows represent a structure that has been operating since the seventeenth century. Some of this trade takes place through barter or market operations and some represents the flow of goods within families. The exchange system relies heavily upon the fact that the most labour-demanding periods of production differ between the zones. For instance, between April and May maize is harvested in the temperate areas, whereas potatoes and wheat need most labour input in June and July. Consequently, both labour within families and the availability of wage labour from different zones allows a flow of labour power to ensure that the overall system flourishes.

The Peruvian example illustrates how the ties between different zones have formed the basic geographical framework for communities within the mountains. In addition this verticality is enshrined in the cultural values of the population, represented by distinct behavioural patterns. One of the most powerful cultural features of the area is the deeply felt division between the Indian and mixed (*mestizo*) populations and this is represented geographically in the belief that, socially at least, higher means lower. The communities at the highest altitudes tend to be poorer, speak predominantly Quechua rather than Spanish, and their main crop, potatoes, is held in less esteem than maize. Symbolically, there are many rituals that are still used to ward off spirits that are believed to occupy the higher slopes. Even as the accoutrements of the 'western' economy encroach, and the rail and road reach these valleys, these beliefs are still important.

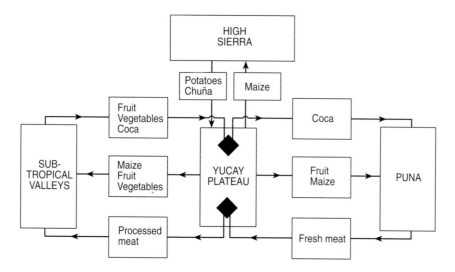

Figure 3.2 Exchange relations between different altitudinal zones in the Andes
Source: adapted from Fioravanti-Molanié (1982)

Not all settlement is permanent, however. Different degrees of settlement mobility exist in certain mountain areas. Ehlers and Kreutzmann (2000) explore this in more detail with respect to pastoral communities in high mountain environments in Asia, emphasising not only the diversity of practices but, additionally, the dynamics of settlement in a situation where the underlying economic framework is changing. For the purposes of exposition however, we identify three main types:

- Nomadic settlement: this refers to transient camps that nevertheless follow annual patterns on long-established routeways – for example, in the Middle Atlas of Morocco.
- Transhumance: in this pattern of livestock management, shepherds have winter and summer settlements. For instance, the once dominant movement of Pyrenean flocks to the lowlands in winter, or, equally, the movement between the main settlements (Yul) and higher locations (Gunsa, Phu) during the summer found in the Khumbu area of Nepal (Stevens 1993). In this latter case, individuals may have several houses scattered at different altitudes throughout the valley.
- Shifting cultivation: in the areas of the Karen in northern Thailand, the villages may be relatively static but the patterns of associated cultivation shift according to a long-term rotation. In some instances, however, the entire village may be relocated, especially in the higher zones where the soils are exhausted after a shorter interval and fallow periods lengthy.

By contrast, in the lower valleys, where soil fertility is higher and fallows shorter, it is more efficient to remain in permanent settlements (Hurni 1983).

Internal structure

The internal spatial organisation of mountain settlements is very varied, more so if traced through a historical perspective. Some could be described as nucleated, giving the impression of a compact village surrounded by its fields. This model might be used to characterise some Alpine settlements, such as Kippel (Friedl 1974) and Torbel (Netting 1981). Similarly, some but not all of the settlements described by Kreutzmann (1993) in the Hunza and by Brower (1991) in the shadows of Mount Everest are relatively compact. Elsewhere, the settlement is much more dispersed; even the typical village may consist of a series of houses where each one is surrounded by fields but the whole is located within a clearing on the mountainside. A good example is that of Melemchi, in northern Nepal, described by Bishop (1998).

The traditional dwellings themselves also show some similarity between regions – in particular, the tendency to house both humans and animals either under one roof or in very close proximity (Plate 3.3). Although the building materials reflect local sources and particular needs, this has led to a basic multilevel design, often with the higher levels being added as the family expands. In the Swiss Alps there are houses dating back at least to the sixteenth century which illustrate this basic architectural design where a stone basement provides the foundation for wooden upper layers. In the Melemchi example, the houses have been built of stone and mud with roofing material of wooden shingles. In the High Atlas, it is common to find the buildings tightly huddled together, each unit interconnected with others and the living accommodation interspersed with pens, especially for dairy cows. Within buildings, the structures reflect the cultural norms of the community. In both Nepal and the Atlas, examples show how originally there would be few rooms, usually with an earth floor. Depending on the wealth and status of the owner, some rooms would be used specifically for entertaining guests whilst others would be family rooms. Gender roles would be reflected in the various tasks within the house, although according to Brower, in her account of the Sherpa in the Khumbu valley, both men and women are involved in the full range of domestic tasks, including cooking. In the past, most of the dwellings would have been dark and smoky, with the open fire hearth as the source of heat for all the functions.

In tropical mountain zones such as those of Thailand and East Africa, traditional construction was of timber (bamboo or teak) with roofing of vegetable matter, for instance banana leaves. Unlike the Alps, animals may

Plate 3.3 House in the Hunza valley

be kept in open pens adjacent to, or sometimes under, the supporting stilts of a house.

In most settlements there are other important functions. In Nepal, the Karakoram and the Atlas Mountains for instance, there is usually some form of religious building. In the case of Melemchi, the local lamas looked after the gomba; in the High Atlas, it is the mark of status of a settlement when an imposing mosque can be constructed. Miller (1984), in his description of the Ait Mizane already discussed above, points out that it is a shrine located several kilometres above Arremdt, the highest village, that gives the area an important religious significance. As a result, the earliest local mosque was built in this village. Later, another was built at Imlil, which itself was replaced by a another new one after only a few years. Elsewhere, in the Anti-Atlas, renowned for its high rate of emigration, most families try to contribute something from their earnings to the renovation/rebuilding of mosques. In Ladakh, many villages are associated with monasteries that may

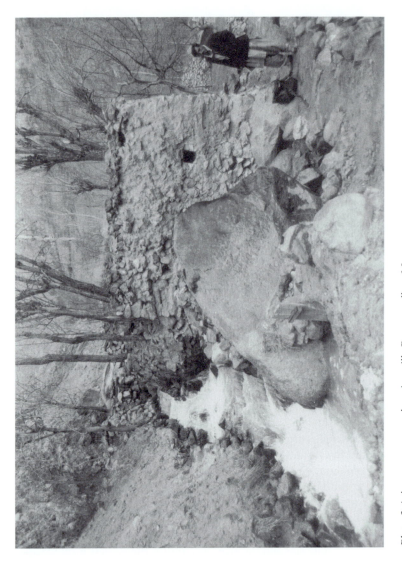

Plate 3.4 A water-powered grain mill, Reraya valley, Morocco

own the land and provide the focus of religious life. As many unmarried males from the local villages become monks the monastery has an important social role in the demography of the villages.

Another important communal function is the village grain mill. In many mountain zones this would be powered by water from a local stream. In the Atlas Mountains, a small hut perched over a stream with perhaps a donkey waiting outside would indicate the presence of a grain mill, where water power would be used to turn the grinding stones (Plate 3.4). These mills would be owned by the community, and often worked by any individual who wanted grain milled. This technology is still used in many mountain areas but is slowly giving way to diesel generators and electrically powered mills. Other important activities in the traditional mountain village would include, especially in the European Alps, storage barns for the hay. Often these were attached to a private house, but some were communal. Also, as part of the grain harvesting process, there may well be particular sites used for threshing, such as the *andrair* in the Atlas.

The site of a particular settlement is very closely associated with the local topography and the prevailing socio-political model of the community. The importance of defended villages has already been noted in the description provided by Kreutzmann and similar features can be found in some settlements in the High Atlas. Equally, many villages are located to minimise the damage from avalanches and landslides (see Chapter 4). On the other hand, whilst proximity to agricultural land is very important, most old villages are not to be found in the floodplain floor, although there are numerous examples of settlements located on apparently ancient floodplains. There are, equally, many examples of such sites being badly damaged when the channel suddenly becomes active. The high-energy physical events of the mountains therefore make the question of precise location important.

Settlement and refuge

One of the most common assertions about settlement in mountain areas is that it represents the final retreat of marginalised civilisations. The Berbers for instance, were once a powerful force on the lowlands of North Africa. However, their influence was reduced by the arabisation of the plains and today Berber (Tamazight) culture (especially language) is found concentrated in the Atlas Mountains. In many parts of the world, particular groups have retreated to mountain zones in time of conflict. A good example is the Druze of Syria and the Lebanon who still occupy the highlands as a strategic site. In the history of Spain, it was the Sierra Nevada that represented the last outpost of the Muslim occupation of the peninsula. More recently, mountain zones have become the places of retreat for rebel movements. In south-east Asia, the many post-1945 wars have been fought against a background of resistance or revolutionary movements based in highlands, and refugees from Laos, Cambodia and Myanmar moved to the mountains of northern Thailand. The Afghanistan conflict has involved military penetration and defence of mountain strongholds.

However, this model of mountain settlement, where the population remains at the fringe of society, is incomplete. In Ethiopia, large areas of the Andes and many other locations – especially in the tropics – mountains form the heartland of the present-day states. Often the areas represent the best (well-watered, etc.) available land, and it is the lowlands that have only slowly been settled. In addition, the role of mountains in civil society changes. In Moroccan history, for example, the mountain Berbers have often resisted overbearing lowland-based regimes (some of which were of Berber origin). Following the imposition of French rule, various campaigns of 'pacification' of the mountains finally provided the colonial state with a degree of control, but only by alliances with some Berber authorities. After Independence, considerable suspicion remained and it is only in the last decade that the central authorities have begun to foster reconciliation.

Associated with this 'sanctuary theory' of mountains, which is critically reviewed by Libiszewski and Bächler (1997), is the defence of cultural values. Whilst language remains perhaps the most important dimension of this process, religion as well as particular social and economic patterns of behaviour are also strongly defended. In Alpine communities, the work of Cole and Wolf (1974) is particularly interesting because it examines two mountain communities close to the Austro-Italian border which have undergone considerable social transformation. The village of Tret has its main cultural links with the Mediterranean, whilst St Felix was settled by German-speaking peoples from the north. Although these links are many hundreds of years old, the differences find expression in attitudes to inheritance. Writers have maintained that, in St Felix at least, the Germanic tradition of impartible inheritance has been retained, despite the fact that only a few kilometres away, in the Italian-speaking villages, it is more common to utilise partible inheritance. Dress styles, music, and literature all play a part in preserving the identity of those mountain communities who fear that their traditions and values are being overwhelmed by 'outside influences'. Where this cultural expression is also seen as a vehicle for nationalism – for example the Basque communities, and also the Berbers – they often suffer from political repression. This is discussed in more detail in Part 3.

Settlement and the role of markets

Markets and fairs play a significant economic and social role in mountain areas. The juxtaposition of different ecological zones, in particular the links between highlands and lowlands, means exchange systems are important and finds expression through specific marketplaces often at the intersection of zones. Any such gathering is also significant for social interaction, including marriage brokering, and some sites are also associated with religious functions so that the market *per se* is linked to a broader celebration as a festival.

Despite their social and economic importance, markets and fairs receive very little attention in the standard studies of mountain regions and little has been written in recent years. Moreover, the general literature on markets and marketing is extremely diffuse, although during the 1970s and 1980s it was a key topic in geographic research. Various reviews from that period still provide the most comprehensive summary of the relevant literature (Bromley 1971, 1974). In mountain areas, studies in the Andean region have been particularly prolific and these have broadened in scope from the analysis of spatial relations of market systems to more development-oriented discussions on the potential role of market institutions in the process of agricultural innovation.

The geography of mountain areas makes the pattern of market provision particularly important. Deeply dissected valleys and relatively poor transport mean that markets enable trade transactions to take place at relatively accessible points within the mountains, or where mountains and lowlands meet. One of the most common characteristics is the fact that most markets are periodic, that is to say they do not operate every day. The market days are established according to a regime that may apply nationally or, more likely, as part of a local custom. Sometimes this is enshrined in statute law, and today the markets may have to be licensed by the state. The theory of periodic markets is well covered in the literature (Bromley 1974). It reflects the level of demand available at a particular location and the accessibility of the site to groups of mobile traders who, through the periodicity, can be present at every market. In many locations, a few traders, often with little shops, may open at the site every day, but the full range of activities only exists on the market days. There are also larger, more specialist markets which operate at a lower frequency, perhaps monthly or, in the case of specific livestock fairs, annually. These tend to be occasions when there are major celebrations, both social and religious and trading functions become subsumed within these other activities.

A good example of a market system in mountain areas is that described by Jackson (1971) for the Gamu Highlands in Ethiopia adjacent to Lake Abaya. A basalt plateau stands high above the rift valley floor and there are rapid ecological changes associated with increasing altitude. The *qolla* zone lies below 1,500 metres where cattle-rearing predominated; the *woina dega* between 1,500–2,500 metres where teff, coffee, soya beans and tobacco can be found; to the *dega*, rising above 2,500 to 4,000 metres where potatoes become particularly important. Within an area of some 600 km², containing about 100,000 people, there were sixteen marketplaces holding twenty-five markets per week in total. Thus some sites had more than one market per week. At the time of this study there were few towns in the uplands but Chenchia, with a population of 2,000, provided an important base for the

Plate 3.5 A market in the Middle Atlas, Morocco

many small traders (*chachari*) who served the rural market centres with trade commodities from outside the region.

According to Jackson, the network of markets included those that served only a local clan area and also those that offered rather greater specialities, so those buyers from the local areas visit them. The bulk of the produce was associated with the agricultural economy. Many local markets in this area locate at sites without a permanent settlement or shops. However, on market days the site would be crowded with people trading in *ensete* (a local grain), chickens, butter, honey, along with a few blacksmiths and the itinerant traders offering soap, salt, razor blades and crockery (see Plate 3.5). These markets varied in size from a few hundred to several thousand participants and the bulk of the sellers were women. By contrast, the town of Chenchia then had sixty permanent shops and a daily local market. However, on two days a week (Tuesday and Saturday), a much larger market occurs, bringing in buyers and sellers from a wide area.

In other mountain regions it was the men who traditionally dominated almost all trades, for instance in the High Atlas (Fogg 1935; Berque 1953), and continue to do so. Market days are very vibrant. In Table 3.1, information collected at a Tuesday market in the foothills of the High Atlas (Amizmiz) illustrates the range of activities.

Table 3.1 Commodities traded in Amizmiz market, Morocco

Fruit and vegetables	Bananas, melons, grenadines, lemons, tomatoes, grapes, beans, onions, celery, apples, lettuce, shallots, radishes, potatoes, carrots, beetroot, chillies, herbs	Fruit from lowlands, except apples
Grains	Wheat, barley (maize in season), fodder	Non-local
Meat	Sheep, goats, chickens, cows (all butchered)	Local slaughter
Livestock	Donkey, ass, sheep, goats, chickens (traded)	Local/regional
Bread etc.	Local (baked in market) and cakes, etc.	Local
Hardware	Wide range of new and secondhand household and farming equipment including sieves, knives, bicycle parts, kitchenware, ropes, hoes, baskets	From cities (Marrakesh)
Clothes	New and secondhand men's and women's plus all additional personal items	From cities
Other services	Barbers, cafés, radio/tapes stores, storytellers (etc.), motor parts, books/papers	Local/regional

Source: University of Sussex, surveys 1994–7

This table emphasises the point that markets play an important role in the interchange between lowlands and highlands. Whilst this has always been the case to some degree, economic development and the impact of the market in mountain regions has meant that this interchange has assumed even greater significance. Many products of the large cities (or even imported goods) now find their way into the mountains. In the other direction, trade involves agricultural goods (crops and animals) and forest products. However, it is often the case that the main exports (apples and walnuts in the case of the Moroccan Atlas) are not actually traded through these markets but are exchanged by individual deal between farmers and traders at the homestead. In addition, some governments have provided specific marketing channels for these cash crops.

At the same time, the periodic market provides an important venue for private and state agencies to offer services. Government extension agencies, banks, and medical services often attend these gatherings as an effective way of linking with the general public. Of course, they are also meeting places where all kinds of discussions and gossip take place and it is possible to observe the wide range of dress styles adopted by different ethnic or clan groups, although this is gradually being displaced by a more uniform dress associated with imported clothing. It is remarkable, however, how the site of a periodic market is transformed from an isolated location into a vibrant mass of people, animals and goods for a few hours on one day.

By afternoon the numbers dwindle until at the end of the day the site is deserted and marked only by rubbish.

As noted earlier there is a profusion of work examining the role of markets in the Andes. Though also becoming dated, studies such as that of Symanski and Webber (1974) examine the links between periodic markets and ecological conditions, whilst work by Long (1975) has explored the development of market entrepreneurship in the mountain context. Long argues that in mountain societies the role of 'brokers' is particularly significant in the exchange process. Their strategic function is to link local, regional and national systems through which surplus may be extracted from a locality as part of the 'development' process. A good example of this are the walnut buyers in the High Atlas valleys (Miller 1984), but almost all areas have such traders. They are also vital mediators in the process by which external goods and services reach remote mountain settlements. This work is important because it provides details on the mechanisms behind these linkages that are crucial for an understanding of development dynamics.

Many studies carried out by anthropologists pinpoint particular people who have established a role as a 'broker' and hence closely manage these relationships. In the context of markets, it is these brokers who often control local trade relationships, and in some circumstances operate both within the confines of particular market systems and also as independent agents outside the market. Long describes how these traders tend to rely very heavily upon kinship links to provide the degree of trust needed in trade. For instance, as was noted in the Ethiopian example, one important set of traders link town and rural areas by providing the channel for the circulation of manufactured consumer goods. In remote areas, this can present problems, even hostility, and one way to secure the appropriate channels is to draw upon kin who have previously migrated to urban centres in order to provide a reliable base for the supply of these goods. Similarly, through kin links, the trader is not regarded as an outsider in a rural market and can be trusted in certain transactions.

One particularly important function of markets in mountain areas today is the exchange of animal fodder. Whereas in past generations summer production of fodder may have been sufficient to allow overwintering of livestock, the more profitable alternative use of the land (for a cash crop) or the fact that increasing demand has led to annual shortages means that larger numbers of farmers purchase fodder. Obviously, this is only viable for those who have transport for this bulky commodity but in some mountain areas (e.g. the Middle Atlas in Morocco), where animal movements take place in lorries between grazing areas etc, this has become a viable proposition.

Another interesting example in mountain areas comes from the Yemen which has an unusually large number of weekly markets given the density of population and the land area (Schweizer 1984, 1985). However, this is in part related to the fact that they still operate as an important medium of distribution. They are also closely connected to tribal areas, each being under the control of a particular tribal council which can dictate opening, closure and other factors in their operation. In the west and central highlands, their actual distribution is closely connected to population density, which is linked to agricultural potential (Schweizer 1984; Swagman 1988). These markets are not attached to settlements and, if in the vicinity of one, do not serve that settlement except in the context of the weekly market. *Suq* sites, therefore, are largely abandoned for six days a week. Some butchers and grocers may trade throughout the week – Swagman (1988) refers to a market in the western central highlands where four or five butchers may trade all week, but this is increased to fifteen on market days. He also notes the proportions of different traders – throughout the week some eighty traders may operate, forty-eight of which provide groceries and dry goods, or tailoring. However, many of these expand their business on market days, and are supplemented by itinerant traders. On market days forty-eight of the ninety-three traders counted were butchers, fresh fruit sellers and *qat* traders in more or less equal proportions. (*Qat*, a privet bush whose leaves are chewed, being a mild narcotic, is a social institution in Yemen, and one of the most significant parts of the economy and social life. Its market is almost entirely internal, lucrative, highly prized, and seems to be funded by external remittance income.) The balance of traders therefore reflects the principal economic role of the market as a fresh food supply point.

The nature of the markets in Yemen has altered as a result of a number of economic changes. The importation of goods (legally or not so legally), especially over the border from Saudi Arabia, has increased in recent years (Schweizer 1984). These include not only staple foods such as sugar, wheat and flour, but also everything from cigarettes and electronic goods to building materials – cement and structural steel – and agricultural machinery. Agricultural trading is considered an 'honourable' sector, due to the high regard that agriculture is held in. This is beginning to be reflected in the changing status of traders – formerly looked down upon, they are increasingly becoming the holders of greater private wealth within the growing cash economy, and many tribesmen are beginning to take up trading as a lucrative and therefore acceptable livelihood. In individual markets, the number of traders dealing in imported goods may be small, but their turnover is probably much greater than for other traders. Craftsmen are still an underclass, but even this has begun to change in a number of instances where the unreliable tourist industry has begun to take effect, and the sale of craft work again provides a lucrative niche.

The Yemen example points to changes that are also found elsewhere. These include the availability of cash through remittance income, which increases the spending power of individuals, and the government construction of roads and highways. The increased mobility of the population has led in some cases to the relocation of markets closer to highways, and the demise of some remoter *suqs*. A second structural element of change is the fact that a gradual shift to a livelihood predominantly based upon trading means that premises are increasingly encouraged to open up each day. They also have access to private or shared transport (for example trucks), which increases the volume of goods brought to, or stored in, the market. The old market buildings become increasingly too small for this growing volume of goods, particularly imported manufactured materials, and the greater permanency of the market leads to further pressures to relocate, rebuild and change the weekly habit to a daily one.

In mountain areas, as elsewhere, the settlement pattern we observe today represents the outcome of a process of continual change in the dominant social and economic factors which influence the location of human activity. In some eras the mountains attract; in others they have little to offer human activity, and the settlement pattern reflects this ebb and flow. However, particular locations are remarkably stable through the influences of inertia. Once-defunct mining settlements may form the basis of a new 'resort location', or the original site of a periodic market becomes the location of a new township. In some instances, a once scattered pattern of farmsteads has been transformed into a nucleated settlement when a new transport route focuses economic activity. Moreover, the pattern and internal structure of settlement reflects the prevailing norms of behaviour, whether of families or of the economic agents locating in the settlement. Thus we turn to the question of culture and its ramifications in mountain societies.

A MOUNTAIN CULTURE?

Culture permeates all aspects of life to differing degrees, and provides the background to the constraints and opportunities under which communities and individuals function. Culture is both the continuously evolving set of ideas and meanings that shapes the community and its behaviour, and also the means by which populations manipulate their environment (Rosman and Rubel 1995). Culture is a peculiarly human thing – whilst all animal groups can have some form of social structure or society, only humans have the more complex rules and forms to which the members of that culture conform. Individual behaviour may violate these rules, and in close-knit traditional societies these violations may result in ostracism. The peer pressure operating within these societies may strengthen conformity, particularly where the need to belong to a particular social group is the critical factor in determining access to environmental resources.

Two aspects of culture are identifiable. First, the content of the culture itself – the cultural roles governing behaviour. Second is the form in which these cultural rules are actually expressed, which may be different to the rules themselves. For example, the penetration of a mainstream religion, or a cash economy into remote mountain regions, either by forceful conversion, opportunistic appropriation or by default, may be expressed in modified ways in these peripheral regions. Culture evolves, often by selecting those behavioural aspects most beneficial or useful to a society at that time; but the society may well not adopt wholesale all the elements that make up the new cultural 'package'. Thus, for example, Islam or Buddhism may sit alongside older animist beliefs of ancestor worship, spirits and sacred places as in Morocco or Thailand, and the cash economy may run alongside a traditional barter economy. Therefore, it is not so much the actual appearance of cultural features, such as a particular religion, but the expression of them in mountain regions which may identify a mountain culture as different from other cultures. In the context of mountains therefore, we are primarily concerned with socio-political institutions, gender and religion, which define the role of individuals in society and the patterns of control and support held over them by the community.

Religion

One of the most important cultural values influencing the use and management of environmental resources is that of religion. For High Asia, Kreutzmann (1993) illustrates how the pattern of religious affiliation cuts across major mountain areas, forming a mosaic of practices that have had important implications for the region's development. Differences in religion and in its expression and relevance for the daily life-world of mountain dwellers to some extent permeate all aspects of mountain economy, social structure, and interaction. For example, it may affect gender roles, as in formal Islam where women tend to be restricted in their activities outside the home. This is actually expressed in much less severe forms in the relatively remote context of mountains when compared to the economic and political centres of an Islamic country. More animist traditions remain important in the same communities – for example in Nagar, Pakistan, women are barred from high mountain peaks because of their 'impurity' (Hewitt 1989). This, of course, affects their ability to perform tasks such as tending cattle on high pastures, or collecting firewood. The descriptions of mountain Berbers by Brett and Fentress (1996) are similar.

In the same way religious values may determine the attitudes to resources, environment, and its use and access rights. Many Himalayan mountain communities are Buddhist, and this has consequences for the use of livestock as a food resource. Inheritance, family sizes and marriage patterns may also be directly or indirectly determined or shaped by religious or equivalent belief structures. The degree of equality between individuals, both in terms of wealth and access to resources and the status of individuals and families within the community, may be influenced by caste systems aligned with religion. For instance, the 'musicians' caste in Hunza was afforded special allowances to settle and gain access to land by the Shiite Hunzakuts. In many areas the role of religion acts as a point of departure for the determination of cultural identity. Thus a particular group, such as the Hunzakuts or the Ladakhis, is 'defined' or socially referenced in part though their religious affiliations. Bertelsen (1997), who describes the genesis of neo-Buddhism in Ladakh during the 1930s, explores this issue in depth. In this mountain region, as in much of the South Asian subcontinent, it was the British colonial authorities that decreed that religion was a critical social reference. Subsequent political actions, which elsewhere may have been associated with other social variables such as class, have traditionally been related to religious affiliation. The issue today centres around how Buddhism itself has been reconstituted into a political rather than a religious force.

The importance of mountains as sacred sites, and as destinations of countless pilgrimages, is also significant. Bernbaum (1997) discusses this issue, suggesting that the sacredness of mountains derives from three elements. In the first instance mountains have always evoked a sense of awe, through their grandeur, their isolation and by their imagery of power. Second, particular cultures identified mountains as playing a part in sacred functions. Gods live there. Mount Olympus is a well-known example in Greek cosmology. Vogt (1990: 19), in his study of the modern Maya in Mexico, states that 'almost all mountains . . . located near Zinacanteco settlement are the homes of ancestral gods'. Finally, many are places of pilgrimage or sites of religious foundations such as monasteries. For Hindus, Badrinath is one of the prime pilgrimage sites of India, receiving something like 450,000 pilgrims each year. Far less accessible, Mount Kailas in Tibet is sacred for both Buddhists and Hindus. Mount Fuji, of course, provides a similar function for many Japanese as part of Shinto beliefs. Today the sanctity of some mountain areas precludes their development in certain directions, and colours the attitudes to mountains of both the local inhabitants and outside visitors or observers. Pilgrimage and tourism have become intermixed in some instances and there are severe problems of damage to the local environment. With changing accessibility and increasing wealth, what was often just a trickle of the hardiest of pilgrims to a remote site has become a flood orchestrated by global travel companies (see Chapter 7).

Other aspects of culture include the language and artistic traditions associated with particular regions. Sometimes particular artistic traditions have developed as a result of specific products available – for example, the abundance of cedar wood in the Swat valley, Pakistan, supported the architectural and carving traditions which were later adopted in other regions (Kalter 1991). Much mountain architecture is similar. An abundance of stone and relatively little timber in many areas, and the need to provide substantial buildings capable of sheltering animals and humans, means that low, stone-built structures abound. Elsewhere, if stone is unsuitable mud brick replaces it, or timber is used if it is available. However, religious, defensive, protective and privacy needs have resulted in buildings which tend to have few windows on the outer walls. Local variations associated with establishing tribal identity, territory and individuality abound. Brett and Fentress (1996) provide an account of the architecture in Berber villages and there are many detailed descriptions of architecture in the European Alps.

Language

Language is an important issue in many mountain areas closely associated with tribal identity. Many different languages and dialects are spoken, not always mutually intelligible. Thus, linguistic and dialectic diversity is often very high – in Hunza, for example, three different languages are spoken, and each dominates in a different part of the valley. Similarly, the different groups occupying the mountains in northern Thailand each has its own language derived from their original homeland. In the High Atlas of Morocco, watersheds and valley systems often coincide with old tribal boundaries and the dialect of Berber spoken by the occupants. In the Caucasus there are twelve official languages in the autonomous republic of Dagestan. However, language affiliation is very dynamic. Mountain communities are renowned for their linguistic ability, often derived from nomadic practices and the need to communicate with neighbouring groups. Today many of these languages are now only spoken in the home and by the women. Male members of the household have been forced to learn 'national languages' as part of their wider employment opportunities, especially if they have been members of the armed forces or other state organisations. In addition, state education in a common language (Urdu in Pakistan, Arabic in Morocco) is often available to the children and men and, more recently, to women.

The survival of either pure or adulterated forms of cultural traditions, such as language or customs, may in some areas be revived as a result of the search for tradition by growing numbers of tourists (Allan 1988; Cohen 1989; Price 1996). Festivals and customs may be laid on especially for visitors – folk dances, music, and crafts are good examples. This may have the effect of preserving culture, but it may also result in stagnation as it becomes dislocated from everyday life. It then exists merely as a spectacle or source of income and thus ceases to evolve as the pressure to maintain a 'pure' cultural 'tradition' for the tourist market precludes assimilation of evolved forms into daily life. With this dislocation comes a break from many of the traditional attitudes and values, and can be a very destabilising influence in the socio-political sphere (see Chapter 7).

Attitudes to resources

The cultural framework indicated in the first instance by religious authority and then through other elements such as wealth, kinship and latterly the state, provides the controls on resource use and maintains the myths that have sanctioned these controls. The attitude to 'ownership' of land is culturally determined. The concept of divine ownership and human stewardship is a fundamental part of attitudes to use, ownership and control of resources such as land, water and forest. Spirits must be placated, summoned or deterred and inauspicious omens may cause the abandonment of land or new clearances, as shown by the case of the Karen in Thailand. These peoples were once shifting cultivators and viewed land as a temporary gift from the gods, the users having occupation but not perpetual ownership. In more recent times, this attitude has changed as a result of both state intervention and the dynamics of the market.

It is instructive to compare attitudes to forest resources between populations in the European Alps in the past and cultivators in Nepal in more recent times. There are clear differences in attitudes to tree felling. In the eyes of the Nepali cultivators, for whom land is scarce and population growing, the cultural tradition does not work to cut birth rates very quickly. In this instance much higher value is placed on clearing forest, and using the land for cereals which support more people per unit area. The careful construction of terraces, and care over cultivation to make the best possible use of this scarce land, results in a productive environment, no more or less susceptible to erosion than the forest but where trees are considered convertible resources – convertible into productive land.

In the Alps since the 1950s there has been much less pressure on the use of land for agricultural production and economic survival because livelihoods can be obtained from other economic activities. In this case, forests are considered a valuable and renewable resource – the value of the land and its stability being greater when covered with forest. The need for the land for agricultural production, the cost of maintaining terraces and the dearth of local population engaged in subsistence farming for their livelihoods means that the management infrastructure needed to maintain geomorphological stability in the landscape is missing, and therefore a different view of forest is appropriate. These views do not, however, translate happily between regions!

Of greater importance is the recognition that prevailing attitudes to resources are dynamic. This is the basis of the use of cultural theory, noted in Part 1. The implicit assumptions of the examples provided above include the fact that whatever cultural norms prevail in traditional society they remain true throughout the period until 'modernisation'. However, it is precisely this static view of cultural values that is attacked by Price and Thompson (1997), who argue from the standpoint of cultural theory that myths have always been fluid and their trajectory remains ultimately uncertain. In their examination of the 'myths' in the context of forest management at Davos in the European Alps, they examine the fluctuations in attitudes towards forest preservation, especially given the role of forest plantations in avalanche control. They note how a combination of specific legal instruments and social/religious control has been used over time to prevent forest removal. However, in some periods this control lapsed and often the community paid the price in terms of increased flooding. The crucial issue was the extent to which a combination of material needs – for example, opportunities to increase wealth through forest removal – and prevailing cultural norms made the exploitation socially acceptable. This example also suggests that the dominating influence behind changing cultural norms was the role of external pressure, in this case the changing demands for resources.

An example of juxtaposed, different but complementary cultural traditions can be found in some areas of the Nepal Himalayas (Fürer-Haimendorf 1964, 1975; Hewitt 1988). The lower altitudes are predominantly occupied by settled Hindu cultivators. They grow grains and keep a few livestock on intensively cultivated terraces. They value large families as a source of security in the future, and follow a form of partible inheritance whereby all sons receive a share of the land. Land is acquired primarily through marriage. Therefore the area is very densely populated, with families having a number of small, widely dispersed plots, and reliant upon the extended family for labour supplies. These are called the 'cautious cultivators'. At a higher altitude there is a predominantly Buddhist population who are mainly pastoralists. They favour impartible inheritance, thus maintaining the capital of the family intact, passed through the male line. Marriage becomes possible only after a man has come into his inheritance, so many men are not married. Many single men traditionally went to the monasteries that were supported by the local community. The paucity of agricultural land, and shorter growing season higher up, meant that crops were not productive, and thus it was necessary for links to be maintained in trade between the Hindus, who needed animal products, and themselves, who needed grain. In addition, the significance of the monasteries to the religion made this region a focus of pilgrimage, and thus trade in supplying pilgrims, and housing them was always an important factor in the economy. Trade also extended across the mountains into China and Tibet, and, having yaks and an intimate knowledge of the mountains, these people were well placed to act as porters for pilgrims and traders, carrying salt southwards and animal products northwards. In the present economy, these 'adventurous traders' found it easy to extend their portering services to mountaineering and later trekking expeditions, thus being opportunistic in outlook and benefiting from change.

INSTITUTIONS AND SOCIETY IN MOUNTAINS

In trying to delineate the various 'cogs' in the mountain landscape, it is important to highlight the role of institutions. Very little work within the conventional mountain literature has focused on this issue and much research remains to be done. Institutions are an essential element in mountain societies, providing the framework through which cultural attitudes and behaviour in the widest sense are mediated, regulated and reinforced. They exist at all scales, from household units through kinship and village organisations to collective labour management and wider political organisations. They influence both individual and collective responsibility, with changing objectives and priorities in space and time. In household units, individuals or groups of various compositions respond to changing conditions and needs by altering their management strategies. As noted above, the work of Price and Thompson (1997) emphasises the dynamic nature of institutions and that they are in a continual state of flux. This might not be apparent from casual observation. Indeed they are often presented as 'timeless', but subtle changes are always taking place. This is particularly important to appreciate because there is a tendency to characterise mountain societies as static and only responsive to powerful external pressures such as those represented by the incursion of state power.

Fisher (1989) makes the distinction between indigenous and traditional community institutions. Traditional implies a degree of antiquity that many resource management strategies in Nepal, for example, do not demonstrate. They may arise from truly old, traditional practices, but their current manifestation is the result of recent changes due to contemporary stresses and conditions. Likewise, Fisher also emphasises that not all institutions in place in Nepal could be considered indigenous – that is, arising from the existing population. Some forms of management may be imported or imposed, and represent a local manifestation of much wider strategies. A good example is the development of the *nawa* system of Khumbu that is assumed to be ancient. This system was designed to ensure, on the one hand, that livestock did not damage village lands, and on the other that winter grazing regimes were preserved. It involved the appointment of officials charged with policing animal use of the lands near the villages. However, Stevens (1993) argues that it may have developed only during the nineteenth century as a response to breaches in restrictions between villages, leading to the modification of older common property regulations.

The key institutions of social and political organisation are the community, kinship (extended family), and family or household and the social relationships within them. Family and kin provide one of the most important building blocks around which social institutions are constructed and this applies as much to mountain communities as those elsewhere. The particular forms of this relationship are expressed through marriage rules and inheritance and hence access to resources and labour. At the community level, families join in social religious and economic activities that sustain livelihoods and ensure the continuation of the prevailing 'myths' (Price and Thompson 1997). These traditional patterns of allegiance are not cast in stone and have always been under some tension, if only that arising from intergenerational differences. However, external factors including war, incursions of the state, market or environmental (climatic) changes provide macro-level forces for transformations, which will be examined more closely in Part 3.

Whilst acknowledging that within the literature of anthropology the interpretations of family, kinship and community are constantly under debate, for the moment the object of this section is to illustrate the various ways in which these terms have been used in the descriptions of traditional mountain societies. At the level of the household and kinship a detailed account is provided by Crook (1994b) of the village of s'Tongde lying at 3,550 metres in Ladakh. In this village of some thirty-two households, each household unit usually consists of two sections: one known as the 'big house' (*khang-chen*), the other (of which there may be several) called the 'little house' (*khang-chung*). The eldest brother whose family still contains young children, along with other brothers who are husbands or co-husbands of a single wife, occupies the big house. In the little house are to be found the parents, along with any relatives (brothers/sisters) of the parents' generation. In the 1980s, the average size of a 'big house' was 6.09 people, for a 'little house' 3.0. This 'typical' household has many variants, reflecting the diverse age structure of a community, but the basic household/kin structure is interwoven with that of inheritance, which is based upon primogeniture. Along with each house are attached an array of plots and also livestock. In addition, the household structure is associated with control over labour, itself a crucial factor in the successful survival of an agrarian economy. While each unit within the household has access to its own plots the ability to provide assistance through the kinship links has been one of the most powerful mechanisms by which such communities cope with problems.

In most traditional mountain communities the kinship links extended into a wider social structure which, formerly, if not today, provided the basis for the political organisation of the area. Returning to the work in the High Atlas Mountains discussed earlier, Hart (1981) argues that in the Atlas social relations were constructed traditionally around segmentary kinship relations. A household is part of a kinship network, which is embedded in a hierarchical organisation of lineages, clans, tribes and confederations. In the Reraya valley, Miller (1984) describes the traditional structure of the Ait Mizane as a tribal group consisting of thirty-six lineages, in four clans (or fractions) occupying twelve villages. The Ait Mizane form part of the confederation of Reraya, which occupies the watershed of the same name.

Conventional historical accounts claim that prior to the early part of the twentieth century, this confederation was one of the 'players' in the continued flux of the political power of this region alongside the better known Glaoua and Goundafa. Miller argues that one of the key elements of the social geography of this area was the close coincidence of the tribal confederation and the watershed. Thus the Ait Mizane occupy a distinctive geographical territory, strategically placed at the head of the Reraya system. The implications of this for the management of an irrigation system include the fact that communities lower down the valley are bound to the Mizane through the interlocking of needs and social obligations.

This hierarchical model of social and spatial relationships provides a neat framework for integrating environmental and anthropological variables. In mountain areas watersheds become more than just a physical entity, they are the home of particular communities whose livelihoods depend upon both the skilful manipulation of the environment and the maintenance of cordial relations with other groups. In the past, if not today therefore, it was the institutions of clan, tribe, and confederation that provided the framework in which life was conducted.

However, despite providing a useful starting point for the social geography of mountains, this neat model is the subject of considerable debate. At the general level, the notion of the 'tribe' has been under attack as more writers examine the problems of 'identification' and 'representation' in colonial regimes. In the Moroccan context the 'tribe' is increasingly viewed as the deliberate result of the colonial obsession with political control and bureaucratic tidiness. In the Himalayas, a similar argument has been put forward for the Ladakh region by van Beek and Bertelsen (1995). An associated issue concerns the claim that in traditional mountain societies there is a close correspondence between watersheds and confederations. This was strongly disputed by some of the earlier writers on Atlas societies (Montagne 1930), primarily because of the constant flux of peoples across mountain areas. In Miller's view, this merely accounts for the inevitable local anomalies within the prevailing institutional framework. This is not the place to examine the general problems of social identification, but, clearly, the claims about territory and tribe begin to look rather shaky if neither can be identified very clearly! In fact, in the Moroccan example, the bureaucratic identification itself to a large extent fossilised the tribal spatial pattern on which later writers have worked. In addition, there remains the important fact that, where neighbouring groups have access to a variety of natural resources, various institutional structures exist to prevent constant conflict. Whereas today such institutions may be based purely on technocratic management demands, traditionally they were an integral part of the society occupying the mountain areas.

In the highlands of Yemen, Swagman (1988) illustrates how different institutional structures can co-exist. He emphasises the importance of the

Bedouin legacy of tribal base and the way in which this has been modified as a result of settlement of nomadic groups. There has been a transition from kinship as the basis of tribal membership to a spatially defined territory with membership by residence. Individual families or households give allegiance to a particular tribe, and whilst this may be relatively inconsequential in everyday life, unity and identity is enhanced during times of conflict. Tribal units may comprise one or two major descent groups, but this may be an abstract relationship and birth into a tribe is still a major source of identity. Tribes are divided into segments that may form close confederations when external pressures and threats arise. Otherwise, interaction may be fairly minimal. There is a difference between tribal identity and identity based purely on lineage, which may be independent of place. Such groups hold common descent through the male line, but again, when the need arises, such groups can coalesce.

In the present day, however, where permanent settlement is more common than nomadism, residence is a more significant factor than kinship in determining membership. This is particularly true of villages where the institutions are more completely based on state administrative units. The state in Yemen has relatively minimal control over many Bedouin and some mountain population groups, but elsewhere it holds the main positions of authority through government-appointed officials. Many state administrative boundaries are based on tribal territories, especially at the provincial and smaller levels.

Equally problematic, but in widespread use in the mountain literature, is the use of the term 'community'. It is a dynamic concept. In some circumstances it may be used to describe a collection of households living in close proximity, thus relating to geographical space, or it may comprise a group of people linked in some social or political sense. Links are extended through the wider family and kinship associations, and by tribal membership. This exemplifies the household, village–valley linkages noted in Part 1. However, affiliations of individuals may change, and in times of stress the identity of the 'community' may be drawn up by different criteria (family membership, tribal affiliation, or by a group which has a common interest in a particular resource). A good example is the 'creation' of a community described as 'Chipko' representing the socio-political movement in the Indian Himalayas, comprising mostly women who campaigned against commercial logging (Shiva and Bandhyopadhyay 1986). These groups have varying degrees of formality, different power and influence structures, and their composition may change with time.

With the increasing importance of external, especially state, agencies the term 'community' has become part of the rhetoric of interventionist management. This has become particularly apparent in recent years with the appeal to 'bottom-up' participation. As was noted earlier in the case of the Ladakh, communities become identified, or even created to suit political or bureaucratic need, and assigned powers that were not inherent in the original constituency. Whereas regulatory powers may have existed within the community controlling such things as access to land, forests and water, the sanctions associated with those powers were themselves mediated by the participating families. Today such sanctions are often enshrined in formal national law and enforced by state power, completely changing the nature of the relationships involved.

LAND TENURE, WEALTH AND POWER

In many mountain communities, the primary source of wealth remains the land and livestock. Without access to land for cultivation, grazing, household construction and forest resources, the livelihood of mountain dwellers is untenable. Cultivable land is relatively scarce, and as it provides the foundation of livelihood every effort is made to retain ownership or tenancy of land. Ownership of land, or control over its access, is therefore not only the basis of wealth but also to some extent of power, although this is less clear in many examples. The full range of tenure relationships can be found in mountains and here we provide a series of examples to indicate the way in which land tenure is closely interwoven with other key social and political institutions.

An interesting example can be drawn from the work of Campbell (1964) in Greece where at least 30 per cent of the population still resides in mountainous regions. Although now dated, it shows how the shepherds in Greek mountains view property as the basis of prestige. Inheritance is subject to the judgement of the family – if the obvious candidate is considered unreliable he can be disinherited. Women are entitled to a dowry. The principal object is to secure family resources in the face of a perceived hostile community, so the maintenance of property within the family ensures continued honour, wealth and status. The head of the household's authority is absolute, but tempered by the understanding that the head's role is that of a trustee for the next generation. Shepherds who did not own land had to rely upon annual renegotiated leases and thus were dependent upon landowners. Power therefore lay with the landowners. However, land reform in 1938 resulted in the state upholding the shepherds' right to continue leases from one year to the next, so making grazing rights more secure. This has allowed shepherds to build up their own wealth by increasing their flocks and has begun to change the balance of power within the community. With the advent of Greece into the EU and the technical and economic changes facing the agricultural sector, particularly in the mountains, the durability of the traditional patterns of tenure is much debated. Some maintain that precisely because of the close links between the tenure system, cultural values and power structures, there is 'no sign that farm villages and farm structures will be dismantled' (Damianos and Hessapoyannos 1997: 304).

A similar process of continual redistribution is recorded in the Swiss Valais (Wiegandt 1977) where, for each generation, partible inheritance occurs, giving rise to fragmentation but also to re-amalgamation as a result of marriage. Partibility is a requirement of the state, but there is evidence here and elsewhere (Netting 1981) that although land ownership records indicate fragmentation, alternative strategies may be in force in order to retain ownership of land. Examples of such strategies are the consolidation of households and their associated estates by retaining celibate siblings within the same household unit, or, by agreement within the family, land titles are not actually transferred to siblings.

The close links between inheritance and land tenure arrangements are a vital ingredient in the dynamics of mountain communities. Later we shall explore the demographic implications of this. In the European Alps there has been extensive research into the problems of fragmentation associated with partible inheritance. Wiegandt points to differences between the formal inheritance patterns and the practice, so that, despite partibility being the 'norm', re-amalgamation occurs often at marriage in order to maintain ownership of land resources. Netting (1981) also indicates that practice is far more flexible than the customary law would appear to suggest.

A number of studies (Stevens 1993; Fricke 1994; Bishop 1998) describe similar patterns of land ownership and control in the high mountain zone of Nepal. All land appears to be owned, and no household is without land. However, in Melemchi, the religious institution that collects the taxes and retains the right to a certain amount of annual labour from each tenant, also claims ultimate ownership. However, tenants operate as if, in effect, they owned the land with respect to inheriting and buying/selling it. The youngest child inherits the family home and some land, and remains there to care for the parents. Fricke (1994) demonstrates the importance of open-ended reciprocity within the kinship, as in this context the exchange of labour and goods is not given a direct economic value. Beyond the immediate family, however, a careful costing is kept of obligations owed, and the acceptance of reciprocal arrangements is considered in the light of the ability to repay the obligation by an equivalent amount. The proportion of land and livestock that is held by each family remains very similar, even though the total wealth of the household may vary substantially across the community. This reflects the allocation of labour and the balance of different resources that provide the livelihood. Informal co-operative groups exist to support some specialist individuals, such as metalworkers, by contributing to their upkeep. Similar conditions exist to support the woodcarvers of the Swat valley, Pakistan (Kalter 1991).

In some mountain regions it is common to find sharecropping arrangements. In the Yemeni mountains the crop may be divided equally between tenant and landlord (Swagman 1988). However, according to Stevens (1993), at least in the Sherpa areas of Nepal, this has fallen into disuse and cash to a value of 25–50 per cent of the crop of a good harvest is payable each year. Moreover, wealth in Khumbu is often associated with the tax collectors. Previously the local *pembu*, a hereditary family head, retained this power and prestige, working as tax collector, arbitrator and patron. In the new state government the local *panchayat* (division) officers are not necessarily these *pembu*. Tax is also only payable on land owned in the immediate area of the main village. In 1965–6 the Nepalese government changed land tenure rules and abolished the entitlement to land cleared and improved by individuals – a measure to counteract the conversion of very marginal land. However, the practice continued without retribution, through the surreptitious extension of boundaries and reclamation of abandoned plots, until the introduction of zealous policing in the mid-1980s following the establishment of the Sagarmatha National Park.

In the Yemen, land ownership falls into four main categories. First is privately owned land; most cultivated land is in this category. This land can be bought and sold, but transactions are almost always kept within the tribe. Serious disputes have arisen as a result of attempts to sell land near boundaries to adjacent tribes, however convenient and rational such a step may be. It is seen as weakening the resource base of a particular tribe and to be avoided at all costs. Such disputes have often led to violent confrontations. The second category of land comprises that assigned to a local mosque on a long lease or as a gift. Such land is administered by the Ministry for Religious Endowments and usually sharecropped by tenants. State land is the third category, and, finally, open tribal land, which officially belongs to the state but is open access and comprises relatively poor land suitable for extensive grazing. Technically title to such land can be claimed by individuals if they take it in, and improve and cultivate it, but in reality it is usually so poor that it is not worth the effort to reclaim.

Swagman also draws an interesting distinction between the tribal-based communities and the state-aligned ones. Following from the divisions in land tenure, there is also a notable difference in the persons appointed to collect tax and hence in the sources of wealth. In a tribal community it is the sheikh who is the main arbiter in disputes and who accumulates wealth through holding offices as judge and tax collector. This is not the case in state-aligned groups because the sheikhs do not necessarily hold these offices, and consequently have seen a notable reduction in their income.

A final example is from the central Andes. Brush (1988) refers to the individual ownership of cultivated plots, but the communal control of planting and fallow schedules. In areas of aseasonal cultivation, planting may occur throughout the year. However, communal co-ordination of fallow-crop cycles is operated in order to ensure that potato crops, a staple food, are separated by seven years, during which one to two years of cereals and four to five years of fallow occur. This ensures that potato disease conditions associated with the nematode do not survive. Intensification of planting regimes, reduction of fallow, and use of chemical fertilisers contribute to the closer spacing of potato crops in some areas, and thus the increased risk of disease.

Although the emphasis so far has been on land, resource rights apply equally to water, in many cases tied to the rights to land. These rights can also be linked to grazing and forestry provision. Vincent (1995) refers to land reforms in the central Andes, which have upset the pattern of land and water rights. Formerly, in parts of Peru and Bolivia, specific water allocations depended upon the quality of land suitable for particular crops. Access to water therefore came with ownership of land. However, with land reform, land tenure has become separated from water rights, permitting not only the privatisation of water supplies but also the possibility that land may be exchanged which might have no associated water resources. In this instance the owner usually has to purchase water rights independently and this has had important repercussions concerning agricultural development. Such reforms are often well-meaning in redistributing land to more people, but often difficulties arise of this nature due to an incomplete understanding of the overall resource provision.

One of the most crucial issues confronting management of resources concerns the regulation of common property resources (CPR). Such resources are extremely common in mountain areas and encompass basic resources such as land and water, and other resources that have multiple uses for the community such as forests, or those that have a sensitive or low productivity such as grazing lands. CPR has become the subject of considerable debate in the last two decades and an extensive literature already exists on these issues (Berkes 1989; Ostrom 1990). Perhaps the most important point to recognise when dealing in a mountain environment is that CPR does not simply entail open, unregulated access. It is this fallacy that has proved so damaging to the debates about mountain degradation and the design of appropriate management strategies which are examined in later chapters. In most instances, it is precisely the role of the institutional structures already noted to ensure that such resources are very closely managed. Thus land tenure institutions constitute one of the most important features of mountain life.

Embedded within the structure of land tenure relationships, and therefore within the overall cultural values of traditional mountain societies, are the political structures through which conflicts are mediated and allocative decisions made. In particular, some political institutions are necessary to ensure a framework for negotiation between local and more regional (and later national scale) authorities. The wide range of political institutions that extend beyond the local community matches the diversity of mountain areas. Power may operate through a formal local ruler, such as the Mir in Hunza, Pakistan before his deposition. This arrangement resembled a feudal system in which the Mir owned much of the land and rented it to tenants. They could count on his protection and judicial action in conflict resolution in return for their labour and rent. Such a system was common throughout a number of Himalayan kingdoms. Alternatively, as has been noted above, power may be administered through a religious institution such as a Buddhist monastery or other temple. Such a situation is common in Nepal and Ladakh. In the Zangskar valley, detailed work by Shakya *et al.* (1994) has attempted to unravel the political economy of monasteries, a task made extremely difficult by the fact that much of the detail of monastic life is locked in the personal memories of the monks. Even when accounts are available, measurement systems need to be interpreted, as the system of weights and measures has to be translated into a common metric. However, on the issue of land we are fortunate as documents, often dating back many generations, still exist. In the case of the monastery at s'Tongde documents show that it owns about twelve hectares of land. This land produces as rent something in the region of 1,500 kilograms of mixed grain per hectare, which is similar to yields on British farms prior to the introduction of high-yielding varieties.

The absolute authority of these systems lies in the feudal landlord or the monastery, and is administered directly. Such relationships require the allegiance of individuals in the community to the ruler or monastery, and conformity to that power structure in order to have access to the benefits and protection they offer. The power and authority within these systems may be nominal, holding the land but not dictating how it is used. In the case of the Mir of Hunza, the landlord or ruler plays an important part in the crop year, setting dates for festivals and being central to rituals. Other institutions may be less formally structured, and may arise and decline according to need. Groups of elected or hereditary elders often constitute a village council that represents the community and makes decisions concerning rights and disputes. This was the case in many mountain regions in the past, for example in the High Atlas and the European Alps. The members may represent the heads of the families of the community who inherit a place on the council – in the Atlas called the *j'maa* – or be elected by the community to represent their interests. Such a council may be the principal authority and jurisdiction over a community's resources and management, and have the right to enforce regulations agreed upon by that community. In either case, the membership of the council may not be exclusive, and other members of the community may attend meetings where matters are of direct interest to them. They may be part-time, when they support themselves but receive payment for services as called upon. Alternatively, they may act full-time and be effectively supported by the community, which provides food, labour assistance with their fields and livestock, or payment in cash. The latter case is the common status of those appointed watchmen over forest and water resources; this is discussed in Chapter 4.

MOUNTAINS AS BORDERLANDS

One of the characteristics of mountain areas is the fact that they form the periphery of states and so national boundaries often pass through their peaks and valleys. Good examples are the Italian Alps, the Ruwenzori in East Africa and segments of the Andes. Sometimes the boundaries run directly through the mountains themselves, in other cases the border lies at the foot of the range. This is a topic that has been extensively explored in the literature on political geography (Prescott 1987).

Historically, political authorities have claimed assiduously that mountain peaks make good boundaries as a glance at many of the nineteenth-century histories will reveal. Whilst there is some logic in this, for example when dealing with watersheds, it presents particular problems for communities who are nomads or where transhumance plays an important part in their livelihood. In many instances, the border zones are politically sensitive and groups who frequently cross and recross are deemed a security risk.

As a result of the closure of borders, grazing areas may suddenly become proscribed. Returning to the work of Campbell (1964) in post-war Greece noted above, he describes how transhumant shepherds have had to use pastures in Greece more intensively as a result of the closure of borders such as that with Albania. The passing of laws by the Greek state giving equal citizenship to transhumant populations and to the settled villagers, together with laws protecting the shepherds' access to summer grazing lands, caused increased pressure and conflict between the two groups. Changes in co-operative measures between the groups, associated in part with the exodus of many young people of the settled villages, enabled co-operation and access to be incorporated. However, population growth and changes in power structures and wealth distribution caused stresses and thereby polarised the identities of the different ethnic groups.

These mobile groups may be incorporated into the wider society in so far as they do not affect the productivity of the original population, or they may make efforts to exclude these people from traditional tribal lands. This often leads to conflict and the intervention of wider political forces to mediate and diffuse these conflicts. It may also lead to imaginative ways of circumventing or interpreting these laws in order to retain maximum resource access for themselves. An example of this is in Khumbu, Nepal where state legislation prevents exclusion, but locals excluded outsiders' flocks on the basis that a foreign shepherd (i.e. one not belonging to even the wider group entitled to access to pastures) led the flock (Stevens 1993).

In addition to the political frontiers noted above, the fact that mountain areas have often been areas where different cultures meet has led to the existence of linguistic frontiers. Among the seven states in the Alps, there are four languages (Italian, German, French and Slovenian) along with a number of surviving dialects including Rhaeto-Romance. A similar complexity can be found in the Andes between Aymara- and Quechua-speaking peoples, and in the Himalayas.

The survival of ancient language groups, as well as the persistence of many distinct dialects (for instance Tyrolese German), reflects one of the salient features of a mountain environment – namely, that often the territory is highly compartmentalised due to the topography. The result is well illustrated in the Alps in the Canton of Graubunden where in the 1980s 35,000 people spoke five dialects between them (Rougier *et al.* 1984).

The Tyrolese example

The name 'Tyrol' goes back to the period when the German-speaking Austrian *Länder*, Romance-speaking Trentino and the Province of Alto Adige all formed a common province. After 1919 Trentino and Alto Adige became Italian, with South Tyrol becoming a German-speaking area south of the Brenner Pass – but only as far as the Salurn (Cole and Wolf 1974).

From the eleventh century the area known generally as the Tyrol, encompassing currently both the Austrian and Italian zones, was settled by Bavarians who occupied the land under the tutelage of the 'Counts of Tyrol'. The mountain communities were relatively independent, however, and continued to fight each other throughout subsequent generations. Even when this area was absorbed into the growing Habsburg Empire during the fifteenth and sixteenth centuries, the Tyrolese were left very much to themselves provided they paid the necessary dues. At the beginning of the nineteenth century the Habsburg Empire stretched over the Alps into northern Italy, but with the gradual rise of Italian nationalism they rebelled against Austrian power, controlling Milan in 1859 and Venice in 1866. The German-speaking Tyrolese began to look towards the rising power to the north. Thus the Tyrol became an 'ethnic battleground' exacerbated by the fact that after the First World War Italy assumed control of the South Tyrol, despite the fact that a plebiscite in 1921 (not recognised) favoured links with Germany. The rise of Italian Fascism, driven by intense nationalism, meant that the subsequent twenty-five years consisted of a process of 'ethnic cleansing' in which many of the traditional cultural practices of German-speaking Tyrolese were forbidden. This included in 1929 the repeal of the law that reinforced the practice of impartible inheritance. In fact the borderlands became a 'pawn' in the formation of the 'Axis' in so far as, once Hitler and Mussolini had agreed to join forces, the Germans were no longer interested in the problems of Tyrolese recognition. Indeed there were suggestions that the German-speaking Tyrolese should be moved out of the area to the Reich. For a brief period the German authorities controlled the area before they were driven out by the allied forces.

In the post-war period an agreement in 1947 fixed the boundaries along the 1918 line. Although the Fascist regime tried to increase the number of

Italian speakers in the area, it was recognised that the South Tyrol would remain a problem area within the new Italian government after 1945. The result was the creation of an autonomous region, which incorporated both German-speaking and Italian-speaking zones so that linguistic partisanship could be neutralised. At the same time, changes in legislation allowed German language TV, schools and media to be established (Minghi 1963).

This example illustrates the fact that a mountain area is playing a multiple role in the geopolitics of a region and needs to be analysed at different scales. At one level, this part of the Alpine chain lies between the lowlands to the south, with their Mediterranean influences, and the Germanic lowlands and Austrian Basin to the north. But it has also functioned as an important crossing point between these areas, with control of the passes being an economic and strategic matter. The communities that established themselves in this area derived from two cultural traditions marked principally by language, but also by other practices such as family structure and inheritance. However, on a day-to-day basis a Tyrolese culture has developed which, whilst obviously reflecting the different cultural traditions, has been built upon the specific characteristics of life in this mountain area. Although not completely autarchic, the communities tended to operate independently of wider politico-social groupings even under the Habsburgs. However, as the linkages between the mountain economy and the wider world increased so did the influence of the various 'cultural' pressures. Surrounding nationalisms, the growth of economic power and the relative decline in the vitality of the mountain economies provided the framework for migration and economic allegiances. Finally, from the end of the nineteenth century the mountains became a political flashpoint with the communities within them largely playing a minor role (as pawns) in the tussles between the powerful neighbours.

This region of the Alps displays many of the facets of boundary zones, both positive and negative. Far from being always marginal 'marchlands', even in mountainous regions, these zones can have important economic attractions. Today, for instance, there are several 'transfrontier regions' where customs duties are normally not applicable. One such area is the valley of the Livino between Switzerland and Italy where a 6-kilometre stretch of the route has 120 places selling items such as perfumes, tobacco, alcohol and petrol all at duty free prices. Another feature of these Alpine boundaries is the fact that several are crossed by large numbers of commuters who work in neighbouring countries. Of particular importance are Geneva, the Rheintal, and also the boundary between Austria and southern Germany. Finally we might also remark on some negative consequences. An oft-cited example is the decline of the town of Gorizia, now on the post-Second World War border between Italy and Slovenia. The loss of the Italian section of its hinterland has had serious consequences, although in recent years there are signs that the area may be experiencing a revival.

CONCLUSION

It is widely recognised that communities occupying mountain areas must face the rigours of an environment in which climatic and topographic extremes play a major role. The settlement pattern is therefore closely bonded to this physical environment, but it is equally necessary to stress that the rationale for settlement itself and the forces which hold a population within the mountains have their own impetus. In many cases this is fundamentally economic: a place to live and derive a living, from agriculture, mining, forestry and, more recently, some industry and service activities such as tourism. It is also cultural in that people's value systems may place a high regard on the specific environment of the mountains. This may be through long association (a tradition) or, equally, through the rise of an ethic which prizes the landscape highly. It is precisely these values, and the institutions and behavioural patterns that they engender, which remain little explored in the literature but which must be at the forefront of any attempt to isolate a 'mountain culture'.

Communities exist in mountains by virtue of the fact that they are able to cope with the specific conditions presented by both the physical and social environments. Expressed in terms of the cog metaphor this means that the various elements interlink in such a fashion that each element is allowed to regenerate. However, this does not imply that the way in which this linkage occurs is uniform in all mountain societies. We cannot simply run the cogs like clockwork because we cannot be sure that a particular combination of cogs will automatically operate like all similar combinations. This is clearly illustrated in the context of how different societies actually resolve the practical problems of survival in the mountains, particularly in the development of appropriate livelihoods. This is the subject of Chapter 4.

4

THE PRODUCTION
ENVIRONMENT

The significance of the physical environment
for livelihoods

INTRODUCTION

High-altitude environments have long been successfully utilised to productive effect by human populations. The various climatological and geomorphological zones have been used for different purposes, and the method
of resource use has, in many cases, allowed sustainable exploitation of these
resources. As a general rule, in the highest zones, the land is effectively
wilderness. On the intermediate slopes, a transition between grazing and
forest use and cultivated terraces occurs, and on lower slopes intensive settlement and agriculture takes place, where every plot of land is carefully
guarded and managed in order to maintain its productivity.

Mountain inhabitants have employed a wide range of technologies to
manipulate the physical environment so that settlement, agriculture and
other activities can take place safely and successfully. These technologies are
often ancient, though each generation adds new experience to form a wealth
of local knowledge. In this section we will examine these in three groups,
although in reality they are not mutually exclusive.

First, there are modifications to the landscape to overcome the problems
of steep slopes, instability, mass movement and landslide hazards. These
adaptations include terracing, irrigation, drainage and water harvesting techniques, along with hazard mitigation and coping. These require both large-
and small-scale engineering technologies. They provide flat land, control the
distribution of available water supplies in time and space, and cope with
problems of flooding, drainage, mass movement and avalanche hazards.

The second group of adaptations has been developed to overcome the
climate and its seasonal or diurnal variation. These comprise cropping
regimes, pastoral cycles and movements of herds and flocks, and choice of
crop and livestock species. On a smaller scale, soil conservation and improvement techniques overcome the perennial problems of fertility maintenance,
length of growing season or diurnal heat and frost in tropical mountains.

Manuring, fallowing and rotation, together with more subtle techniques which adjust the albedo of soil, are used to modify ground heating and evaporation. The timing of irrigation and the different methods of seed propagation allow optimum use of the diverse microenvironment even on a single terrace. This also permits a variety of crops to be grown, fertility to be maintained and diseases to be restricted.

Third, the use of natural vegetation reflects the risk avoidance strategy of the traditional farmer. A wide range of local plant species is cultivated for human consumption, medicinal and fodder use. In addition, grassland and forest ecosystems represent important resources in themselves – for example, for grazing, fodder, and construction timber. The highly diverse genetic resource base of mountain flora has not only been put to good use, but this use has enabled the survival of many species up until very recent times.

The sum total of these adaptations exemplifies the risk avoidance strategies adopted by communities living in marginal environments. Should one crop fail, or diseases or other disasters strike their animals, there are alternative sources of food to be drawn upon. As noted in Chapter 3, a critical element, which binds the activities of groups of individuals together to form a coherent community, is the institutional structures that control the use and distribution of resources such as land and water. They may also control timing of pasture use, rights of access and even times of sowing, harvest and the proportions of crops planted. These institutions provide the framework through which members of the community may be mobilised for engineering and agricultural activities.

Finally, not all activities in traditional mountain societies were associated directly with the agricultural economy. Hunting, fishing and forestry are all important in mountain economies, but so also are activities associated with mining and handicraft production. The chapter ends with a brief account of the way these activities skilfully utilised the natural environment.

LANDSCAPE ADAPTATIONS

In order to overcome the problems of scarce flat land suitable for cultivation, the eternal problem of erosion of mountain slopes, the provision of water for crops and drainage of excess overland flow, a number of indigenous engineering strategies have evolved simultaneously throughout all the major 'old world' mountain systems. These include the construction of terraces and check dams, irrigation systems and various methods of coping with surface runoff. These are considered below in three sections: first, slopes, terraces, check dams, etc.; second, the issue of water and irrigation; and finally, mass movement and avalanche hazards.

Slope engineering

Terracing

Terraces occur in most major mountain regions. In Central and South America, they date at least from Inca times and were abandoned at the time of the Spanish Conquest (Guillet 1987). In Europe and the Mediterranean historical accounts provide relatively good documentation of terrace construction and other slope engineering practices (Foxall 1996). These structures increase the amount of flat land suitable for cultivation, create deeper soils which can be further improved by manure, increase infiltration of scarce rainfall or irrigation water, and prevent erosion by controlling runoff across the slope surface.

The construction of terraces requires the co-operation and co-ordination of large groups of labour, but the terraces tend to be individually owned or tenanted and cultivated intensively throughout the growing season. They require constant maintenance in order to prevent collapse, which affects terraces up- and down-slope in a domino effect. The terraces tend to be quite small in area but can still offer a wide range of microenvironments for exploitation. This applies both in terms of the whole valley slope, where location on the slope is important (see pp. 143–150), but also on the terrace itself, where areas of relative dampness and shade occur against the riser of the next terrace compared with the outer edge of the terrace step. Construction proceeds up-slope in a series of steps. A foundation wall, usually of stones, although in some areas this may be an earth bund, is laid following the contours of the mountain slope (Plate 4.1). This wall is gradually built up to varying heights, depending on the steepness of the slope and subsequent size of terrace area. On steeper slopes, walls need to be higher in order to provide wider strips, but if too high they become unstable. On a very steep slope (about 45°), walls may be one and a half metres high, providing a terrace width of perhaps only one metre. This is, however, sufficient for cultivation and tree planting in areas where shallower slopes are more rare. Soil is carried up the slope to infill the area behind the wall. Vincent (1995) notes that little data exist on the actual labour requirements for terrace construction, despite many studies on terrace construction, use, maintenance and ownership. She quotes work from Peru (Treacy 1989) indicating that 610 worker days are needed for rebuilding a terrace of three hectares, while Guillet (1987) estimates that a 0.03 hectare walled terrace requires forty days labour. Clark (1986) suggests one hectare of terracing in eastern Bhutan requires 1,320 worker days. These all stress the importance of hired or co-operative labour. Certainly, the investment of time and effort is significant – the figures from Bhutan might indicate that a team of ten workers would take four to five months to construct their hectare of terraces, assuming no delays. This could be accomplished in slack growing

Plate 4.1 Terracing and agricultural activity, Hunza, Pakistan

seasons, given labour availability, and certainly the whole hectare is unlikely to belong to one family given the much smaller average holding size.

Variations in construction techniques occur in response to local topographic or climatic conditions. For example, in the Himalayan regions affected by the monsoon, two types of terraces are built. The *khet* terraces have a bund or lip on the outer edge, allowing the ponding of water on terraces used for rice paddy. The *bari* terraces are gently sloping outwards, which allows water to drain off the surface. The former are built on shallower slopes better able to support the weight of saturated soils, whereas the latter are built on steeper slopes, where draining the soils is important in order to prevent collapse. The terms *khet* and *bari* are also used to distinguish between wet and dry season crops respectively.

Water harvesting in arid regions

In more arid regions, such as the Anti-Atlas Mountains, farmers use more complex systems of water harvesting and water capture practices (Kutsch 1982). On the surface of the land this amounts to the construction of low walls or bunds, behind which soil collects. This serves to slow the flow of runoff across the surface and increase infiltration. The resultant crop growth often resembles stripes, with trees and healthy growth located adjacent to the bunds, whilst intervening areas have much more sparse growth. These bunds, although only perhaps 30–50 centimetres high, and scattered systematically across hillslopes, may appear relatively insignificant but are a vital and effective way of capturing runoff resulting from the relatively heavy and infrequent rainstorms such areas experience. They are, however, subject to damage by these very heavy rainstorms, as are all artificial constructions on slopes and the slopes themselves in their natural state.

Whiteman (1988) refers to the common alternative approach to water harvesting which occurs particularly in areas of bimodal annual rainfall, such as in the Yemen and Ethiopia. Where the rainfall from one season is insufficient, it may be 'stored' in the soil and crops planted after the second rains, as the cumulative amount of water is then more likely to support crop growth. Water can be collected over several seasons' rainfall. Long-maturing plants, such as sorghum, which need both rainy periods to mature are therefore particularly vulnerable if one of the rains fails, although when the rains fall they are able to survive the intervening dry period. Water from flash floods in wadis may also be collected and diverted to cultivated ground or stored underground in cisterns. The main problem of using flash flood waters is their sudden appearance, high sediment loads and destructive energies.

Rainwater harvesting involves the collection of surface runoff and infiltrated groundwater using methods that impede surface flow by bunds, ditches and obstacles following the contour. Water is diverted into underground cisterns, surface ponds and reservoirs. Where impermeable strata lie below the surface, dipping parallel to the slope angle, subsurface water can also be collected at the base of the slope. In some cases a wider area is used as a catchment to supply a limited cultivated plot (Pacey and Cullis 1986). In other cases, a 'net'-like system of low bunds with cultivated plots in between is constructed on gently sloping land. The greater the slope, the more difficult is the retaining of walls and the more forceful the runoff flow. Bunds are constructed of earth, mounded vegetation covered in soil, or stone walls. All may be overtopped, and stone walls allow some infiltration of water through the wall. In Morocco, trees are planted on the downslope side of the walls where water is more concentrated and where they may assist in strengthening the wall.

Erosion control

Check dams may be built in gullies in order to arrest erosion. Walls of stone, which are often semi-permeable, are built across gullies to catch sediment carried down and slow the rate of water flow (see Plate 4.2). A series of these may be built down each gully. In time, the gully may become partly infilled, and each dam may form a pseudo terrace. However, the most effective means of controlling erosion is the careful maintenance of terrace systems.

Plate 4.2 A check dam in a ravine, the High Atlas

Where rainfed agriculture is practised, and especially where surface collection of runoff takes place, there is a risk of concentrated erosion. The construction of slipways, channels and chutes is effective in assisting the flow of water, but also reduces friction and thus increases the erosive capacity of the water. Erosion can therefore be concentrated and increased adjacent to such structures, and at the ends of discontinuous cross-slope bunds and walls. Erosion by sheetwash, however, on slopes crossed by a network of low walls and bunds, can help the process of landscape alteration for agriculture. As the water is halted by the wall, the sediment is deposited, gradually accumulating to form low, flat 'pseudo' terraces. These encourage infiltration by the lower slope angle, and result in the distribution of infiltrated water more evenly over the slope, thus increasing the effective cropping area.

The maintenance of terrace walls and channels ensures effective stability in most conditions, but gullying tends to occur, particularly along pathways. Much of this can be mended, and waters diverted into channels, although the channels will need clearing of the accumulated sediment load. Some natural gullies and small valleys are kept free of structures in order to ensure some escape for water. In the monsoon regions of the Himalayas, the different types of terraces have already been mentioned. *Bari* terraces, which gently slope outwards, may have a channel on their inner edge, so water draining off the terrace above can be directed off the lower terraces. These require constant vigilance, and small areas of damage can cause large-scale collapse if left unattended.

The Ethiopian highlands have been deeply dissected by centuries of erosion, producing deep canyons. The soils comprise heavy clays derived from basalt. These have a low infiltration capacity and rapidly become water-logged after heavy rainfall, leading to high rates of overland flow. The tendency to plough across the contour has contributed to the high erosion rates (Whiteman 1988). Planting takes place after the rains have fallen, when farmers know if sufficient soil moisture has accumulated to support the crop. This means that rains fall onto bare ground or prepared seedbeds, offering largely bare surfaces and enhancing soil removal. Different methods of ploughing, ridging, or blocking furrows at intervals have helped slow the rates of runoff and thus increased infiltration. This has the combined effect of reducing soil erosion and increasing amounts of soil moisture, similar to the effects of bunds and stone lines in arid regions for harvesting water (Hurni 1988).

Efficient drainage systems are necessary in order to limit erosion damage during periods of heavy rainfall. In areas where the flow of water is controlled by sluices these can be shut, allowing storm water to drain through the irrigation network. With flexible construction of the offtakes from channels onto the fields, the volume of water through different parts of the system can also be controlled. Elsewhere, especially if runoff provides a significant proportion of water for crops, runoff control plays a major role in cultivation practices. In many tropical regions, such as in the mountains of Zimbabwe, storm water control is a crucial element in erosion control strategy.

Maintenance and abandonment of slope engineering constructions

Maintenance activities require suitable physical conditions and the availability of labour. Thus in temperate areas, maintenance work can only take place once the snow has cleared. In other areas, without significant snow cover, it is often the labour constraint that is the major problem. There is a significant responsibility attached to the maintenance of terraces by individual farmers, as breaches in the walls of one terrace can quickly weaken those above and below and channel runoff waters, rapidly leading to gullying. Whilst unusually heavy or persistent rain can cause the collapse of even the best-maintained terraces, it is exaggerated by the poor maintenance of terrace walls. Gerrard (personal communication) has observed in the Himalayas that numerous small slips occurred during the monsoon season, which were immediately and effectively mended. Some of these slips were caused by leaking irrigation channels, others by the saturation of soils and slides. Breaches were rapidly mended before they could spread, using labour provided by the whole community. Larger slips involving many terraces were left to stabilise for a year or more, before being terraced over again.

Major problems occur as a result of poor maintenance of terraces, through, for example, abandonment. At the end of the Roman Empire the Mediterranean suffered extensive rural abandonment and a return to pastoral farming. The dense and complex systems of terracing gradually collapsed, resulting in erosion of all the accumulated soils exposed through the collapse of terrace walls. This sediment accumulated in valley floors throughout the Mediterranean as a result of climatic changes coinciding with large-scale abandonment, and is known as the 'younger fill' (Vita-Finzi 1969).

WATER AND IRRIGATION

Sources of water

Water for mountain farming comes primarily from precipitation of snow or rain, but in some areas, such as the Karakoram, may be derived from glacier melt. In the case of snow, the water may be stored for several months or possibly years until melted and thus available for use. In the same way that the total amount of rainfall provides a finite supply to the valley, so too does the quantity of snow melt. Other sources of water include direct rainfall, which may be seasonal (for example, the monsoon). As the monsoon has a well-defined season, the availability of water is concentrated in a few months of the year, and becomes the critical factor determining the timing of crop growth. In the case of snow melt, this occurs in spring and early summer, and this is therefore the time of peak water availability. Glacier melt is at its peak in the summer and can be particularly valuable to areas having little summer rain.

In valleys supplied by one predominant source, the growing season may be severely constrained by water availability. In valleys supplied by more than one source – for example, a mixture of snow and glacial melt – the availability of water is spread over a longer period of time and, other factors permitting, a longer growing season may prevail. In some regions, such as the southern Atlas, only very little rain falls on these arid, rainshadow lands, hence the need for the water capture techniques mentioned above. Even this is often insufficient, rendering the livelihood extremely marginal and vulnerable to failure. In the Hunza valley similar low levels of rainfall (150 mm) occur on the valley floor region (Kreutzmann 1991), but here, with the altitude reaching over 7,000 metres for many peaks, snow falls in winter, and the occurrence of glaciers provides water, rendering it highly productive. Other sources such as springs tend to provide a relatively minor contribution.

The source of water has a bearing on its quality. Snow melt tends to be relatively pure and clear, but glacial melt may be several degrees colder, and also contains a high proportion of fine rock flour. The combination of ice cold water and the fine dusting of silt and clay deposited on the leaves of young plants may slow their development by retarding growth or reduce photosynthesis.

As a result of the finite supply to many mountain valleys, and the critical need of water for crops during the summer when evaporation rates may be high from local winds and hot sunshine, the conservation and efficient use of water is a critical factor in the maintenance of livelihoods. Water, whether from snow or glacier melt, is concentrated in river channels, flowing below the level of many terraced field systems on the slopes. It is necessary, there-fore, to direct this water onto the fields. In the case of rainfall, the water arrives directly on the fields but must be carefully directed off the fields without inducing runoff, even during high intensity monsoon rainstorms. The resultant management of irrigation and drainage systems is therefore a fundamental component of the mountain landscape and its livelihood.

Irrigation systems

In association with terraces, the irrigation of mountain valley slopes is perhaps the most astonishing factor testifying to the ingenuity and efficacy of the mountain farmer. The extension of terraced land requires the provi-sion of irrigation water in many areas, and therefore is an integral part of the development of a cultivable landscape. Various forms of irrigation systems have been identified in mountains and are summarised by Vincent (1995). She classifies eight types of system – offtake, underground, spate, collection, storage, lift, combined and wetland. The gravity led offtake system is the commonest, but often occurs in combination with other types to supplement supplies during periods of shortage. Different types may also predominate, according not only to supply but also to topography and access to rights from different sources. In Yemen, for example, irrigation using the flood or spate waters (*sayl*) is an ephemeral feature, depending on the rare flash floods of the wadis. Water is captured and diverted directly onto the fields. By contrast, the hillslopes are supplied with more regular *sawagi* irrigation, which is runoff captured on the hillslopes and diverted onto the fields (Vogel 1987, 1988). This is similar to systems of water harvesting in the Anti-Atlas of Morocco (Kutsch 1982) and means that cultivation is only possible during times of flood. Because of the ephemeral nature of flood supplies in Yemen, these waters are not subject to the same water rights and regulations, being essentially freely available to those having the land in the path of the flood, and the means to capture the water.

Other systems use underground collection and transport. The existence of *qanat* systems in the Middle East is well-documented (English 1968;

Plate 4.3 A small reservoir in the High Atlas

Joffe 1992). These also occur in many other semi-arid regions, such as Morocco, where they are referred to as *khettara*. They consist of subsurface galleries tapping water from strata in the foothills and carrying it underground some distance, usually to a cistern. Many of these are now defunct as diesel pumping requires less maintenance and is a much faster way of acquiring water.

Storage systems may comprise tanks for temporary storage of water if, for example, the allocation for an individual village is not used up immediately. The water is collected into a reservoir tank and stored until needed. This allows a more even distribution over time of supplies, particularly in cases where the village may not receive water for several days in rotation, or where their turn for water comes at night when it may not be feasible to apply it to the fields immediately. Allowing the water to stand also permits sediment to settle and the water to warm up – this is advantageous in systems supplied by glacial meltwater. In semi-arid regions, surface runoff water may be collected in cisterns and stored, carefully locked and guarded as in southern Morocco. An example of local water storage is provided in Plate 4.3.

Simple streamflow diversion techniques apply in areas where flow is more reliable. In the Swiss Alps, for example, Netting (1974) describes the system

Plate 4.4 A complicated irrigation offtake with stone check gates in position, Hunza valley, Pakistan

of irrigating meadows, which may increase productivity by four or five times. The stream is simply dammed by a stone plate and the water spills over the meadow. The length of time permitted to irrigate each meadow in this instance was the time taken for the water to reach the lower end of the meadow as runoff, when a person calls back up the meadow to stop the flow at the top. Such procedures often take one to four hours to complete, including time to travel to the plot and to irrigate it, but the increases in productivity made it worth while.

Most traditional offtake irrigation systems follow the same pattern of construction and management, being essentially gravity led, communally managed, and constructed in close association with terrace building. The offtake generally comprises a diversion from the channel using simple stone, earth and timber dams, or a collecting pond high up in the valley, located against the valley side at a level a little above the highest terraces it will serve (see Plate 4.4).

It must be located where, each year, it can capture the meltwaters into the pond, and thus feed the system, but without incurring excessive damage as a result of being located in the path of floodwaters bearing heavy bedloads. Obviously damage often occurs, as discussed below. Water flows under gravity from the pond along channels cut into rock faces or built onto slopes, similar to terraces. Stone walls and earth banks channel the water along the slope contours, gradually dropping height. As the channel reaches the fields, sub-channels take water to groups of terraces and smaller channels onto individual terraces. On the terraces, the water is channelled through furrows, ensuring even coverage. All these channels are fed by gravity, and each has a sluice gate, allowing water to enter each subsystem or to be cut off. This permits accurate allocation of water and also reconstruction and repair of channels without waste of water. In Morocco, the *seguias* of the High Atlas have an average gradient of about 3 per cent (Miller 1984). This is, however, often sufficient to drive the flour mills described earlier.

Traditionally, the whole community contributes labour, materials, or nowadays often cash, to the construction of irrigation networks, thus each individual who contributes may benefit from the completed system. The work may be overseen, or initiated by a ruler, landlord, or group of village elders. Design and construction occurs without sophisticated surveying, but with an intimate knowledge of slope character and gradient, location of common avalanche, landslide hazard zones, and altitudes. In some cases, as in Netting's example of Törbel, special engineers exist who have knowledge and experience of construction and maintenance of such systems. As the system is fed by gravity, the offtake of the channel must be higher than the highest terraces or irrigated meadows. To achieve this, a channel may have to be established high up the feeder valley, and run for many kilometres – in Hunza, for example, some channels run for 8 kilometres.

The location of the offtake to capture meltwaters is critical, but also subject to problems. The high potential energy and abundant sediment carried by mountain waters in spate can wreak considerable damage, and the offtake needs to capture the water without excessive sediment. Settling ponds located at the head of channels and at intervals throughout the system permit finer particles to settle. These tanks, and often the larger channels, need regular cleaning throughout the year. The difference in gradient between the very gently sloping irrigation channels and the usually steep natural channel means that heavier material is not carried far, but constant vigilance and removal of this material is needed throughout the flow period in order to prevent blockages or reduced capacity of channels

The maintenance of the main feeder channels is also carried out by the community, often in spring before the first meltwaters arrive. During this time, channels are cleaned and repaired. Channels leading onto individual plots are the responsibility of individual farmers. The common need for water unites the community and brings to bear a common responsibility for the upkeep of channels passing alongside or through privately owned land. Technically, this is to prevent a tailender problem arising so that those down-channel do not lose out due to carelessness on the part of any one farmer. In practice, this is never perfect and tailender issues are one of the common complaints of irrigation systems. However, labour demands to build and maintain such systems are very high. An estimate of the labour requirement for *seguias* in Morocco is 10,000 man-days to construct and another 32,000 before fully operational (Pascon 1977). In Nepal, Martin and Yoder (1987) estimate the labour requirements for maintenance to run to an average of ten days per hectare of land irrigated, but this may rise to over a hundred worker days for rectifying damage. One problem that has affected irrigation systems in many areas, for example the European Alps and the Himalayas, is the effect of glacier retreat and advance on these offtakes. These may either be stranded by retreat, or overrun by advance, and therefore rendered useless or in need of modification.

Water allocation

In the discussion of institutions in Chapter 3 there were several references to the management of water. The allocation of water to clans, families and households is critical to livelihoods. The annual pattern of water supply is often marked by periods of abundance and scarcity. This means that the allocative mechanisms and their management become political issues. This process may be controlled by a community of users, or by a local ruler. In many cases a village council appoints a guardian to oversee the working of the water flow and monitor channel maintenance. This occurs, for example, in Hunza (the *chowkidar*) (Kreutzmann 1988; Sidky 1993) and in Morocco (Mahdi 1986). The guardian may be full-time, supported by contributions from the other families, and elected by a group of community representatives for a period of perhaps a year, so each family will have a turn at this responsibility. Alternatively he may be a permanent appointment, making his livelihood from it. In the Swiss Alps, for example, traditionally one man was appointed to oversee the whole valley system for a year (Netting 1974). Such guardians may be paid in crops or livestock, but in recent years cash is more commonly given in payment. In addition, where water from major channels is distributed to many villages, guardians from each village may be appointed to ensure the fair and accurate allocation of rights to each village, and to share the responsibility of maintenance.

The allocation of water may involve the lands of more than one village or community. In these cases, guardianship responsibilities may be shared between beneficiary groups to ensure that the rules are adhered to; such is the case in Hunza. Here, a sector may comprise the lands of a particular village, part of a slope or, on a smaller scale, the plots of a particular family. Each sector supplied by the channel may have its sluices opened to allow water through for a specified period of time – perhaps two to three days. The villagers then subdivide the water to different plots, so that within each allocation period water is fed to every plot. During the night, water may be stored in reservoirs for distribution during daylight hours. Each family knows and guards well its allocation times, as it may be that in times of scarcity, water is only available for their land perhaps once a week, or once every ten days (Kreutzmann 1988; Vander Velde 1992).

Water allocation is based on a variety of procedures and measuring systems. One Moroccan example uses the time taken to fill a container of a given size as the basic unit of account (Mahdi 1986). In the Swiss Alps it is the time taken for water to flow from the top to base of a strip of meadow (Netting 1974). In the Zangskar valley (Ladakh) the system is based upon the rule that each household has 24 hours of water within a cycle made up of several households. Much of this irrigation may be carried out at night and so it is common to see lights and hear people busy in the fields well after dusk as they ensure their full allocation of water. Other methods include the 'shadow' method where timing is based on the movement of a shadow between given markers. The rigidity of these water allocation rules also varies according to the volume available. When the water is plentiful, the arrangements are flexible, but when the supply becomes scarce, rules are very strictly enforced.

In Yemen, there is a dynamic and flexible system of irrigation according to the economic needs of *qat* cultivation, and to the variation in source of supply of irrigation water throughout the year in some areas. *Qat* is a privet-related bush which is cultivated on terraces. The young branches are cut throughout the year and the leaves chewed fresh as a mild narcotic. The crop is worthless if too much time elapses between cutting and consumption. It is therefore necessary to maintain growth throughout the year to meet demands. The standard, possibly ancient, entitlement was one measure of water (*tasah*) to one hundred of land (*libnah*), but this is now blurred by the demands of *qat* and alternative supplies. Different rights apply to new sources, such as wells, reflecting these new economic demands (Mundy 1989). Under Islamic law, which holds in Yemen, water can only be rented and sold if a known, reliable quantity can be guaranteed (Varisco 1983). Flood and spate water, therefore, is exempt from these transactions because it is unreliable in time and space.

Rights to water in Yemen are tied to land, and transactions of land ownership carry with them rights, at least for traditional water supplies. This has broken down in some cases where new land is taken into cultivation, or new sources of water such as wells and diesel-pumped river supplies are exploited. As noted earlier, in Peru, Guillet (1991) reports that the separation of water from land rights is the result of state intervention. Grants for water were made separately to land, and made to territories within which the communities concerned oversee the allocation of water. Some of this distribution occurs by auction.

Water is also allocated on a priority basis, which comes into operation particularly in times of scarcity – for example, in spring in Hunza before glacier melt; in summer in the Atlas after snow melt. The order of priority in these cases is usually domestic use, vegetables, grain, fruit trees and finally fodder. In principle, no one should be permitted to irrigate fodder land or tree crops if someone is without water for vegetables. This ensures an equitable distribution, and no family is prevented from growing subsistence crops whilst others grow cash crops. Violations of water rights, particularly in times of scarcity, are still a major cause of disputes within and between villages in mountain regions. The ownership of land is normally the means by which access to water is gained, and ownership of land is by heredity, kinship, clan membership, or permitted by village councils.

ENVIRONMENTAL HAZARDS

A number of environmental hazards are part of the way of life of mountain dwellers. Hewitt (1997) tabulated the major natural disasters of mountain regions between 1953 and 1988. There were nineteen volcanic eruptions, 154 earthquakes, seventy major landslides, eighteen avalanches and 118 floods in the major mountain regions. Landslides of various sorts, triggered by heavy rains, earthquakes, deforestation and the construction of roads are a common feature. Many mountain areas are inherently unstable due to their youth, and the active uplift they experience. Other mountains are volcanic cones which may be in a state of dormancy or currently active. The isolated peaks of East Africa such as Kilimanjaro, and the Hawaiian islands (Mauna Loa) are volcanic cones. Historically the volcanic mountains of Italy (Vesuvius, Etna) and the island of Santorini are all associated with widespread destruction of civilisations. Volcanic hazards, however, remain less frequent than earthquakes.

Steep slopes, unstable regolith accumulations and the natural variability of precipitation make mountains particularly susceptible to mass movement events. Other hazards arise from flooding caused by unusually rapid or extensive snow melt, or summer storms such as those which have caused severe floods in the Alps and North Africa in recent years. Glaciers are also a source of hazards. The movement of the lower parts of glaciers can block valleys, impounding water that may break through by sheer force or by the stagnation of the glacier snout later on. Advance and retreat of glaciers caused significant problems in the construction of the Karakorum highway in Pakistan (Kreutzmann 1994). Finally, avalanches have always been a hazard of many seasonally snow-covered Alpine regions, but the hazard has become more critical with the growing use of these areas during the winter season for skiing and other winter sports.

These hazards are not mutually exclusive – they are more often interactive. It is also the case that although many of these hazards are not new, the perceived and actual threat is much greater due to changes in the distribution and nature of human activities in mountains. The construction upon and development of valley floors in the Alps, for example, has constrained the floodplains, so the natural capacity of the valleys to absorb the energy of floods is much reduced and so severe damage will occur. In the High Atlas valleys, during the floods of 1995, water neatly scythed through properties newly built on the floodplain (Johnstone 1997). The construction of roads on the sides of unstable mountain slopes, for example along the Karakorum highway or the Llasa–Kathmandu road, not only places a structure in the path of landslide but in the act of construction, and in changing the nature of the slope profile, actually increases the hazard.

Over the centuries, local populations have often been able to cope with unusually severe hazards in a number of ways. Their intimate knowledge of the pattern of avalanche tracks, many often marked by different tree species or damaged specimens, has enabled them to avoid these areas during the spring and early summer when avalanches may be more frequent. The various uses to which valley floors may be put tend to be determined by the nature of soils and the propensity to flood. So although relatively abundant flat land may be available, these areas are usually free of constructions such as permanent walls and dwellings, and the land (which is often damp and with a tendency to frost due to cold air subsidence) is used for grazing or hay pastures. Such is the case in Morocco where the valley bottomlands are divided into plots by low stone walls and used for hay and pasture. This leaves the channel and floodplain free to absorb floods, and minimises damage. There are, however, still occasions when floods damage even old structures due to highly unusual rainfall events, as in Morocco in 1995 (Johnstone 1997). In Pakistan, settlement is concentrated on the alluvial fans and glacial terraces where soils are deeper, land is flatter, and where the land lies above the flood levels of the Indus river.

Although hazards are always considered primarily in the light of potential damage to humans and their property, the increasing human use of mountain areas has resulted in increased exposure to environmental hazards. Changes in population density and distribution, expansion onto new land arising from the increase in communications and tourist development have often occurred on land that was once considered risky in the face of landslide danger.

Volcanic and earthquake hazards

Several mountains are active volcanoes but still support dense populations. Volcanic ash provides a fertile, well-drained soil which may be attractive to sedentary agricultural populations, particularly if the surrounding region is relatively less attractive (too dry, or even too wet and cloudy). The upper part of many recent cones tend to have very steep sides, which limit the practicality of constructing terraces and therefore the potential populations that might be supported. Such is the case above 1,500 metres in some Indonesian mountains (Uhlig 1978). The risk attached to eruptions does not seem to prevent their settlement. History is littered with examples of destruction of settlements and landscapes as a result of eruptions – Mount Etna, Sicily, and Vesuvius in Italy which destroyed Pompeii and Herculaneum; Mount Pelée destroyed St Pierre on Martinique some 8 kilometres distant by a '*nuée ardente*' in 1902. In more recent times, Mount St Helens in the USA erupted in 1980, although this area was not densely populated as is the case of most 'old world' mountains. In 1993, Lascar in Chile erupted causing significant fallout, and also Mt Pinatubo in the Philippines to which global changes in precipitation and sunshine were attributed.

Perhaps the most famous study of volcanic islands was that of Krakatoa which erupted in 1883, triggering tsunamis which drowned an estimated 36,000 people in the neighbouring Indonesian islands. The dust and ash cloud was projected some 80 kilometres into the atmosphere, having significant effects on global climate in subsequent years (Smith 1992). Detailed studies have been made of the recolonisation of the island by fauna and flora, and of soil development and processes operating to create a new living landscape from the volcanic debris.

Most major mountain chains are associated with active plate boundaries, and these mountains owe their existence to collision and uplift. The Andes, Rockies and Himalayan chains are actively being uplifted at the present time, and therefore are also the foci of much earthquake activity. Some 97 per cent of the world's most destructive earthquakes occur in mountain regions (Hewitt 1997). The most numerous occur around the Pacific Rim (about 80 per cent), along with the most abundant active volcanoes. Kreutzmann (1994) enumerates forty-two events between 1876 and 1911 in the Northern Areas of Pakistan, and 102 between 1912 and 1971. In the latter period there was less damage, as the epicentres of most twentieth-century activity seem to have shifted to north of Kabul in the Afghan Hindu Kush. The tectonic processes not only give these areas high hazard risk but also are the source of the abundant mineral resources for which mountains are renowned. The earthquakes themselves may not be the greatest cause of damage. Most lives are lost as a result of the consequent landslides, mudflows, dam bursts and fires in settled areas. The greatest problem in coping lies in the unpredictability of the earthquake event, and there are the additional problems of access due to poor weather, few places to land aircraft bringing supplies, and the frequency with which the single road in and out may be blocked or destroyed.

The recent earthquakes in Afghanistan illustrate these problems. In the north of the country near Rostag, an earthquake measuring 6.1 on the Richter scale struck at around 6 a.m. on 3 February 1998. About 2,300 were killed and at least 8,000 made homeless. A few months later, on 30 May, another earthquake registering 6.9 on the Richter scale struck mountainous Afghanistan, this time killing 5,000 persons and making 45,000 homeless. According to the relief agencies help was delayed not only by blocked roads, bad weather, broken communications but also by problems arising from the political disruptions of the area (Oxfam 1998). The scale of displaced refugees is generally much greater than the death toll and more difficult to accommodate, particularly in countries with limited resources. Allan (1987) examined the impact of 3.5 million Afghan refugees fleeing (from civil conflict in this case) into Pakistan upon the vegetation resources. He documents the degree of degradation, particularly in forested lands settled by people originating from barren semi-desert lands. Such degradation is difficult to control, given the needs of the refugee population, and on a scale too large for agencies to cope with. Similar effects apply to the location of refugees from natural disasters.

It is obvious, therefore, that whilst an earthquake itself is mainly a 'natural event' the nature of its impact depends considerably upon both the population density of the affected zone and the ability of the society to cope. Perhaps this is best observed in the mountainous western USA where earthquake events are extremely common but where fatalities and serious dislocation are relatively rare. The Afghan quake referred to above was of a similar magnitude to that of the 1989 Loma Prieta quake in the San Francisco region. However, whilst the damage to property and infrastructure was considerable in the US case, the number of lives lost was a fraction of that in Afghanistan. An enormous disaster management infrastructure exists in the US context, including FEMA (Federal Emergency Management Agency) to cope with the event itself, whilst considerable resources are devoted to research, monitoring and predicting earthquake events. Seismic stations, such as that at Menlo Park in California, provide the data (including effectively real-time maps) that allow emergency agencies to watch the patterns of tremors to see if any large-scale build-up is imminent, and hence alert the population. Even if equivalent warnings can be offered to those in other parts of the world (the seismic service can do this) many countries simply lack the resources to mobilise appropriate assistance, and provide long-term relief.

Mass movement hazards

Landslides, rockfalls and mudflows are annual events in steep mountain terrain. They are often triggered by heavy rain when soils become saturated and may liquefy (mudflows), slump or slide (see Plate 4.5). The construction of terraces can stabilise slopes by reducing surface runoff and erosion, but the increase in infiltration rates can cause saturation and increase the risk of failure. Farmer adaptations of terrace type, according to rainfall, slope angle and crop type, are finely tuned to the risk potential, a factor discussed further below. In addition, the greatest number of failures identified in

Plate 4.5 Aftermath of a slope collapse in the Hunza valley, Pakistan

studies such as the Nepal mountain hazard mapping project (Kienholz *et al.* 1983, 1984; Ives 1988) were relatively small and quickly repaired by farmers before they spread further than one or two terraces. Larger failures tended to be left to stabilise and later re-terraced and reclaimed.

Ives (1988) made a series of calculations of landslide hazards based on extrapolation of average observed conditions in the Nepal mapping project region, which make alarming reading. Given an areal density of two land-slides per kilometre, and mean expansion in area of 60m^2/year, an estimated loss of land of 120m^2/km^2/yr for landslide results. This was doubled in order to account for new locations, estimated at 1 per 6 km^2. Given a mean depth per landslide of 4 metres, the annual soil loss was estimated as 1,000 m^3/km^2 from landslides alone. However, both temporal and spatial variability is masked by such extrapolations. Although there is a real problem of soil loss by landslides in cultivated areas, it is often the torrential, un-usually heavy rains and major landslides triggered by dam bursts and earthquakes which displace the greatest quantities of soil. As indicated earlier, the mountain sediment system represents a sporadic redistribution of soil from storage to storage – that is, from slope to valley floor in a time-less, uncontrollable way in the long term. Such losses therefore are accepted as part of the risks of cultivating steep slopes, and as suggested, many indige-nous management systems have coped with these for centuries.

Out-of-season rains cause frequent problems on the Karakorum high-way – for example, the heavy rains in September 1992 (Kreutzmann 1994) and again in 1997 (Parish 1999) both caused numerous blockages, cutting off the higher villages for several days. This was taken philosophically by the local populations, who are accustomed to being temporarily cut off on several occasions during the winter. They consider it a small price to pay for the much greater advantage of road communication. Similar events of flash floods and movement of sediment causing widespread destruction are a common feature of the Himalayas, as high magnitude events such as the 1968 flooding in Darjeeling demonstrate.

The dynamic nature of natural and anthropological interactions means that the distribution of mass movement events, their magnitude, frequency and cause are difficult to predict. The afforestation of slopes increases surface stability by reducing saturation of the soils, and root binding effects. However, whilst forests may reduce the number of shallow, small failures of the type readily rectified in terraced localities, they make little difference to deep-seated failures where the slip plane may lie below the level to which roots bind the thin mountain soils. These deep-seated slides are more likely to be triggered by earthquake or possibly road blasting and construction activities, and facilitated by strata of rocks such as slates and shales dipping parallel to the slope surface. Such slopes are also vulnerable to undercutting at the foot by swollen or meandering rivers. Nevertheless, arguments have been put forward which maintain that the magnitude and frequency of the landslide problem in regions such as the Himalayas is significantly increased by deforestation. This issue will be examined in detail in Chapter 8.

Avalanches

Mountain populations have long accepted the hazards presented by avalanches. Many mountain dwellers have a detailed knowledge of the tracks regularly followed by avalanches. They can adjust their settlement location and activities accordingly to keep out of the way. In seasonal climates, the high pastures are inaccessible during winter, and the hazard is confined to the spring thaw period. Once the snows have melted in early or mid-summer these zones can be used for grazing. Such tracks are identified in summer by the presence of flattened or damaged low trees, often of contrasting species. Whereas landslide scars involve the removal of the surface regolith, avalanches tend to strip the surface vegetation, or damage it, but do not necessarily involve disturbance of meadow turf or soils. In the central European mountains avalanche tracks have resulted in the long-term tree-lessness of some locations during the Middle Holocene when warmer conditions would otherwise be expected to have favoured tree growth (Jeník 1961). This represents a continuing threat to biodiversity in these regions (Jeník 1997). Avalanches also tend to occur in similar areas in successive years, as it is the combination of particular local conditions that triggers them. For example, concave slopes rather than convex slopes help compact the snow and reduce stress. Windward slopes tend to have snow that has been compacted by wind action, so lee slopes are more likely to fail. South-facing slopes or those receiving more warming also fail more often. Smooth surfaces, such as grassy slopes, offer less resistance than boulder-strewn slopes with shrubby or wooded vegetation cover, however sparse.

Avalanches occur either as loose snow drifts down very steep slopes, or as slab avalanches which are much deeper, larger and potentially more damaging. These are most common on slopes between 30–45°. Shallower slopes do not build up sufficient stress to fail, and steeper slopes are less likely to accumulate sufficient depths. Slab slides may be triggered by heavy snowfall on frozen surfaces. The new snow does not bind to the frozen surface and therefore may fail under its own weight, and may be more than 10 metres deep (Smith 1992). The runout path, or track, is the key element of concern for avalanches, and the force of gravity operating on large masses moving down steep slopes under lubricated conditions is enormous.

In the Swiss Alps, settlements such as Davos have demonstrated the effect of development impinging upon hazard zones. Early settlement in the twelfth to thirteenth centuries was located out of the way of such hazards, but during the sixteenth to eighteenth centuries the population increased, pushing settlement and land use into avalanche zones. Local industrial development (the iron industry in the fourteenth century for example) led to deforestation (Price and Thompson 1997). The pressures for development, especially in the arena of winter sports, means that not only do settlements expand into the threatened zone but they are also at their most active during the period of greatest hazard. During the winter of 1999 there were widespread and devastating avalanches throughout the Alps. The media reported snowfalls of 2 metres or more on rooftops and slopes. This was the most snow for 400 years in Chamonix, and subsequent rapid melting caused avalanches that destroyed not only new tourist devel-opments such as skiing chalets but also fifty-year-old houses. It was thought that these events had a return period of at least one hundred years and the destruction of older buildings, which had not been affected previously, supports this.

Many issues have arisen as a result of these avalanches, including the anger felt by those tourists who had to be airlifted out. With considerable media coverage the tourists indicated that not enough protective measures were in operation, and that the chalets should never have been built there. There were also insurance issues – for example, travellers' policies can cover the eventuality of too little snow, but do not cover loss and damage to life or possessions caused by too much snow. These issues bring to reality the problems of coping with the economic as well as the physical effects of unprecedented natural events where human activity is affected. This is the nature of manufactured risk in today's environment. The effects of global climate change on the frequency of such events cannot be predicted, and so it is really impossible to plan accurately for all such eventualities.

In such regions, therefore, there are a number of mitigation activities, which either influence the avalanche itself or the location and timing of settlements and activities. Passive measures of controlling the avalanche track include fencing crossing the slope, boulder mounds, etc. which break up the mass of snow and dissipate its energy. Planting forest belts and clumps reduces the surface snow accumulation depth and also breaks up any snow masses arriving from upslope. Alternatively the track may be diverted or deflected by chutes or fences away from major settlement locations. This is illustrated in Figure 4.1. More active measures include triggering avalanches in particularly prone areas so small, frequent avalanches can reduce the risk of any larger, more destructive events from occurring. Hazard zoning is a common feature of the Alps, whereby different constructions and activities may be excluded from high risk areas either totally or at high risk seasons. The evaluation of the degree of hazard is based on the size, frequency and nature of the threat, and of the economic or other importance of the threatened installation. Since the 1970s avalanche hazard maps have been produced for the Alps. The hazard maps were originally produced for strategic planning, and planning zones have been designated permitting or excluding different land use intensities (Frutinger 1980). Red Zones comprise areas where hazard events have a return period of less than thirty years, or larger events with a return period of up to 300 years. Blue Zones represent potential hazards where public buildings should not be constructed, and White Zones where avalanches rarely occur. One of the key issues surrounding the Chamonix avalanches of 1999 noted above concerned the fact that they included avalanches in White Zones.

Although originally fairly limited in its technical scope mapping has been the subject of considerable research and development, especially with the introduction of GIS-based systems linked to automatic meteorological reporting systems (Margreth and Funk 1999). For example, most countries now have online avalanche warning maps and so the data is used not just for long-term planning based upon prior experience of tracks but on current

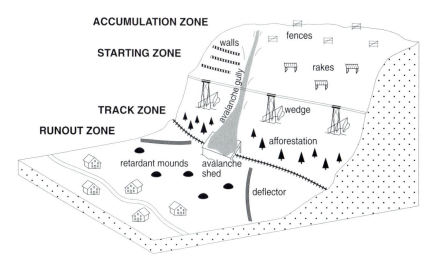

Figure 4.1 Methods of protection from avalanche damage (adapted from Smith 1992)

meteorological information. At the same time, there have been alterations to the designations of avalanche zones (Pietri 1993).

As most of the financial risk from avalanches lies in the loss of tourist income and capital, much of the cost is borne by insurance companies. The costs of insurance will vary according to the risk of a particular location and type of structure. Regional or national planning controls or the willingness of the company to bear the high cost of insurance, given the greater risk, will determine whether a construction or activity takes place. Where the high incomes generated by investors in winter sports in the Alps have attracted high densities of development, the pressures on tour companies and those providing the infrastructure to capitalise on the attractions of a particular location may result in development occurring in very high risk locations. It is, therefore, not surprising in the sophisticated technological era we live in, that the attractions of economic gains will inevitably result in occasional disaster.

Floods and glaciers

As noted earlier, flooding is another common characteristic of mountain environments. Seasonal melting of snow in spring and glaciers in summer creates annually fluctuating regimes. Superimposed on this are the effects of rainstorms or seasonally concentrated precipitation such as monsoons. The problem of flooding in mountain valleys lies in the 'flashiness' of the regime. In other words, the response to storm rainfall is very rapid. In arid regions, where surface vegetation is minimal, there is little to impede the flow of surface water to channels, so responses can be spectacular as in the case of flash floods in wadis. Similar conditions occur in high mountains where steep slopes, minimal vegetation cover and large expanses of bedrock or scree do little to impede runoff. For this reason, traditional mountain cultures located their settlements in the valley slopes as we discussed in Chapter 3. This preserved the flatter land for pasture once the spring floods were passed, and avoided cold air drainage zones.

Glacial hazards cause severe flooding as a result of the impounding of meltwaters behind glacial ice or debris. Kreutzmann (1994) describes such hazards in the northern areas of Pakistan. The numerous glaciers in the Karakorum mountains provide ample water for agriculture but also are the cause of frequent floods where glacial advance creates a natural dam which later collapses under conditions of retreat or by the pressure of impounded waters. During the 1960s the Shimshal glacier was the source of such floods. The Karambar glacier in the north-east Ishkoman valley gave rise to three major floods in the nineteenth century, documented by writers such as Biddulph (1880). The oldest settlements in this valley are all located above the levels of these floods, as lower settlements were destroyed. Glacier floods are notable for the fact that they are perhaps the most extensive hazard in area, causing floods in distant downstream locations in rapid response to upstream events.

Again, as for avalanches, concentrations of development in valleys such as in the Alps have increased pressures to build. The ease of building on flat valley floors rather than on slopes, even where these have been abandoned in terms of productive agriculture, has resulted in the constriction of river-beds, and dense construction of hotels and car parks. The concrete surfaces serve only to increase storm runoff, and the straightened and constricted channels reduce friction and energy absorption, thus increasing the rate of flow and erosive capacity of the river. Storm rainfall can therefore have devastating effects, particularly where the storms are unusual for a given time of year – i.e. the high magnitude, low frequency events.

In the Alps severe floods resulting from a combination of late snow melt and thunderstorms occurred in 1987 (Schwarzl 1990). A heavy, late fall on top of the usual winter accumulation fell in June, and within two to three weeks the temperature had risen some 15°C. A significant temperature and pressure gradient developed over the Alps causing thunderstorms. The result was a flood combining sudden and rapid thaw augmented by heavy rainfall. Many Alpine valleys were severely affected, with damage to roads, communications, bridges, new buildings, power stations, etc. These developments had all encroached onto the natural floodplain, on flat land. As there had been no such severe floods for a long period, insufficient consideration was given to the potential risk of floodplain development. During the previous fifty years the river had deepened its channel by 3.5

to 4 metres as a result of canalised, straightened sections increasing velocity whilst reducing friction and bedload. In the Otzal region, for example, all the damage affected constructions post-dating 1915 when significant expansion occurred.

Similar events have occurred in all mountain regions. In Darjeeling in the Himalayas a one in 100 year event occurred in 1968 (Ives 1988), where 500–1,000 millimetres of rain fell in three days at the end of the monsoon. The saturated condition of the soils resulted in some 20,000 landslides, and along 60 kilometres of highway some ninety-two cuts and blocks. Schelling (1991) recounts a similar event in 1987 in eastern Nepal where 18 kilometres of road was destroyed and severe flooding occurred in Sun Kosi Bazaar. Damage to a power station, bridges and other structures also resulted. Other major floods have occurred on the three major rivers of eastern Nepal in the last two and a half years.

The flooding in the Reraya valley, Morocco has already been briefly mentioned. Johnstone (1997) describes how, in an unusually heavy storm in August 1995, about 70 millimetres fell in two and half hours, increasing the discharge of the river by twenty-seven times. A tributary stream brought enormous quantities of gravel and coarse debris down, which was deposited in a huge fan, covering large areas of cultivated land. This caused a shift in the channel of the river and severe undercutting of the banks, drowning of terraces, and destruction and burial of trees. A number of vehicles, the road and several lives were lost. The effects stretched a considerable distance down valley and are illustrated in Plate 4.6. Compensation payments have been made between $20–200 per household depending on size, and the government instigated a four-month project to reinstate the road. The tourist season was in mid-swing, and effectively ended some two weeks early. Given the short season, this represented significant losses in revenue.

AGRICULTURAL ADAPTATIONS

The adaptations undertaken by indigenous farming populations demonstrate a deep and applied knowledge of the details of the microenvironmental mosaic offered by mountains, and which they exploit to the full. Long-term experiments by trial and error of different species of crops and techniques of cultivation, harvesting and storage testify to the readiness of farmers to test new ideas and crops, but in such a way as not to threaten their livelihoods. This overturns more dated concepts of the conservatism of mountain populations and their resistance to change. Hewitt's (1988) critique of Price (1981) argues for the increased recognition of the complexities of adaptations manifest in mountain communities. They are not resistant to change if they can see how it can help them in their constant search for improvement in livelihoods. Equally, farmers are constantly fine-tuning techniques,

Plate 4.6 The aftermath of a flood: the buried base of trees, Imlil 1995, the High Atlas

and testing new crops in small areas. However, they are unlikely to turn over all their fields to some new wonder crop offered by a development agency until they can see for themselves how it copes with what they know to be a vulnerable, diverse and marginal environment.

Farmers have also been shown to balance the returns from a number of different crops, seeking to improve yields by easing the most limiting factors to growth, such as water or nutrients, but not necessarily making the maximum yield from one crop at the expense of others. They know full well the need to have a range of sources of food and fodder in order to minimise the risks associated with failure of one principal crop. The example of the practices of farmers in the Hopar region of Hunza, Pakistan by Butz (1994) mentioned below illustrates this.

Agricultural adaptations comprise those strategies adopted by farmers to overcome the physical constraints of mountain environments. These will be considered in two categories: first, those relating to soil fertility and productivity; second, techniques used to deal with climatic conditions such as seasons, temperature, precipitation, etc. The combined effect of these micro-environmental conditions is to favour a land tenure system that supports the fragmented distribution of plots, a common feature in mountain regions. Households often work a number of small plots distributed across several altitudinal zones and each with a different aspect. The farmer maximises exposure to different radiation conditions, both diurnally and seasonally, permitting cultivation of a range of crops. Whilst farmers may have some distance to travel to some plots this drawback is outweighed by the advantages gained in spreading risk.

Soil conditions

The friable, poorly structured nature of mountain soils, and the problems arising concerning stability and fertility, have already been discussed. Both these issues are critical to sustainable agriculture, and mountain soils generally need a considerable degree of improvement. Continual tilling and manuring have a cumulative effect on soil and thus after long periods of cultivation well-tended soils can be significantly improved in structure, texture and nutrient status. This in turn will increase water-holding capacity and improve the conditions for crop growth. The maintenance of soil cover, by slope engineering to reduce slope angle or by maintenance of vegetation cover, is critical as the steep slopes and (in many areas) precipitation patterns enhance the inherent propensity for erosion.

In the Andes, archaeological investigations have examined a raised bed system of cultivation, and local farmers are currently reintroducing this through experiments. In Peru and Bolivia (Erickson 1992) prehistoric land management systems, dating back to 3,000 years, involve narrow raised beds with intervening canals. These raised fields, or earth platforms, had multiple functions – enhancing drainage, channelling water, providing irrigation water, increasing fertility and nutrient cycling, soil moisture, etc. Phases of abandonment (AD 300) and reuse (AD 1000, 1450) have been identified. Erickson estimates that around 37.5 persons per hectare could be supported. Some 82,000 hectares have been found, thus indicating a population of 1.5 million. This raised bed system produced high yields, and the other advantages outlined above were demonstrated during field trials, so that the system is being reintroduced as part of development projects.

Other soil conservation measures include embankments, mounds, ridges and check dams in valleys. Allen (1988) and Sillitoe (1998) have described the use of mounds to facilitate the production of potatoes within a zone of significant frost risk in Papua New Guinea. Immigration and population growth was supported by the arrival of the sweet potato in the sixteenth century. Permanent settlement became possible up to 2,300 metres using a process of composted mounds where green manure (crop residues) is heaped up and covered with soil. Potato runners are placed in the upper part of these mounds, above the level of ground frost and also above cold air that drains off between the mounds. Harvesting and planting is continuous, as there are no seasonal constraints to growth. Above 2,700 metres mounds are smaller in diameter but taller and have a thicker sod covering the green compost in order to protect against the greater frost risk at this altitude. The adoption of these mounds came about by the movement into higher altitude zones of farmers who had designed this strategy to counteract the friable, nutrient-poor condition of volcanic soils. These mounds have proved a successful adaptation to periods of hard or frequent frosts.

Maintenance of fertility and improvement of terrace soils by the addition of manure mean that livestock remain an integral part of the mixed farming economy. Rotation systems of crops ensure that exhaustion of the soil is kept to a minimum, and the inclusion of fallow periods, where grazing and thus manuring by livestock is permitted, helps restore the nutrient status.

In the Andes, the aseasonal nature of the growing period means that cultivation can occur continuously. However, the high altitude of many fields prevents the slow breakdown of organic matter and thus the release of nutrients into the soil. The use of rotations, therefore, is critical in retaining productivity in the traditional cycle. Grain is grown first, followed by potatoes and several years of fallow to allow the soil to recover. According to Sarmiento *et al.* (1993), the land use system of the Venezuelan *paramo* comprises cultivation phases of one to four years of potatoes followed by one year under cereals, after which the field is left to rest. This is followed by a succession–restoration phase of 7–20 years, ending in a restored field ready for cultivation again. The vegetation biomass accumulated during restoration cycles is sufficient to restore the fertility of the soil. Fields that are abandoned during regeneration also are subject to pest control, as the fallow period exceeds the survival of most diseases such as potato nematodes. Leaving such fields fallow doesn't increase erosion rates as fallow fields at different stages are randomly scattered throughout the mosaic of cultivated fields, so only small areas are left bare at any one time. The aseasonal climate and adaptation to continuous, staggered agricultural cycles means this is always the case. The land surface at any one time may comprise 10 per cent cropped area, 65 per cent natural vegetation and 25 per cent in succession (transition between crop and fallow). Of the cropped area, 24 per cent is given over to cereals and 76 per cent to potatoes.

This pattern of land use represents an effective rotation and fallow allowing the long time necessary to regenerate tired soils under high-altitude conditions, with slow rates of breakdown of organic matter. It also operated without the need for artificial fertilisers. It is in these conditions that communal control of crop planting cycles operates effectively to maintain fertility, reduce disease and minimise erosion. Such systems are at risk where the increasing demand for more production leads to reduction in fallow, compensated for by the increased use of artificial fertilisers. Although this allows more cropping and less fallow, it also leads to monoculture of specific crops, such as the potato, and increased risk of disease. In other communities, such as in Hunza where seasonal planting occurs, land is too scarce for significant areas to be left fallow, and the risk of disease is reduced by the exchange of potato cultivars between farmers (Parish 1999). Farmers who had not participated in such an exchange reported declining yields from their fields.

Climatic conditions

The nature of climatic interactions, operating on different scales with soils, topography and vegetation, produces a complex and diverse microenvironmental mosaic of growing environments. These are known and exploited by farmers in order to maximise the range and the total productivity of crops grown, balancing between pastoral activity and different cultivated crops. As farmers have learnt to adapt to and cope with the geomorphological and edaphic conditions of mountains, so too have they adapted to cultivation in a marginal environment where many of their crops are grown at the limits of their climatic tolerances. The use of local varieties of crop plants and of livestock is important, as these are more tolerant of the extreme conditions of mountain climate, and of the low-nutrient status of the soils. They can be productive with minimum additional inputs from the farmer, in marked contrast to the greedy and more specific demands of many new hybrid varieties. A small change in one climatic variable can cause widespread failure of any one or several crops in a growing season and, certainly in previous times, famine. Many farmers in mountain areas operate on the principle of addressing the most limiting factor for a given crop in a particular location, thus rationing the scarce resource available to them.

The critical components of climate that affect the growing environment of crops involve solar radiation, temperature, and precipitation. The greatest climatic risks faced by mountain farmers include drought or even too much rain, early or late frosts, and local wind patterns affecting evaporation and creating cold stresses. These are no different from many other agricultural systems, but it is the severity of the risks, their frequency and the longer time needed to recover, which marks out the mountain environment as different. In other words, it is the fragility and sensitivity to change of mountain conditions which increases the need to spread risk over different crops, different ecological zones and to balance productivity and inputs when resources are scarce.

One of the most significant factors in mountain agriculture is seasonality. In Chapter 2 we have already discussed the effects of latitude on climatic factors, and how this is combined with altitude to increase the stresses of high mountain living. Climatic seasonality determines the duration of the growing season in which agriculture can take place and, given sufficient conditions for part of the year at least, a livelihood can be sustained. However, it depends on the productivity of the growing season rather than its length alone, and this is intimately connected with altitude. An example given below cites the greater productivity of irrigated rice at higher altitudes in the subtropical, seasonal Himalayas rather than the aseasonal humid tropics of south-east Asia.

Variations may also occur within seasons, such as the timing of rainfall, or the rapidity of snow melt. These may significantly affect the capacity to grow more than one crop if, for example, there is an unexpected delay in the onset of rains. Many agricultural cycles in mountain areas have little 'space' to extend the growing season (see pp. 162–170). A good example of this concerns the timing of the pre-monsoon rains in the Himalayan region, which are vital in 'priming' the soil and initiating germination.

Seasonality and frost

Growing seasons are determined by either winter–summer dichotomies, determined by latitude, or, alternatively, the season is constrained by wet or dry periods. Cultivation is therefore dependent upon sufficient rains and appropriate growing conditions during or after this time. In many high-latitude mountains rain either falls throughout the year, as in the Alps, or there is a distinct winter maximum when it falls as snow. This is stored until it melts with the warmth of spring and is then a usable form of water for cultivation – for example, in the semi-arid parts of the north-western Himalayas. In seasonally wet and dry mountains, the precipitation may be unimodal (occurring once a year as in the case of the monsoon) or it may be bimodal (occurring twice a year as in many Mediterranean and Middle Eastern mountains).

In the tropics, the difference between seasons is negligible but diurnal fluctuations are considerable. There is a risk of frost at any time of the year above 3,500 metres on Mount Kenya on the equator. This is relatively low, due to the fact that there is no mountain mass effect on this isolated peak. In the Andes, frost occurs for 100–250 days per year at 4,600 metres, and for 320 days above this in the periglacial belt of Ecuador. This zone is characterised by *paramo* grassland vegetation (Lauer 1981). Frost occurs all year above 4,500 metres in the seasonally arid mountains of the Hindu Kush. In the seasonal latitudes (beginning 15–20° N and S) frost is restricted to the winter period when little cultivation can take place due to snow cover and cold. Frost retards growth, kills new buds and the blossoms of fruit trees, so can have a devastating effect on many crops.

In Ecuador, the cultivation of potatoes occurs continuously, producing a chequerboard landscape of fields at different stages of growth (Lauer 1981). In the case of Papua New Guinea cited earlier, where mounds protect tubers from frost, extraordinary frosts associated with cyclic events such as El Niño, cause greater problems than are solved by the compost mounds. The El Niño effect, which occurs about every decade (although more frequently in recent years), causes reduced rainfall, clear skies and increased frost risk – up to three or four successive nights, two to three times over six months. This causes a substantial lowering of temperature in the mounds, killing the potato runners. Families then migrate to lower altitudes, residing with households with which they have connection through kinship or marriage links. They take pigs with them to contribute to their keep. Allen (1988) describes them returning immediately to plant new runners, but residing at lower altitudes until they are ready for a new harvest (in twelve months' time) when they can then return to their villages. This need for kinship links is therefore critical in order to survive in times of higher altitude famine.

During the winter, plants become dormant. Dormancy is determined not only by temperature but also by day length, which affects the balance between photosynthesis and the processing of carbohydrates into grain and root storage. The importance of topographic shading effects becomes more significant. Growing seasons can be extended by nurturing seedlings in protected beds, or indoors, and planting out when frost has ceased, thus giving the crop a head start.

In Peru and Bolivia, maize is restricted to the wet season when the frost risk is reduced above 3,000 metres. At 3,800 metres in the Titicaca region, cultivation of a local cereal, *Chenopodium quinoa,* alongside potatoes, provides the staple crops. Potatoes can only be grown during the frost-free season, and so in order to preserve sufficient to support the population a technique of freeze-drying has been developed. This involves exposing the tubers at night to frost, and to running water during the day. Eventually a pure starch substance is produced (Troll 1988). Today, barley is grown up to 4,100 metres, above which are pastures grazed by llama and alpaca within the permanent night-frost zone.

In Pakistan, despite the very low rainfall in the valleys, there is abundant meltwater from snow and glaciers. Frost ends in April after quite severe winters, but until the frost returns in October the long, hot summer provides excellent growing conditions for a variety of crops, including potatoes. Whiteman (1988) cites particularly the world record potato production of 90 tonnes/hectare, and apricots, which have a 20 per cent sugar content and are therefore particularly succulent. Cooling at night under clear skies facilitates the latter. Seasonality, the relatively low precipitation, along with the dependence on meltwater irrigation, have not restricted the growing conditions in this region as much as abundant precipitation and cloud cover have in the tropics.

Solar radiation: sunshine and clouds

Precipitation requires the presence of clouds, which shade the sun's rays. As discussed in Chapter 2, the heat energy in mountain regions comes directly from the incident rays of the sun. Any shading, or the presence of clouds, can therefore severely restrict the heating of the soil and the effective growing conditions of crops. Shading effects are also caused by topography (i.e. aspect), tall vegetation and constructions such as buildings, terrace walls, etc. Where the growing season is dependent on rainfall, as in dry farming regions, the presence of clouds coincides with the rain and thus continuous cloud cover during the time of germination and early growth can severely retard development. The risk here is that maturation may be delayed too long, creating drought stress for the plants in the later stages of growth when water is needed to swell fruit and grain.

A clear example of the interaction between cloud cover and precipitation is provided by the altitudinal limits at which rice is cultivated in south-east Asia and the Himalayas (Uhlig 1978). Cloud cover, and therefore net radiation receipts, is the limiting factor here, rather than the availability of water. In the subtropical Himalayas, rice is cultivated at 2,200 metres in Kashmir and at 2,700 metres in western Nepal. The mountain mass effect increases heat availability and the lower precipitation ensures high radiation receipts, as discussed earlier. Meltwater provides sufficient moisture to permit rice to grow. In the Outer Himalayas, the altitudinal limit of rice cultivation is much lower – 2,200 metres in Darjeeling and Simla, where the monsoon provides the rainfall but also the clouds. Above the cloud level, skies may be clear but precipitation insufficient to support rice. The highest altitude at which rice is grown falls to below 1,500 metres in equatorial Indonesia, despite sufficient land occurring above 3,800 metres and population pressures. This is again due to cloudiness, but also due to the isolated nature of the volcanic peaks, which do not benefit from the mountain mass effect. In addition, the steep sides of the upper part of the cones, and the nature of the rocks, make it very difficult to construct viable terraces. The limit for double cropping of rice in this area is at 800–900 metres. Above this altitude the crop takes seven months or longer to mature.

Solar radiation has other important effects. Water that is kept ponded before flooding onto rice paddy not only settles its sediment load but also warms up in the sun. By flooding paddies with warmer water in the afternoon, the plants can be protected from the night's cooling effects. Whiteman (1988) notes that this can keep the paddy above 15°C for an extra five hours (an average of 3.8°C above paddies not so treated), which can hasten the growing season by several days. The working in of darker organic matter into field surfaces can also increase the soil's absorption of heat during the day, but although this may reduce night cooling it may also increase moisture and heat stress to the roots of plants.

Shading by terraces and trees also has important effects on plant growth. Even the loss of one hour of sunshine a day can retard maturation of a crop by several days or even weeks, and make the difference between successful double or single cropping. Where terrace walls are of similar height and aspect, most farmers would be similarly affected. However, raising walls to create bunds, or planting trees, may create problems for a neighbour's crops. In the Hunza valley, rules apply to the planting of trees on terrace walls. They may only be planted on the southern end of terraces, thus casting the noonday shadow over the farmer's own field, not his neighbour's.

Topographic shading on a larger scale, involving the aspect of slopes and orientation of the valley, has been explained above in the Climate section. Whiteman (1988) gives an example of its significance for agriculture in Nepal, where on north-facing slopes at 2,800 metres wheat takes a full 12 months to mature, whereas on the south-facing slopes at the same altitude it needs only 9 months.

Temperature and precipitation

These two variables interact closely, providing fine distinctions in soil moisture, evaporation and humidity conditions on the level of the individual plot. Variations in these conditions occur at different altitudes in the valley, according to the position on the slope (with respect to local winds and inversion conditions), slope orientation (with respect to solar radiation) and clouds, etc. Farmers can control the water supply to irrigated plots, but must respond to precipitation events – planting, for example, may not take place until the rains have fallen. Waterlogging and runoff must also be controlled. Temperature conditions are primarily dependent on incident solar radiation and therefore beyond the control of farmers to alter significantly. However, they can and do vary the locations of particular crops in order to avoid frost hollows, and make the best use of night-time temperature inversions that create warmer conditions mid-slope than on valley floors. They also alter the surface albedo by adding dark-coloured matter to increase heat absorption, and lay mats over tender seedlings to reduce evaporation during the day or frost damage at night.

Perhaps the most complex strategy is that of the location of crops and controlling scarce inputs – water, manure, labour. The object is not necessarily the maximisation of the yield of one crop, but to balance the total productivity, being the sum of the yields of a number of crops. They will identify the most limiting factor, such as nutrients or water, and concentrate inputs on that. The crop therefore can benefit from extra inputs of this most limiting variable, without wasting resources on less limiting variables in that particular case. Butz (1994) studied the distribution and cultivation strategies of potatoes, wheat, beans and alfalfa in the Hopar settlement area on the south banks of the Hunza river, north Pakistan, with respect to identifying limiting factors and selection of inputs. Cultivation here takes place between 2,500 and 3,000 metres altitude, on terraces and gently sloping lands. The distribution of crops in relation to specific conditions of radiation, access to water, nutrient status and soil temperature and moisture conditions reflects this limiting factor approach and the priorities of the farmers to retain balanced production. Potatoes are a high income cash crop and given the best land, in terms of fertility and sunshine, but were the least frequently irrigated (at 10–11 day intervals), being sensitive to overwatering. The response of potatoes to increased nutrient inputs is much better than that of wheat, so manure is concentrated in potato fields, which show good response with increased yields.

The wheat cultivated in Hopar is predominantly of local varieties adapted to poorer soils, temperature fluctuations and sunshine conditions. Although it is less productive than potatoes, it remains the preferred crop (70 per cent of terraced area). The limiting factor is water, and this is reflected in the increased frequency of irrigation (6.5 day intervals), but less fertiliser is applied as yield increases in response to manuring are not as great as for potatoes. Beans are recognised as the crop that is grown closest to its ecological tolerances and therefore requires more care. Beans are therefore grown on the plots receiving most sunshine and are irrigated most frequently, but also for a longer period of time at each application (although this may be a function of plot size; as beans are grown on the largest plots, these also take longest to irrigate). It is also significant that the bean plots are furthest from the source of the water, which enables it to warm up and deposit its sediment load before reaching the bean plots. As colder water and sediment loads both cause stress in plant growth, by reducing both these effects the beans are again not subjected to additional ecological stress, which is important for such marginal crops. Finally, alfalfa was either not irrigated, or infrequently. Where it was undersown in orchards, the trees might be irrigated by water flowing into circular bunds around the base of the trunk, but the alfalfa would not be watered.

The pattern of the careful allocation of resources therefore reflects not only the priorities of the farmers to increase all yields, not favouring one crop above another, but also an intimate knowledge of the requirements and limiting factors of individual crops. Resources such as manure are limited by the number of livestock kept, controlled by the capabilities of the household to feed and tend these animals. The availability of water is limited by the summer glacial melt and water distribution controlled in space and time by a rotation system. Each farmer must therefore be able to distribute these resources on his own land, between the different uses, in the most effective way possible. The ability to identify a constraining factor quickly and take appropriate action, is one of the key factors for survival in mountain areas.

In many semi-arid regions precipitation is the critical factor limiting plant growth. With rainfed cropping, farmers must plant in response to expected or actual rainfall, and they cannot control the application of water to plots as irrigation systems allow. Planting occurs when the first rains fall onto the prepared seedbeds. The strategy of temporarily storing water in the soil from one rainy season to the next in regions where it falls twice a year has already been mentioned. In the case of the monsoon, however, the tendency is for abundant rain to fall in a relatively short time, and, coinciding with cloudiness, this introduces radiation limits to growth rather than water limits. Rainfall derived from local orographic effects may be highly localised, falling on one valley whilst the adjacent valley remains dry. Such rainfall can also give rise to highly erosive runoff and flash floods, which not only damage the soil, wash away seeds and plants, but also is difficult to control and collect, and much is therefore lost down valley.

The combined negative effects of temperature and precipitation tend to retard growth but, depending on crop type, may not limit the actual yield of each plant. For example, maize grown in Kenya takes 69 days to mature at 1,268 metres but 96 days at 2,250 metres for similar weight of cob (Cooper 1979). Potatoes at 2,700 metres in Papua New Guinea take 12 months to mature, but only between 5 and 7 months at lower altitudes (Allen 1988). The upper limit of double cropping in the Hunza valley is at 2,300 metres (Sidky 1993), with only single cropping possible above this limit. This altitudinal pattern is further modified by aspect and shading effects mentioned above. In north-western Nepal, between 2,000 and 2,400 metres on south-facing slopes, there is sufficient heat for double cropping but insufficient water; this is reversed on north-facing slopes where sunshine becomes the factor limiting double cropping, and reduced evaporation ensures sufficient water supplies. In semi-arid regions, farmers adjust the planting density according to expected rainfall or soil moisture resources. If the rainfall is lower than normal, fewer plants per unit area will be planted in order to reduce stress. For example, in the Anti-Atlas, higher densities are planted (90–120 plants per m^2) in topographic hollows than on slopes (70–90 plants) due to the damper conditions (Kutsch 1982).

At the other extreme where rainfall is abundant, albeit seasonally, alternative strategies of 'hydro-agriculture', such as paddy cultivation, have been developed. In these systems of inundation agriculture, control is concentrated on the flow of water to maintain appropriate levels for crop development. The distribution between *khet* terraces, where paddy cultivation occurs, and the *bari* terraces has been mentioned earlier. Alternatively, raised bed systems can be used. The control of water on paddy is critical to prevent any collapse of the terraces. So too is the control of the temperature of water, which may be modified by shading or by warming (e.g. using darker soils and organic matter to increase heat absorption, or by topping up in the afternoon) (Whiteman 1988). This can speed up the rate of crop development. Paddy cultivation may also be fed directly by rainfall or by using ponded supplies derived from precipitation or snow melt. In addition, standing water around the roots of young plants reduces the frost risk, but consistently higher temperatures of paddy water can increase the risk of disease to the plants.

Wetland cultivation can occur on the hills of the monsoon Himalayas, the mountains of south-east Asia, and is also associated with swampy valley bottoms, for instance the Lake Titicaca region of Peru (Erickson and Candler 1989), and Rwanda (Loevinsohn *et al.* 1992). In the latter case drainage furrows can be altered to allow different crops to be grown at different periods according to need.

In northern Thailand, intermontane basins such as Chang Mai are filled with rich alluvium and climatically suited to irrigated rice. Fields are flooded in late spring by both rainwater and irrigated supply. Increasingly, ploughing takes place using diesel-powered machines with paddles, these having replaced buffaloes. Seedlings are transferred after 30 days, and are weeded by 'treading the weeds into the mud'. After harvest, the fields may be resown with soya bean (Walker 1992).

Ecological resources

A final consideration of the resources available to mountain dwellers in terms of subsistence livelihoods, and increasingly commercial enterprises, is that of the role of natural vegetation and animals. Both grasslands and forests produce abundant and different natural resources for the agroeconomy of traditional livelihoods. In addition, the diversity of plant life provides a rich resource of medicinal and commercial products. It is only in recent years that official agencies have begun to recognise the diversity of these resources and the depth of indigenous knowledge among mountain communities.

Ploeg (1993) describes the indigenous strategies of Andean farmers with respect to their knowledge of the environment, plants and the application of this knowledge to improving productivity. Cultivars of potato are distributed over a range of plots. Some plots are planted with only one variety, others intercropped with several varieties. Thus the productivity of cultivars under different plot conditions can be tested with the minimum risk of crop failure. The selection of cultivars is not a hit and miss affair, but based on the concept of '*art de la localité*' – that is, a detailed understanding of the variable plot conditions and cultivar performance. It is stressed that farmers are not averse to experimentation, or to a certain degree of risk in trying new varieties and combinations of variety, timing and environmental conditions. To do otherwise is to induce stagnation into the farming system. Such indigenous experimentation is immediately accessible, comprehensible and demonstrable to the community, more so than in the case for introduced hybrids and western-based scientific research.

Meadows and pastures

Livestock are an integral part of all mountain agricultural systems, and the grassland ecosystems at high altitudes have provided a valuable, although often a low productivity resource. In North America, for instance, the grassland of the mountain ranges has been an important element in the development of the ranching industry. In the Andes a variety of different *puna* and *paramo* environments are grazed by indigenous camelids and by more recently introduced sheep and goats. In the tropics these grasslands are accessible all year, but complex systems of pasture control along the transhumance routes are needed to prevent overgrazing and degeneration. These grasslands may take longer to recover than lowland grasslands, and thus fallow or resting periods are necessary. In middle and high latitudes alpine grasslands are generally only accessible once the snows have melted in spring and until the snow falls again in autumn. Such grasslands can be a very rich resource, but can be quickly damaged by continual overgrazing.

Mountains such as the Alps, Pyrenees and the Himalayas have many examples of an annual cycle of movement of herds and flocks to successively higher pastures during the summer (Plate 4.7). These may return to the village for stall feeding during the winter, or alternatively the animals might overwinter in the lowlands during periods of fallow, as is shown in Figure 4.2. They can provide manure whilst feeding off the crop residues. Some of these highland–lowland reciprocal linkages are established through kinship and tribal networks over several generations. The movement of flocks and herds involves the removal of labour from the village for most of the summer, and the operation of such a system requires sufficient labour and fodder to operate effectively. Where long distances are involved, one ethnic group may be concerned almost entirely with herding, whilst settled agriculturalists live permanently at lower altitudes (Puigdefabregas and Fillat 1986). Reciprocal trade networks operate to fulfil each group's needs. Such a situation occurs in the Andes (Guillet 1987), Nepal and Papua New

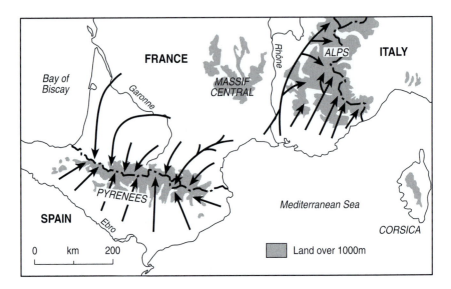

Figure 4.2 Transhumance routes in the mountains of the western Mediterranean (After Puigdefabrigas and Fillat 1986)

Guinea, and further case studies are noted below. The sensitivity of many alpine pastures to overgrazing, and also the need to identify particular grazing rights over a given tract of land by tribes, communities and companies, has created complex systems of land management. In a similar way to the communal management of water as a critical resource, relatively extensive low productivity resources such as pastures are often managed as common property resources. Rules and regulations are enforced to conserve grazing land to ensure that fodder is available each year. To overcome difficulties of labour shortages, sharing and renting may occur. In the latter case, a shepherd may tend the flocks or herds of several households as an optimal use of labour. Alternatively, one household may rent out their animals to a shepherd who makes his living from them, in the same way as a cultivator rents land.

Grassland productivity is constrained by the growing season, and a combination of climatic variables. Turf is an efficient slope-stabilising agent, but is easily damaged by trampling and takes a long time to recover. The resilience of these pastures, therefore, is dependent upon not damaging the turf cover and not overgrazing. The pasture management systems, which have been developed to prevent this occurring, are diverse, ranging from pastures where no controls operate to closed or reserved pastures. Bishop (1998) reports examples of zomo and yak herding in Nepal, where grazing is not permitted near the village during the summer. This removes the stock from nearby crops, and reduces damage. In addition, these lands are reserved

Plate 4.7 High-altitude pastures, the Pyrenees

for sick, pregnant animals or for the early spring and late autumn grazing when higher pastures are inaccessible.

Elsewhere, as in Yemen, there is increased reconsideration of the traditional utilisation of resources (Kessler 1995). The nature of the land resource and access by the community changes during the seasonal cycle of agricultural activity. The stretch of land surrounding the cropped area, about 5 hectares, is temporarily classed as private and protected from livestock until after the harvest of the cultivated crops. Then, it reverts to communal grazing land, and animals of the whole community can graze on the stubble and add their manure. The appropriation of such common land by landlords, its enclosure, and the development of irrigation supplies enabling an extension of the growing season have excluded the community from this grazing resource.

A more formal, rigid system of use rights and controls is illustrated by Gilles *et al.* (1986) in the High Atlas of Morocco. They examined an area of reserved grazing or *agdal* in the Oukaimeden zone above the Ourika valley. This pasture consists of an area of approximately 4 km^2 at an altitude between 2,600 and 3,260 metres. The grassed valley has springs, lower hay meadows, slopes and upper rangelands. Access is restricted to those with use rights that are determined by the ownership of a corral on the *agdal*. Ownership of these rights is inherited, but if a family does not use it, the rights can be rented out to another household (although within the same tribal confederation group). Actual use is limited primarily to those who have sufficient labour resources for the size of the herd and, to some extent, by distance from their home village. Those from very distant villages with only a few animals may not consider it worth while sending many people up to the pasture, unless flocks are amalgamated. In turn, herd size is constrained by labour resources. This arises from the fact that although sheep are permitted to graze freely on this *agdal* pasture, cattle must be kept in a corral and fed grass cut from the meadows. Thus the number of animals that can be kept is constrained by the number of people who can be spared from the crop harvest on the lower terraces. The cut and feed policy enables the maximum use to be made from the fodder by eliminating trampling losses and it is not permitted to take hay or grass cut from the *agdal* to sites outside its boundaries.

The *agdal* is actually closed between 15 March and 10 August. Allegedly, these dates were set by an eighteenth-century saint and are rigidly observed. This late opening ensures a good, ripe grass crop, allowing for later snow lying at these altitudes. Disputes primarily associated with rights of access occur, sometimes arising when a corral is unused for several years and slowly dismembered by neighbours. When the owner finally returns, they find it has disappeared, or been taken over, and thus they have effectively lost their rights of access. Alternatively, during times of fodder scarcity, flocks from adjacent tribal confederations may encroach on the area.

A different system is discussed by Netting (1972, 1974), who studied the uses of meadowland in the Swiss Alps. These pastures may be privately owned, as hay meadows or grazed lands. Barns are constructed to store hay, and during the winter cattle are moved from barn to barn. South-facing meadows are often irrigated. This may take 6–7 hours for a large meadow, but is worth while due to the importance of the dairy component of production. It has been the meadows, however, that were first abandoned as a result of labour shortages after the Second World War, although they have taken on a new role in summer and winter recreation and tourism.

Plant resources

Mountain agricultural systems tend to retain a larger variety of crop species and a significant proportion of native, indigenous species than lowland systems (Brush 1986). This is, in part, a result of the marginality of mountain environments and thus the need for a variety of crops to cope with possible failure due to environmental stresses. Added to this, the physical and ecological difficulties of mountain farming and the peripheral location of many mountains have often protected these areas from being absorbed by large-scale, commercial, externally (e.g. state) run farming enterprises. Nonetheless, more and more mountain areas have been subject to monocropping practices. These range from the long-standing activity of companies producing tea in the Himalayas and coffee in the Andes, to the spread of opium in south-east Asia, *kif* (cannabis) in the Rif (McNeill 1992; Maurer 1992), and more recently potatoes in Pakistan. These production systems are designed to exploit the particular ecological and market niches available and are often a vitally important part of the local economy (see Chapter 7).

As noted above, many traditional crop varieties are derived from local species of plants. In Central and South America, there is an extremely diverse genetic 'bank' of plant varieties (Zimmerer 1992). The local inhabitants, including children, exhibit a deep knowledge of these resources, their potential and actual uses, and differences in productivity. In the subtropical montane forest of north-western Argentina 181 species of plants have been identified and categorised by local people. This list contains seventy-two cultivated species (with twelve medicinal plants), and 109 wild species, including forty wild medicinal herbs. This includes seven varieties of maize (*Zea mays*) and multiple species of potatoes, sweet potatoes, beans, nuts, fruits, etc. (Levy Hynes *et al.* 1997).

In the Andean Titicaca region, the traditional staple cereal has long been a local *Chenopodium* variety, which was well adapted to the wet and dry season and the frost risk during dry seasons (Dollfus 1982). In every mountain community, similar examples occur of local varieties of cereals, often selected and derived from local grassland species. Tubers, vegetables and fruit species are also well adapted to local conditions and offer a comparative advantage of greater ecological tolerances than many modern imported higher yielding varieties. These latter are often less able to cope with poor soils and require high inputs of nutrients which may not be readily available. The number of different varieties grown enables microenvironmental differences to be exploited. However, commercialisation of mountain farming and the introduction of new species mean that although yields can increase, often in response to increased nutrient inputs, the diversity and resilience offered by a range of traditional varieties can become eroded. One example of this has already been noted in the Hunza valley where the introduction of new potato species has required new methods of disease control (Parish 1999).

An additional point to note is that traditionally cultivated species often performed functions beyond simply producing food. Crop residues may be equally important as the crop itself. Local varieties of cereals might produce less grain, but may be suitable as fodder; longer straw may be used for several purposes, such as weaving mats, reducing surface evaporation or thatching. Changes in crops grown, even in the variety used, involved not only a consideration of the part that the crop has as a food resource, but also its other roles in the domestic economy.

A final point relates to the experimental nature of many farmers. This may arise out of curiosity, the need to solve a particular problem, or in response to the dissemination of new technology or varieties (Rhoades and Bebbington 1995). For example, in Peru in the eastern high jungle potatoes are grown in the less humid zones above 2,500 metres. Below 2,000 metres the conditions are too humid. Farmers discovered that green aphids attacked green sprouting potatoes, so they used red and purple sprouting varieties instead. Also, diffuse light storage prevented tuber moth attack where dark storage did not. Both these 'discoveries' were later scientifically explained and 'proven'. Such experimentation, along with the testing of new varieties, takes place on a small scale, intercropping with old tested varieties until the farmer is satisfied with productivity and performance over several growing seasons. The risk of failure is of too great a consequence for farmers to abandon their tried and tested technology and varieties wholesale in order to follow a new one. Such instances as these have increased the awareness of western scientists to the great practical and applied knowledge developed by generations of these farmers (Rhoades and Bebbington 1995).

The increasing recognition of the highly diverse and rich biological resources of mountains has meant that the role of genetic resource protection has become a primary conservation issue. This extended from local species richness in traditional crop varieties such as potatoes, cereals, local rice, wheat and barley, to the unknown medicinal plants of forest ecosystems.

Livestock

The livestock species, which are suitable for domestic use in high-altitude environments, are derived from local wild stock, and adopted over the centuries for a variety of purposes. The yak, llama and indigenous species of mountain sheep and goat are well able to cope with poor quality forage, long winters or high diurnal temperature ranges. They provide meat, milk, wool and live animals for domestic use and for trade with communities down valley, or further afield, who may lack the pastures for grazing. An example already cited was that of New Guinea where during times of potato famine, higher altitude communities decamped to lower altitudes, bringing their pigs with them. Pigs are not kept in these lower villages, so this contribution is considered good payment for their keep during times when the potato crop is replanted and matures in the higher zones.

Interbreeding of yak and domestic cattle in the Himalayas over time has produced hybrids that have productivity advantages over traditional species but tend to be less tolerant of extreme conditions at higher altitudes. For example, the domestic cow is unable to survive above 3,000 metres, whereas the yak can survive all year round up to 6,000 metres. Cross breeding between these related species produces a variety of hybrids such as the zomo, which combines the milk and meat producing capabilities and qualities of lowland cattle with some of the altitude tolerances of yak. As these hybrids can only be produced in the zone between the lower limits of yak tolerances at 3,660 metres and the upper limits of cattle tolerances at 2,100 metres (the hybrids are sterile), this gives a comparative advantage in production in this altitudinally constrained niche. Trade networks operating laterally between communities and valleys can therefore be strengthened, as in Nepal (Bishop 1998).

Many of the craft techniques such as weaving have developed in response to the wool or hair type available. In the Andes, weaving techniques are locally adapted to the long staple hair of llama. The introduction of the shorter staple breeds of animal resulted in a change in the trade and craft networks. Shorter staple wool is exported to lowland factories for weaving, when the traditional long staple is unsuitable. However, local trade markets still favour long wool woven textiles, so some of the original stock is kept to supply local demands.

FORESTS

Forests are a critical part of sustainable mountain livelihoods, both in the past and at present. Forests not only supply timber for construction and fuel, but fodder for livestock, which produce manure used as fertiliser. Gurung (1992) estimates that some 50 per cent of fodder comes from grass

Plate 4.8 Slash and burn agriculture, northern Thailand

and leaves gathered from a forest or wooded landscape. In addition, systems of agroforestry and silviculture are significant components of the productive economy, concentrating on dual use of land: grazing, or growing fodder beneath tree crops, and by selective planting of productive species for cash crops (for example fruit trees). In slash and burn economies, forests provide land which is used on an intermittent or cyclical basis; clearance of hillslopes in Thailand, for example, leaves stumps and selected individual trees on cleared plots to promote regeneration of forest once cropping has finished (see Plate 4.8)

The primary resources of forests are first (by volume at least), timber and fuelwood (Table 4.1). Timber production has resulted in modification of the diversity of species, either by overcutting of selected species, resulting in a depleted forest in productivity terms, or by removal of unwanted species or replanting with selected species in order to increase productivity. In both cases the forest structure is substantially altered, as is the case for most European and American forests. The emphasis on management here is either the sustainable use of plantations, or the regeneration of diverse forest stands to restore the habitat and ecosystem which formerly prevailed. In the Mediterranean, a long history of exploitation has resulted in great impoverishment of the current forest stands in most cases. Agricultural activity,

Table 4.1 Forest resource use

Scale	Products/use
Local	Grazing, fodder (leaves, branches, litter) Timber (construction) Fuel (deadwood, lopped branches, litter) Medicinal and food (herbs, fungi, fruits, nuts and other plants) Cultural (e.g. sacred groves and tree species)
Regional/national	Recreation (wilderness, tourism, summer and winter sports) Timber (for national use and for export) Water (quality and quantity affected by presence or absence of forests)
Global	Biodiversity (fauna and flora) Global climate (carbon 'sink') Medicinal plants (as yet unidentified or to be commercially exploited)

and prior to that some forest-based gathering and hunting activities, was established in the eastern basin by 10,000 years ago, and spread to the west some 4,500 years later. A significant expansion of use during the Roman era was followed by periods of regeneration as pastoralism expanded, and periods of extensive cutting of selected species such as juniper, cedar and pine for construction of buildings and ships, especially by the Byzantines and Venetians (Meiggs 1982). In more recent years the trend in southern Europe and the Mediterranean towards abandonment has resulted in an increase in forest fires – a doubling since 1970 is estimated (Alexandrian *et al.* 1999). In France and Italy, new forest legislation introduced penalties for firing forest, and obligations to clear deadwood and litter in privately owned forests. The cessation of collections of litter and deadwood is a major factor in enhancing the spread of fires. In France, some 6,000–10,000 hectares a year undergo prescribed burning in order to reduce this risk. In Colorado, similar proposals to introduce prescribed burning met with opposition from local communities on the grounds of air pollution and damage to the aesthetics of the forests (Price 1991).

Fuelwood is one of the other main resources for which forests are exploited. Sharma *et al.* (1992) estimate that globally some 3,000 million cubic metres are used each year, forming two-thirds of the world's wood consumption. Some 3 billion people are entirely dependent upon wood as a source of fuel. The growth in population and the visible degradation of forests sparked a huge interest globally in the state of forests, and assumptions of linkages in a cycle embracing poverty, population growth, deforestation, degradation of mountains and flooding of lowlands. This 'Theory of Himalayan Degradation' is discussed at length in Chapter 8, and is viewed rather less hysterically in the light of more recent research.

Forests in south-east Asia have suffered from colonial exploitation of prized species such as ebony, teak and mahogany. These slow-growing hardwoods continue to be selectively felled in present day commercial logging enterprises. Other products from forests include cork from the cork oak (*Quercus suber*). This is limited to the western Mediterranean, particularly Spain and Portugal. A recent study (Varela 1999) estimates that some 2 million hectares of cork oak remain. Although cork production is under pressure from synthetic replacements, some 60 per cent of the production is used for bottle stoppers. Other uses include floor tiling and insulation. In many areas, a greater proportion of younger trees is being harvested, as the older stock dies off. Trees tend to be stripped around every ten years, and unless constant replanting occurs, the supply of older trees diminishes. Other products which are increasingly being commercially exploited from Mediterranean forests have resulted in selection of particular preferred species such as the stone pine for kernels and maritime pine for resin, and thus promotion of less diverse stands.

In more traditional societies, such as in south-east Asia, a highly diverse range of non-timber forest products is collected. In Thailand, for example, the Akha people collect bamboo shoots, stems and poles, ginger, medicinal plants for human and animal use, insects, fruits, nuts, seeds and vegetables, as well as hunting the available fauna (Bragg 1992). The children and women display a detailed, practical knowledge of the forest's resources, and much of the cultural identity and spiritual practice is embedded in forest lore and ritual. This is typical of many forest-based traditions. A similar range of products was identified by the people of Mount Elgon, Uganda (Lampietti and Dixon 1995), including honey, mushrooms, fibres for ropes, resin, etc. Such traditions survive in Europe – for example, the truffle hunting rituals of France. This detailed knowledge of the resources and their uses is now recognised by biotechnology companies as well as by development agencies. There is an increasing need not only to integrate such knowledge into developing solutions to the current pressures upon these resources, but also to protect the indigenous peoples from exploitation (particularly with regard to biotechnology research), and to ensure they receive an appropriate share in the potential profits from the utilisation of this knowledge. Whilst the recognition of the value of non-timber products has benefited some communities, these high-value products are often scarce and large-scale exploitation has resulted in selective depletion.

Finally, forests have other less tangible functions, such as protection of hillslopes from erosion, and reducing (although not eliminating) avalanche, flooding and landslide hazards. This protective function has been emphasised in forest protection strategies, such as in the Swiss Alps during the nineteenth century in the wake of devastating floods. Similarly, the reaction to flooding in 1988 in Thailand instituted a ban on logging. Forests also have a growing amenity value, as the development of tourism in many mountain regions becomes the main source of income. Whilst ski developments in the European Alps, for example, tend to concentrate on the high Alpine pastures, the maintenance of forests for summer tourism, and for their protective function remains important. It is not sufficient to leave these forests to their own devices, and careful management is required to prevent diseases and fire risk. Planned growing is necessary to maintain a healthy, balanced and stable standing stock. This promotes forests as a wildlife habitat, and protection of individual species such as the endangered gorillas and spectacled bear has given an incentive to protect forests as habitats, as well as to maintain populations for game hunting.

Forests have long been held sacred in many cultures – Hindu, Buddhist and animist traditions identify with forest deities and spirits. Even the Judaeo-Christian tradition revered forests and maintained sacred groves previously dedicated to pagan deities. Such groves often form the only relict stands of ancient forests. In the present era, forests have again become important as growing numbers of people seek spiritual solace in areas of wilderness and natural beauty. When linked with more earthy amenity values, these intangible assets of forests become an important force for their protection. In a global context as well, the realisation of the potential of forests to counteract global warming has promoted international action to conserve them, and this is especially valuable when such use can be combined with sustainable use of varied products, amenity and other protective values.

Traditional management of forest resources

Traditionally, most forest areas would be held under some form of communal property system. The nature of common property resource (CPR) management is varied both in time and space. The impacts of constraints in access due to actual or permitted availability affect all aspects of life – fuelwood collection is dependent on the resources being sufficiently near for available labour to exploit. There are also preferences for type of wood used – in the Himalayas, birch is preferred as it is slow burning and hot, but scarcity has meant that fir and juniper is burnt instead. In Khumbu, Nepal an individual house was estimated to require fourteen trees for beams and joists, twenty to twenty-five for rafters to support a slate roof and additional timber for internal divisions, door and the like (Stevens 1993). Changes in house size, fashions for internal panelling and other fittings place considerable strain on available resources. Juniper is preferred as it is strong and dense, but increasingly local fir and imported pine have been used due to local scarcity.

As in many other mountain regions controls have operated on forest resources in Nepal for a long time. The origins of such systems are difficult to establish as the rules, regulations, penalties and enforcement methods appear to have evolved into local forms, which are not necessarily traditional, in the sense of being long established, but are certainly indigenous (Fisher 1989). These systems operate whether by controlling the users by establishing rights of access and central control by some institution (a village council, monastery or temple), or by controlling what products are taken and in what quantities. The various local manifestations of these systems have been recorded by a number of authors (Fürer-Haimendorf 1964; Campbell 1978; Molnar 1981; Campbell *et al.* 1987; Messerchmidt 1987). Special conditions may apply in selected areas. For instance, in Nepal there are sacred groves, 'lama forests' where any cutting is forbidden but gathering is permitted, and 'bridge forests' which are set aside to provide construction timber for the repair of bridges. However, the distinctions are sometimes blurred. Some protected forests may be exploited by certain groups by licence, whilst others may remain untouched. Some forests have become severely degraded, now consisting largely of shrubland as a result of overexploitation. Sometimes this may be the result of attempts by the state to exclude users from other forest resources, so increasing the pressure on particular patches of the forest. There are examples of this in the Indian Himalaya (Rangan 1997) and the highlands of Thailand (Cooper 1984; Tapp 1989). The implementation of use rights or harvesting 'quotas' is varied, but most systems appoint forest guardians similar to those appointed to monitor irrigation water distribution. The guardians are entrusted with the task of policing the reserves. When appointed by the government, they may either turn a blind eye to local 'nibbling' at the edges of state reserves, or find themselves resorting to arms. In either case, it can be very difficult to monitor the different rights for different people at different times. It is becoming increasingly common practice to attempt to reinstate some form of community control over forest resources, to retain a feeling of involvement, interest and benefit and thereby responsibility for forests. They have been the source of considerable dispute and contention, and this is discussed further in Chapter 8.

AGRO-ECOSYSTEMS

Up to this point the discussion has focused mainly on the technical practices which are associated with rural production in mountain areas. This is insufficient, however, as it must be recognised that farming systems also involve social and economic issues, particularly with regard to labour use. A useful concept is that of an agro-ecosystem (Conway 1987) which provides a framework linking ecological, economic and social dimensions of rural production. Is there, or has there been a distinctive mountain agro-ecosystem? Once again it is important to emphasise that at this point we are dealing mainly with 'traditional' systems that have been subject to 'relatively' little change resulting from government intervention or the full-scale impact of capitalism. This helps set out a benchmark against which the impact of the 'development ethos' of Chapter 7 can be evaluated. In practice, of course, both traditional and modern systems often operate side by side in the same area, even on the same farm, but the division assists in highlighting key elements of each system.

Agriculture, pastoralism and silviculture have provided the principal sources of livelihoods in most mountain areas. Although the last century has seen major changes in some parts of the world, these still remain dominant elements in others. It is therefore rather difficult to generalise across the diversity of environments and societies involved and pinpoint a truly distinctive mountain agriculture. In traditional mountain societies, the choice of management techniques for crops, animals and trees is clearly closely related to the particular social demands and environment conditions of the particular locality. Fundamentally, in higher latitude regions agricultural activity is determined mainly by seasonal cycles whereas in lower latitudes more complex continuous activity is possible within an aseasonal environment. This is most apparent when observing mountain ranges such as the Andes which traverse many degrees of latitude. However, as has been made clear earlier, the 'verticality' component of mountain systems means that the particular crop/livestock mix utilised will be partly a function of altitude and so the net effect is a mosaic of systems, each operating in a 'niche' provided by the local conditions of production. With this variety in mind, much of what follows draws upon a number of detailed examples.

We have already discussed the notion of verticality in Chapter 1, but it is important to reiterate that this concept has been widely used to describe farming systems operating in mountain regions. As noted earlier, Stevens (1993) argues that the verticality concept and the production zone concept can be combined to highlight what he calls the strategy and tactics of agro-ecosytems in these areas. The following diagram (Figure 4.3) is adapted from Stevens and illustrates that there have been a variety of agro-ecosystems operating in the Nepal Himalayas, ranging from mixed farming by settled agriculturalist to nomadic agropastoralism. As the diagram shows, each one of these systems also has a vertical component, particularly what he describes as middle-altitude and high-altitude agropastoralism. However, particular communities may find themselves specialising in a production system tied to a specific altitude whereas other groups utilise the full range of production systems in a mixed farming strategy that encompasses different altitudinal levels. This is illustrated in the various studies of Uhlig (1973, 1978, 1995). The altitudinal gradient associated with agriculture is based,

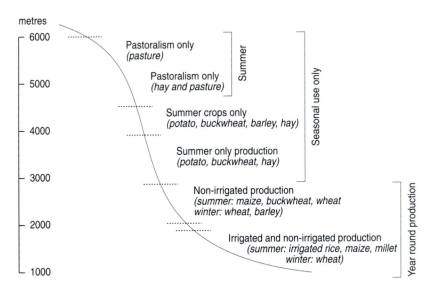

Figure 4.3 Verticality and land use in the Himalayas (adapted from Stevens 1993)

in part, on the fact that sections of this area experience water deficit and suitable production levels demand irrigation, whereas in other areas production under humid conditions is possible.

In Kashmir, eight distinct social groups exploit the mountain resources. At the lowest altitude, Kashmiri rice farmers utilise wetlands on the valley floor. At the other end of the spectrum, Ladakh mountain farmers occupy the highest level cultivating wheat and barley in irrigated oases, along with herding. Various degrees of co-operation and interchange exist between these groups. In the Jaunsar-Bawar zone of northern India, one group utilises several altitudinal zones. In this case the communities work at rice cultivation in the valley bottoms and also manage the cultivation of temperate crops such as wheat and potatoes at the higher levels. At the same time animals are moved with the changing availability of suitable grazing. There is a striking similarity between this example and that described earlier for the Andes.

Whilst the 'verticality' concept provides an overall framework for examining traditional farming systems in mountains it is important that we examine in more detail some of the principal factors that influence the way in which farmers utilise their environment. One way to illustrate this is through the examination of a particular system from which we can then extract some of the key points about traditional mountain farming. A number of detailed descriptions of mountain systems are available, for instance the work by Netting (1981), Miller (1984) and Bishop (1998) but

Table 4.2 s'Tongde cropping pattern

Field crop	Percentage
Cereals (barley/wheat)	73
Peas	22
Fallow	5

we will develop the account by Crook and Osmaston (1994) of a village in the Zangskar valley in Ladakh, which was mentioned briefly in Chapter 3. This focuses on s'Tongde and its locality, situated at 3,550 metres and which has remarkably good yield levels. It is a mixed farming system, with the close integration of cultivation based upon irrigation and livestock, with herd movement to summer pastures typical of the *alpwirtschaft* model. The main crops are six-row barley, wheat, peas and lucerne. At the time of this survey there were no potatoes cultivated, although they can be found in some other villages of the area. Rotational practices vary but the basic sequence is peas followed by barley and then wheat.

Cultivated land is owned either privately or by the nearby monastery that leases the land to local farmers. Grazing land is operated under communal arrangements. The cultivated area comprises about 57 hectares, much of which is irrigated, and another 57 hectares of lucerne and grasses mainly on strips between fields. The average size of cultivated land per holding is difficult to determine due to the multiple households, although a figure of 2.7 hectares for the bigger units is provided. Individual fields, however, range in size from 0.02 hectares to 1.2 hectares. The cropping pattern of a typical year is set out in Table 4.2; the annual crop cycle is shown in Figure 4.4.

The Tibetan New Year occurs in March (spring) and is usually associated with snow melt. In most years the winter snow cover would be several metres in depth and, until a substantial proportion of this melts, the cattle cannot be moved from their stalls to the fields. One of the first jobs in spring is to clean out the domestic excreta and move this to the fields to join the piles of animal dung left over winter. This is then spread over the fields to be used for cereals. Usually sometime in April, most of the livestock, other than ploughing oxen and newborn small stock, are taken up to the grazing areas. Thus by May the cultivation cycle has begun.

As we have seen in Chapter 2, the mountainous terrain results in marked differences of insolation and day length between one locality and another. Therefore the timing of important agricultural activities varies even between adjacent villages. In this instance farmers may take the opportunity to share equipment such as ploughs and oxen. This is an important factor given the heavy labour/livestock demands for this task. Usually the land is ploughed twice for wheat but only once for other crops, the typical

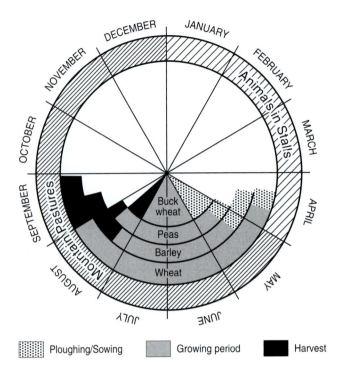

| Ploughing/Sowing | Growing period | Harvest |

Figure 4.4 Crop cycle in Ladakh (derived from Crook and Osmaston 1994)

two-oxen plough taking about six days to cover the 1–2 hectares of the larger holdings. Seed is usually broadcast and covered by either the plough or a special hand implement. The length of the growing season is three months for peas, three and a half months for barley, and four to four and a half months for wheat. The summer period is one in which the burden of crop husbandry falls primarily on the women, who weed the fields and tend the milch cows. Osmaston makes the point that these cultivated areas are exceptionally 'clean' in the sense that the women's hard labour in the fields keeps the weeds down to an extent normally only found in fields subjected to high doses of chemical weed killer. This means that the crops themselves tend to be of good quality at harvest-time. It is also during this period that irrigation takes place, in a manner very similar to that already described earlier. Both men and women are involved and if necessary the work continues through the night.

Harvesting would normally begin in mid-July with the peas, followed by barley, and most wheat would be cleared by the end of September. In addition to harvesting cereals and peas, lucerne and grasses would be cut and haymaking takes place. Most of the harvesting is carried out by hand and the women play a major part. The crop is taken to a threshing area and piled onto a hardened surface of about 8 metres diameter. Animals are brought from the higher grazing areas and tethered in a line to a central pole and then driven around this pole, so breaking up the grain stalks. This may continue all day for several days, depending on the size of the harvest, but usually after about three hours a floor full of stalks is reduced to small segments ready for winnowing. This process separates the grains from the chaff. The grain is then placed in sacks and stored in special rooms within each family's house.

During the period from September to early December, the cattle, yak and horses may return to the grazing areas at least until the first snow falls. Osmaston points out that depending on the altitude of the grazing areas, some villages bring all their stock down. In others the hardier animals such as yak spend winter on more distant, usually drier, high grazing areas where the snowfall may not be too deep.

Winter in the village sees the stock living in stalls that are located around the outer walls of the houses. In this particular area, the house design is such that, in winter, the animals do not occupy a ground floor and the humans live above (as in the Alps). By housing the animals in stalls on the outer walls heat is preserved in the interior, and so the family occupy an inner chamber. During this period the animals are fed from the fodder accumulated during the summer, and by day animals may be let out onto the local fields.

One interesting feature of this study is the finding that yields of the main crops are high. Unlike many reports of farming in mountain areas, this case study emphasises the difficulties of accurate yield assessment. Nonetheless, estimates range from 1–8 tons per hectare for barley and wheat, with an average around 3 tons. This is almost as good as yields in more advanced countries with their panoply of inputs, and Osmaston attributes the results to abundant water, warm days and cool nights, lack of weeds and pests, and careful husbandry. The result is that this village – and, Osmaston asserts, many others in the area – regularly produces well over the minimum dietary requirements set by the FAO. At least during the 1980s when this work was undertaken, there were no signs of a shortage of staple food.

The picture painted of this Himalayan village serves to indicate some of the key points of traditional mountain farming that find resonance in many other areas: the significance of livestock providing dung and draught power, the crucial role of winter fodder and the particular importance of seasonality with the relatively precise timing of the key farming operations designed to maximise the use of the growing season. Equally, at least in this case the research suggests no significant shortage of staples. However, the vulnerability of the system can also be recognised stemming primarily from the vagaries of the environment. The impact of altitude is not revealed through average conditions but on the increasing frequency of extreme events. Thus late snow melt or early snowfall, sudden floods or avalanche can easily disrupt the tight cropping schedule and radically affect the level of farm output. It is then that the various 'coping strategies' become vital. Some families have very large grain stores and can handle a seasonal failure. The monasteries may also help, although their existence is closely bound with the success of agriculture. Animals might be sold but the slow rate of herd replacement means that such an act is likely to impoverish a family for a generation. For recent generations it has been the state that has provided food security, indicating one of the critical changes that such villages have experienced.

As we examined in Chapter 3, pastoralism has usually played an important role in traditional mountain farming systems. Not only is it an integral part of mixed farming systems but in some instances provides the principal form of livelihood for mountain communities. In the earlier discussion the different types of pastoral economy were noted, and all these can be found in mountain areas. Some sheep and goats along with cattle can be found in most mountain zones, although the relative importance of each species does vary, often as a function of altitude. More noticeable is the fact that the most important animal species does vary according to location. In the Himalayas at higher altitudes, yaks thrive on poor pasture and harsh climates. Donkeys are also kept for draught power. Camels are important in the Middle East and central Asian regions where they perform the role that donkeys play elsewhere.

In the Andes, camelids provide the basis of the agropastoral system, being adapted to the harsh diurnal ranges of temperature and the climate of the dry *paramo* grasslands. Llama and alpaca are found alongside the ubiquitous sheep and goats (Browman 1990). Breeding is a highly skilled process to try to produce the most valuable animals, which depends upon location and business context. Commercial buyers usually prefer shorthaired animals with a white fleece because this can be easily dyed, whereas local buyers prefer long staple, coloured wool because this can be used for hand spinning and reveals less dirt. Herd management involves maintaining the right balance of males and females to ensure that breeding animals and followers are provided with appropriate pasture and that all animals are protected from predators. The alpaca likes a wet environment whereas the llama easily contracts foot rot and so prefers the drier areas. Equally, whilst male llamas are good as pack animals, the females often abort young if they are placed in stressful situations that may arise on a long trek. Given the complications of herd management and its labour demands, a herd may be handled under contract or reciprocal arrangements which can help keep the different herds separate. In these communities, fodder is rarely grown and so they are particularly dependent upon open grazing resources. Consequently, many of the institutional controls noted earlier which seek to control herd size and grazing quality are still in operation.

An examination of other farming systems in the Himalayas, the Atlas, the Andes and European Alps suggests that all have very similar characteristics depending upon the balance between crop production and pastoral activities. The importance of seasonality is also common though less so at low latitudes where much of the production is aseasonal. As has already been emphasised, mountain farmers are adept at utilising the many micro-environments that mountains present and they are highly adaptable and experimental in their outlook to new ideas. Most systems are based upon the integration of livestock and cultivation, the exploitation of different production zones and the broad range of activities that extend beyond farming to include forest management and handicrafts. Thus any mountain agroecosystem produces a wide range of products, mostly for staples but also, depending upon location, for exchange.

In the same way, the use of distant high summer grazing pastures allows livestock to be well fed during the summer, and also kept away from the ripening grain fields, thus limiting damage by trampling. In Nepal, strict laws apply to the removal of livestock at given times in the year in order to do this. The distance of these high pastures, however, means that some of the family are absent at the time of harvest, so often flocks and herds are amalgamated within the wider community, thus releasing more labour for the terraced fields and leaving the animals in the care of selected shepherds. As in the *agdal* example earlier (see p. 154), distance from these pastures, insufficient labour within a family, the labour cost of taking only a few animals up to the pastures, and some control over stocking densities can be incorporated into this system.

Communal labour is also called upon for infrastructural tasks – roads, terraces, irrigation systems, houses, etc. all require pooled labour. In the same way, farmers with adjacent fields help each other at peak times – ensuring harvests can be rapidly gathered and fields replough and planted for a second crop within the limits of the short season. Community responsibility is also extended to the access of resources beyond those officially communally managed. In Morocco and Pakistan, the second hay cut from valley floor fields is left for those members of the community who have live-stock but no such hay lands. This ensures some equality and provision of basic needs for all.

Farming systems respond to ecological patterns in various ways. In Chapter 1, the altitudinal impact on settlement was noted and this is closely related to the operation of the farming system. In many valley systems, such as those in the Atlas and Alps, individual households may own or exploit land in all altitudinal zones. In others, where distances are much greater, such as the Karakorum and Andes, different communities may be more specialised in living and working at different zones. In the former case, each household directly benefits from all zones, but in the latter complex systems of interaction and trade develop, producing a pattern of interdependence between the communities. Networks of interaction develop on all scales – from household, pooled labour to major tribal interactions. The spatial scale involved coincides with temporal scales – on the former, interactions are daily, but major intertribal actions may only occur intermittently, or annually, or in response to particular events or crises, such as disputes over boundaries or during times of conflict.

However, it is not just the spatial variation in farming activities that characterises mountain systems, but the fact that most households are also integrated into a wider net of activities beyond agriculture. This characteristic is not just a recent feature of mountain societies. In many areas, what today is known as pluri-activity has been perhaps the most significant characteristic of livelihoods for a very long time.

NON-AGRICULTURAL ACTIVITIES

Many of the traditional non-agricultural activities in mountain societies are inexorably linked to the physical environment. For instance, although often overlooked one of the most time-consuming tasks is that of labour associated with household reproduction. In most, but not all, traditional mountain societies this would be the domain of women, and their access to these resources has become increasingly contentious. Cooking, cleaning, firewood collection, clothing repair and the myriad tasks of house maintenance would draw heavily on the local resources of food, water, timber and fibres. The gender division of labour has been well documented in some areas, but this pattern is far from immutable, as Hewitt (1989) has demonstrated. It is also important to emphasise that these duties require considerable knowledge, as well as hard labour in order that the properties of natural products can be employed. For instance, the conservation of foodstuffs requires considerable skill and an intimate knowledge both of the climate and of the properties of the foods.

Many handicraft activities have been an important component of mountain life, and often remain so today. Men and women can be found who are skilled weavers, potters, dye makers, wood or stone carvers, and jewellers. In those regions where the mountains experience seasons of harsh weather, many of these handicraft tasks would be confined to the winter months. Whilst initially these tasks have been oriented to domestic needs, some mountain areas have long exported products to towns. Indirectly, the water (for power) and timber resources (for charcoal etc.) have provided critical resources for these processes, and in some cases have been the foundation for more advanced industrial activity (see Chapter 7).

The development of trade relations between mountains areas and elsewhere is closely associated with the emergence of 'cottage industry'. Historically this meant that families would be integrated into a production system based upon their contribution of labour (sometimes raw materials) provided by an entrepreneur, usually from outside the locality. The family would carry out the necessary tasks, such as weaving, within their homes and the finished goods would be collected. The effectiveness of this process was limited in many mountain areas by the problems of inaccessibility, both through remoteness and through the physical barriers of winter snow. However, some of the mountain zones of Europe, for example in Germany and Austria, participated in these activities, as did groups in the Andes and Himalayas where long-distance exchange, especially of animal products, was an important occupation.

Mining

In many areas, however, mountains became famous for their mineral wealth and it was mining that generated the most important non-agricultural activity. By virtue of often complex and highly convoluted geology mountains tend to be a prime location for the exposure of important minerals. In the Andes, after the Spanish subdued the Inca civilisation in 1532, there was systematic exploitation of gold and silver resources. A similar picture could be related for almost every mountain area of the world, whether we refer to the Roman exploitation of mines in the Alps, or, nearly 2,000 years later, the first wave of colonial exploitation of small mines in the High Atlas. At this point in the discussion it is sufficient to indicate some of the early linkages of the mining industry: its subsequent development is described in Chapter 7.

A particularly clear example of the way in which the mining industry in mountains emerged is that of the early developments in the Rocky Mountains of North America. This has already been mentioned in Chapter 3, but the focus here is on the technical aspects to illustrate the linkage with the physical environment. The mining booms of the late 1840s and then 1850s/60s brought large numbers of people to the mountain states of California, Colorado and Nevada. In the first instance it was the prospectors who rushed to seek their fortune and gain a place in American folklore. They could do this because of the fact that in the mountains gold in particular would be washed out of seams and deposited in creeks. This allowed 'placer mining' that encouraged the 'stampedes' across the US to such famous places as Cherry Creek in Colorado or Alder Gulch. The process involved the separation of loose gold from the river deposits by a process called panning in which the gravels were washed in a pan that was swirled by hand so that the heavier metal sank to the bottom. Improvements on these simple techniques relied upon the local topography where, just like irrigation development, miners built simple conduits to bring water to sites away from river banks and to more elaborate washing apparatus. Similarly, water power was used in the development of rock crushing mills which enabled exploitation of the more substantial exposed deposits themselves.

The topography of mountain slopes has played a very important part in the exploitation of the mineral lodes themselves. Metalliferous seams were exploited by digging adits into the seams and then extracting the ores. This meant that more and more of the operations occurred underground and could continue throughout the year, whereas placer mining could not be carried out in the severe winter months. Soon the individual, low capital operations were replaced by various corporate operations as the landscape became littered with crushing mills, waste dumps and criss-crossed with water conduits. In one site, Gregory Gulch, a relatively untouched landscape was first filled with prospectors' pits and then, in the space of 15 months about sixty big crushing mills (Dick 1964).

Elsewhere in the USA, the Appalachian Mountains were the site of mining activity in the nineteenth century (Dunaway 1996), whilst Europe provides us with examples from a much earlier period. Mineral wealth in the mountains was exploited to serve distant markets. In the Gastein valley of the Austrian Alps the fifteenth century saw a spectacular rise in mining enterprises, especially for silver. In each case it was the attraction of the mineral wealth that led to the considerable effort to construct appropriate roads, railways and aerial ropeways, etc. that characterise many of these exploits. Later, these tracks have provided some of the first 'improved routes' into mountain areas.

Although the individual mines may have been on a small scale in these early examples, the landscape nevertheless suffered, as there were often many mines in a small area. If any smelting took place on site, then large numbers of trees were removed, although it has been argued that mine-masters were well aware of the effects of excessive removal of trees and coppicing was regularly applied. However, the damage to the landscape through the physical process of extraction and the deposit of mining tailings remains even today as evidence of these early activities.

Despite the negative effects on the landscape, mining has offered some mountain communities a new and relatively higher source of income than farming. In many mountain regions it has been the only reason why settlements have occurred. When a mine is opened one of the first sources of labour may be from the farming community in the locality, but this is very variable for several reasons. First, farmers may be unwilling to take part in mining because it is considered both a dangerous and low status job. In some cases the employers require men with previous skills, and this has become more the case as the technological sophistication of mining has increased. In this instance, premium wages are paid to attract labour from elsewhere. In turn this has important social implications. In the Tyrol, in the fifteenth century for example, it was the migration of large numbers of miners from southern Germany that boosted its population, which doubled over a century and a half. Elsewhere in the Alps mines were opened in the sixteenth and seventeenth centuries in the Italian regions, and immigrants, usually bearing different linguistic and cultural characteristics, settled in these areas (Viazzo 1989).

The mining activities are interesting precisely because of their combination of local resource, external markets and a variety of sources of capital. From the medieval period onwards Europe's highlands have been the location of the enterprises, with the abundant sources of wood and water associated with mining, smelting and other processes. In the highlands of Brescia, in northern Italy, descriptions from the seventeenth century report that there were, in one locality, '84 mines, 11 blast furnaces and 100 forgers. Production consisted of nails, blades and cutting arms and agricultural tools' (Belfanti 1993: 261). In the Austrian Alps and the mountains of Poland, Moldavia and Bohemia, similar clusters of early industry were also to be found (Myska 1996).

Whilst mining activities may have brought some wealth to mountain areas, perhaps the more abiding fact is that mountains throughout the world are littered with the debris of closed enterprises. In some cases this may have been the direct result of the exhaustion of supplies or collapse of demand. More often, it is the result of comparative disadvantage; that is, some other site can produce cheaper or higher quality output. Mountains are particularly prone to this given that the often harsh conditions and remoteness often demand high wage rates and hence costly production. The relative instability of mining has meant that not only are there unsightly physical remains in the form of dumps but often old settlements serving as a monument to a bygone period in the economic and social life of a mountain area. Good examples of this can be found in the Rocky Mountains. Ghost towns such as Vicksburg or St Elmo in Colorado were once a hive of activity following the boom period of the 1860s/70s, but many were deserted by the beginning of the twentieth century, others lingering on to the period of the Depression. Today, some form the basis of an increasingly popular tourist circuit (see Chapter 7).

Proto-industrialisation

The mining sector has had a long history of links with mountain areas, but it is other sections of manufacturing that provide us with an important clue to the role that mountains have played in the process of social transformation. This is associated with proto-industrialisation, a concept largely developed by Mendels (1972) and examined mainly in the European and North American context. This concept refers to a period of the development of rural manufacturing activity which, depending upon locality, emerged in the late Middle Ages. In Europe, the mountain areas played an important part in this process which has rarely been acknowledged (Belfanti

1996) and owes much to the work produced in specialist studies of upland areas – for example that by Mezario (1989). According to Belfanti, the process is part of a wider change in the relationship between the mountains and lowlands, and has two elements. In the mountains, the relatively poor agricultural prospects, often coupled with land ownership rules that left many adults without access to land in sufficient quantity to farm an adequate living, meant that in times of hardship there was a surplus of labour.

On the other hand, in the urban centres of the lowlands an increasing number of entrepreneurs sought to free themselves from the costly restrictions imposed by urban-based guilds. Together with the availability of resources, this led to the development of rural manufactures in some mountain valley areas. Of particular importance were the textile trades involving cotton, wool, along with flax, and in some areas silk. In Switzerland this transition was most noticeable between the fifteenth and seventeenth centuries (Pfister 1996) when some of the Alpine areas became important producing zones. Given the problems of accessibility in mountain regions, the organisation of this production process developed along various pathways. Here the process was relatively simple, such as spinning cotton, and an artisanal trade pattern developed with itinerant traders selling the raw cotton to individual households and buying the yarn. This process fitted neatly into the labour demands of the mountain economy. In other cases, usually where a more reliable flow of the processed product was demanded, and therefore usually confined to the lower foothills, manufacturers used the 'putting out' system, with tight controls of turnover time and standards of processing. Pfister also argues that the rise in the number of households using this source of income occurred in the period between the fifteenth and sixteenth centuries. Many mountain farmers switched from the cultivation of crops to the management of dairy or beef cattle under a ley farming system.

Furthermore, the development of these 'cottage industries' within mountain areas links neatly with the discussion on demographic characteristics of the communities and the changes experienced over time. For example, in some of the northern foothill areas of the Alps, proto-industrialisation is associated with lower levels of mortality, through higher levels of income. Similarly, there is some evidence to suggest that as more women secured an income through involvement in production, so nuptiality increased. Nonetheless, there was enormous variation between one valley and another.

SUMMARY

This chapter has provided an account of the particular ways in which populations traditionally extracted a livelihood from mountain environments. In many instances this has been a hard process, bearing in mind the short growing season, often relatively poor soils and the problems posed by steep terrain. On the other hand it is quite wrong to visualise mountains in a somewhat negative way, even when examining traditional economies. Whilst it may be folklore to envisage settlers in the North American ranges as 'pioneers', elsewhere, long-established communities had produced highly sophisticated habitats using complex irrigation structures, well-integrated agro-ecosystems and closely linked settlement and exchange patterns. In some cases, the mountain resources themselves such as water, timber or minerals were of specific value to lowland economies and so mountain communities were able to exploit these advantages.

Nonetheless, to retain a relatively stable landscape is not easy in mountainous terrain. The hazards described above place all human activities at some risk, and in some cases the outcome is disastrous. There are increasing efforts to understand and monitor the hazards, and to develop production systems and forms of habitat that reduce the danger. In many cases, however, for example where settlement is located on an active volcano or in earthquake zones, the inevitable happens. The scale of devastation, particularly the numbers of fatalities, is, however, as much a function of the available resources and institutional capacity to cope with the disaster as a direct output of the scale of the physical event itself. These dramatic disasters should not conceal the fact that, at least in traditional mountain communities where agriculture has been the mainstay of the economy, long-term declines in productivity and economic returns, and other less dramatic events, have had equally deleterious effects.

Part 3

MOUNTAINS IN TRANSITION

The next group of chapters explores the dynamic changes or transformation process experienced by mountain regions. This necessarily requires a historical dimension, which can be seen particularly in Chapter 5 which looks at the relatively recent environmental changes experienced by mountains, and in Chapter 6 which is concerned with demographic factors. Chapter 7 provides the framework for an analysis of economic and political development. In terms of the 'cog model' this represents a situation in which key elements are recognised and functioning, but this very process brings about dynamic and evolutionary change. In turn this means that new factors, or old ones operating in a different context, come to play a major role in the transformation of mountain regions.

5

ENVIRONMENTAL
CHANGE

Change in any environment is the norm. There is a degree of underlying stability in climatic, ecological and geomorphological conditions which is overlain by constant change and operating on different timescales. Cyclicity in environmental response to external change (especially climate and human activity) implies a degree of resilience within the system. This is manifest as resistance – a capacity to absorb change without altering structure (constancy) or an ability to return to a previous state (elasticity). However, most change is to some degree irreversible so cycles of change actually behave in a spiral manner with a longer-term trend of cumulative change. Spirals may be 'downward' (for example, deforestation leading to degradation) or 'upward' with terracing, which if maintained will increase the stability of the surface. The fundamental problem of studying environmental change lies in the difficulties involved in separating out the longer-term trends from the shorter-term impacts within this 'spiralling' structure. Consequently, depending upon the spatial and temporal scale of any set of observations, a particular 'event' may be interpreted as positive or negative and the process stable or unstable (Gigon 1983).

The marginality of mountain environments with respect to ecological limits, topography, and high-magnitude, low-frequency events makes them particularly sensitive or vulnerable to climatic or other drastic change (Glaser 1983). This vulnerability can be useful as an indicator of climatic change and environmental response, but in a livelihood context there is a relatively narrow margin of error between sustainable and unsustainable activities. Therefore in mountain regions, concepts of stability and fragility are especially pertinent in any assessment of sustainable development (Messerli 1983; Winiger 1983).

This chapter examines climate as one of the key agents of environmental change in mountain areas. In addition, aspects of geomorphological processes are examined along with the impact of both climate and geomorphology on the natural vegetation.

CLIMATE CHANGE

A changing climate, such as the current process of global warming, induces an environmental response. The sensitivity of a mountain system means that small changes in climate can produce significant or large-scale effects. Marginal environments are under a high degree of constant stress. Small changes to water availability, occurrence of floods, droughts, landslides and late frosts can have drastic effects on the agricultural economy. The diversity of the economy, however, not only spreads the risk but also allows more rapid and potentially greater adaptation to new conditions. Climate has changed significantly in the past, and fluctuates on a regular basis, diurnally, seasonally, and from year to year. Climate affects erosion and weathering processes and rates, hydrological conditions, rivers and channels and thus sediment transport, deposition and erosion and water supplies. It also affects the type, quantity, quality and stability of vegetation cover and thereby the associated wildlife. Climate is therefore a critical factor for inducing change and its effects are not easy to predict or reconstruct.

There are several problems in the study of climate change in mountains. The first is the dearth of long and reliable climatic records on timescales beyond the last century or so. Such records as exist are rarely sufficiently evenly and densely distributed on spatial scales appropriate to account for and describe inter- and intra-valley diversity. A partial exception is the European Alps, with relatively long sequences of data. The harsh weather conditions, inaccessibility of high mountain areas and the lack of skilled researchers to maintain and operate weather monitoring equipment mean that even now only a sparse network of recording stations exists in most mountain areas.

A specific regional project developed to examine the particular characteristics of mountain areas is the Mesoscale Alpine Programme. This involves the weather forecasting agencies of six Alpine countries and also weather bureaux of other adjacent territories in a research project which is examining heavy precipitation and local föhn windstorms. These give rise to seriously damaging impacts on human activity as they are closely associated with flooding and landslides. In particular the research seeks to discover the principal factors involved in the generation of precipitation in areas of complex topography. This research involves the collection of high-resolution (fine spatial scale) data to input into numerical models for forecasting. The programme is described in detail along with some of the data at: http://www.map.ethz.ch/proposal.htm

Second, the uniqueness of each valley, and the spatial variability of microclimates described earlier, mean that it is difficult to account for and determine trends and responses within a valley (Slaymaker 1990). Average figures for mountain ranges and continents may therefore bear little relation to the actual conditions for crop growing due to the importance of factors such as aspect, local winds and temperature inversions. Local farmers build up a detailed knowledge of such factors and exploit them on a daily basis in making decisions about where and what to plant.

Haslett (1997) has tackled the problem of spatial heterogeneity by describing complex nested mosaics of landscapes. He also emphasised the need for a landscape approach where the scale of the study is important (a point discussed further in Chapter 8). The physiological response of plants to increased CO_2 and photosynthesis, their response to changes in humidity, temperature, light intensity and quality all operate at the micro-scale. Plant response is the basis of all habitat mosaics and has knock-on effects for animals and human land use. Diversity is the key to resilience, and the ability of plants to recolonise and adapt is critical to maintaining surface plant cover, and to successful cultivation strategies. The fragmented nature of mountain landscapes and the limited mobility of plants leave little room for recovery if the effects of change are dramatic. Insects, by contrast, some of which favour specific habitats and others that are more catholic, demonstrate a more mobile adaptability, which increases with a species' ability to live in a variety of habitats.

In considering livelihoods, it is not only important to confront the problems of describing current climate but also the diversity and timescales of change. The response of environment and farmer is inherently unpredictable, and timescales over which some climate change is predicted are too long for farmers to comprehend in their annual strategies for production. They work from the basis of known environment and conditions, forwards in time, not as western scientists tend to approach planning – predicting the future and working backwards in time to adopt planting strategies which attempt to pre-empt this future environment. Farmers usually adapt by responding to changes, a little at a time. By contrast, modern planning assumes comprehensive knowledge, both of the problems and the likely outcome, and implements programmes with little regard to their underlying experimental nature. Western society has become adept at handling this form of 'command approach' but elsewhere the effects can be particularly damaging. Much depends upon the sensitivity or vulnerability of both the environment itself and the society. Consequently it is the adaptability of societies to the processes of climate change that provides the key to long-term survival.

This view does not deny that climate is changing, or claim that scientific endeavour should not proceed to improve our ability to understand and predict the outcomes. The problem is to make sense of these changes in a way that can be appreciated by the small farmer operating within relatively short timescales and in the context of the specific conditions that apply to each locality. This is particularly true in mountain areas where each valley generates its own microclimate and consequently will respond slightly differently to the global changes discussed by scientists.

Modelling problems

The predominant way of predicting climate is to model it. Global Climatic Modelling has been responsible for predicting changes on a continental scale and for the debates and disparities that surround these predictions. In order to refine the models and apply them to more regional or even valley scale studies, a variety of other modelling techniques are employed. The use of a series of nested models superimposes predictions and models at different geographical scales in order to provide suitable scenarios that can be used by policy-makers. With all models, however, problems of topographic smoothing occur, whereby the actual height of mountains and their complex topography are simplified. Unfortunately, we have already noted that these factors are critical at the valley scale. The large size of grid used may also provide insufficient detail. Models based on grid squares of 100 km^2, for which a mean figure of change is evaluated, may be too coarse for individual valleys, and do not differentiate between slopes of different aspect and altitude which are so critical to the exploitative strategies of farmers. The quality of data has already been mentioned with respect to the existence of spatial and temporal records. This is being addressed, to some extent, by nested hierarchical modelling techniques for shorter timescales.

The MAP project mentioned earlier examines local and regional weather predictions for the European Alps. At present spatial resolutions between 14 and 80 kilometres are typically used in short-term predictive forecasting, but some experimental models have been run at a scale of a few kilometres. At the same time, whilst there have been some useful results regarding the impact of topography modelled in simple two-dimensional form the next generation of work will involve three-dimensional landscapes which will allow forecasters to understand the flows of air masses around complex topography typical of mountain regions.

The project also relates to global change issues due to the importance of mountains in modifying air masses, capturing moisture and influencing the distribution of climatic zones. Upscaling of mesoscale models is an increasingly important element in the development of global modelling. One such simulation used the Atlantic airstream over the Alpine system and reproduced well-known topographic effects, but what was so surprising from the work was the fact that the Siberian climate was shown to be heavily dependent upon the presence of the Alps (Broccoli and Manabe 1992). Upscaling models can increase the accuracy of topographic and regional detail, which is very important in analysing the impact of energy exchanges. This detail is lacking in the current approach, which uses downscaling of global models to attempt to tackle local differentiation.

Both up and downscaling, however, are complex and subject to considerable margins of error. Although progress in this field is rapid uncertainty is still a major problem, together with the wide disparities between predicted scenarios. It is therefore necessary to remain cautious in relying on these models for accurate predictions of future conditions. At the moment only very broad trends can be considered valuable as an input to policy discussions.

One approach to the modelling of future climate uses a combination of analogues, GCMs (Global Climate Models), and nested hierarchical models. It is possible to achieve a resolution of 10–100 km² for limited areas. Studies of climatic change and environmental response tend to be limited to specifics, for example the glacial–hydrological system. A glacier–climate model study in Austria suggests that fifty years of increasing greenhouse gases could induce a rise of 11 per cent in glacial runoff. But this study did not take into account other changes in conditions. A GCM study in California concluded that there would be less winter snow, less spring runoff and consequently less soil moisture. There would also be an increased winter runoff from rain and from earlier snow melt. This has impli-cations for water availability in relation to temperature-defined growing seasons. In a study of five mountain regions in the Canadian Rockies Slaymaker (1990) demonstrates distinctly different detailed responses to climate change, but all produce trends which suggest warming and an increase in precipitation.

The nature of past changes

The most significant effect of climatic change is the upward and downward shift of altitudinal zones, which causes compression or expansion of some vegetation belts and the disappearance of others due to the reduced area available higher up the mountain slopes, or the increased area lower down. Halpin's (1994) study of simulating change in altitudinal belts is discussed later (see Figure 5.1). Reconstructions of past climate change, particularly of the last glaciation and the current Holocene period, are an important source of evidence of the impact of cooling and warming of global climate on mountain environments. The general trend during cooling includes the expansion of glaciers and the movement downslope of vegetation zones. Dresch (1941) estimated that a cooling in mean annual temperatures of 0.5°C resulted in the displacement downslope of vegetation zones by 100 metres in the High Atlas of Morocco. Some transitional zones were compressed whilst others disappeared. Forests extended to much lower altitudes, which can be demonstrated by soil pollen analysis. Subsequent warming led to an increase again in the altitude of treelines and other vegetation belts, along with an expansion of zones and the reinstatement of transitional zones by speciation and adaptation.

The legacy of past glaciation is most clearly demonstrated by landforms of glacial and periglacial erosion and deposition, and the continuing read-justment of slopes, debris and sediments to current conditions by reworking and stabilisation. The evidence of past climatic changes is derived primarily from the landforms and sediments associated with glacier fluctuations, from the study of ice cores, palaeoecology (pollen, beetles, charcoal and the like) and using present-day analogues where they exist. Absolute and relative dating, for example by dendrochronology and radiocarbon, supports this work. The reconstruction of past environments is subject to considerable errors and also to poor spatial and temporal resolution. For example, local winds mixing pollen lead to inaccurate determination of local plant associations and locations of altitudinal limits of trees or other species. Slaymaker (1990) advocates caution in the use of proxy data sources, as it is unclear if such analogues are realistic for the period of the next fifty years or so.

The use of historical records, such as documented catastrophic events, tax and farm revenue information, diaries and accounts of travellers can be useful, and have been important in adding details of the impact of the Little Ice Age (fourteenth to nineteenth centuries) in Europe. This information gives indications of changes in yields, abandonment of land, access over glacier passes and irrigation infrastructure (Lamb 1982). However, such records are most complete in the Alps of Europe, but do not exist for many other mountain areas. They are also subject to inaccuracy, contradictions between accounts and misleading descriptions. During more recent years, the application of remote sensing techniques has been applied to evaluations of the recent changes of glaciers and snow cover. However, again the problem of accurate measurements applies here for models where topographic differences are not fully represented. In addition, the problems of clouds and poor visibility limit the use of such techniques, either seasonally or at all in some regions.

Past conditions

Barry (1992) has made a comprehensive study of mountain climate, including climatic change. The last glacial maximum provides the best evidence of substantial change, as previous landforms and sediments have been destroyed by this last major event. Around 18,000 years ago, ocean surface temperatures were around 2.8°C less than at present. Glacier and geomorphological evidence, however, suggests a much greater cooling of 5–6°C, indicated by a 1,000-metre drop in altitude zones (for example in Papua New Guinea and Hawaii). Steeper atmospheric temperature lapse rates prevailed, together with more snow.

Messerli and Winiger's (1992) review synthesises climatic changes in Africa. There are no glaciers in the Atlas Mountains at present, but during the last glacial period they occurred at 2,600–2,700 metres. Periglacial activity extended down to 2,000 metres and evidence of frost weathering activity to 1,000 metres. Dresch (1941) estimated a total depression of 800 metres in the altitudinal zones, corresponding to a cooling of 4°C. In the mountains of the Sahara, nivation and periglacial activity occurred at 2,000 metres, and frost action in the Hoggar Mountains between 1,100–1,400 metres. This indicates cooling to −10°C in winter and −6 to −8°C in summer. Precipitation changes implied drier summers and wetter winters due to the southward extension of humid polar air. In Ethiopia there was an increase in the size and number of ice caps. Terminal moraines have been found at 4,000–4,200 metres in East Africa, at 3,100 metres on Mt Kenya, and at 3,300–3,600 metres on Mt Kilimanjaro.

The environmental response is marked by a shift in the upper timberline during the last cold maximum. On Mt Kenya at 2,400 metres open Artemisia steppe prevailed up until 16,600 BP, where present conditions are more humid. A depression of timberlines by 1,000–1,100 metres occurred, equivalent to the Equilibrium Line Altitude (ELA) of the glaciers (indicating a cooling of 6–8°C). During the Holocene humid phase (12,000–10,000 years) wetter conditions prevailed in Saharan mountains, resulting in more fluvial activity and occurrence of evergreen oak. This was followed by the Holocene 'optimum' (8,500–6,700 years) when conditions were warmer and wetter than at present. Oak and juniper forest was abundant in the Atlas of North Africa and neolithic settlement occurred in the Sahara that was then savanna. This northward extension of savanna is marked by buried soils and was facilitated by the northward shift of the ITCZ causing reduced seasonal differences and greater similarity between highland and lowland climates. Climate then turned more arid with deflation and dune formation from around 6,000 years.

The mid- to late Holocene period is increasingly marked by human activity, particularly with respect to changes in the vegetation cover and effects on local erosion patterns. In the Middle Atlas of Morocco, for example, Lamb et al. (1989) record anthropogenic forest degradation dating from around 2,250 BP, with a recovery of Cedrus atlantica forest approximately 450 years ago. Establishment of the forest occurred around 4,000 years ago as a result of wetter conditions. This sequence is supported by evidence from nearby lake sediments (Lamb et al. 1989). Periodic clearance of pine occurred about 2,000 BP, accompanied by soil erosion which intensified approximately 1,500 years ago as pastoral use was replaced by arable farming. Evergreen oak become the predominant species primarily due to its ability to reproduce after burning and cutting, resulting in a gradual reduction in the diversity of the forest (Lamb et al. 1991).

In other areas such as the now densely populated Khumbu area of Nepal the sedimentary record is far more difficult to interpret, in part because of the highly dynamic nature of the landscape which disturbs surface layers. As noted in Chapter 3, settlement and clearance appears to date from the fifteenth or sixteenth century.

During historical times oscillations varied in space and time and human impacts became increasingly significant. More recent cyclic dry periods occurred from 1969–74 and 1981–5. Desertification arose due to intensive cultivation during the Roman era in North Africa, and the impacts of colonisation and ploughing in recent history have become more important than natural climatic changes in changing the nature of landscape and resources. Soil conservation and vegetation regeneration phases occurred during medieval nomadic Arab domination, leading to abandonment of terraced cultivation and soil erosion in the Mediterranean. A second phase of desertification followed European colonial expansion and also arises from present-day intensive use. Climatic change has intervened or coincided with degradation, but it is impossible to attribute degradation to solely natural or anthropogenic causes. A complex interaction of long-term natural processes and short-term economic and political processes has coloured the state of the mountain livelihoods and environment today (Reij 1988). For example, out-migration has led to underuse and collapse of traditional land uses – a 'developed country' model of change (Messerli 1983) – but also immigration and population growth have led to overuse and collapse in many developing nations. The continuation of traditional agropastoral systems under which landscapes have evolved is the key to successful maintenance (Naveh 1982). However, under changing global pressures which mountains and their populations experience, such traditional institutions and mechanisms are inadequate.

The Little Ice Age

A particularly well-documented phase of climatic change is that of the Little Ice Age in Europe during the fourteenth to nineteenth centuries. Broadly similar fluctuations in climate occurred globally. Advances of glaciers are recorded in Central Asia, New Zealand and the European Alps. The general climatic deterioration is also marked in the human historical record, by increases in diseases, famine, land abandonment and other disasters.

In the European Alps settlement spread from the fifth century, and by AD 1200 irrigation using glacial meltwaters was established in areas such as the upper Valais of Switzerland (where today there are only 40–90 days of rain or snow per year). These structures were developed when the glaciers were at their minimum extent. By 1300 these systems had fallen into disuse despite growing populations. Climate had turned cooler and wetter and glaciers advanced, destroying the structures. This was true of much of the northern Alps, although the Italian Alps continued to build irrigation systems until much later (Lamb 1988). Climate continued to fluctuate until the nineteenth century, with phases of cooling and glacier advance and warming and glacier stagnation. These fluctuations after AD 1200 are widely documented (Le Roy Ladurie 1972; Bradley and Jones 1992).

Periods of glacier advance not only overran irrigation systems, but also land and settlement, and affected communications between valleys. Routes across the passes had to be diverted, although in some instances it was possible to cross the glaciers from valley to valley. Floods, avalanches and other natural phenomena led to wide-scale land abandonment. Losses of agricultural land are recorded as reduced tithes and taxes – for example, it is estimated that one-third of Central Europe's agricultural land was lost due to these events.

Crop failure and land abandonment was enhanced by the prevalence of diseases affecting humans (the Black Death) that gave rise to labour scarcity, and also to crop diseases (such as mildews) and pathogens. The blocking of the westerly winds just preceding the Little Ice Age permitted locusts to swarm much further north than their southern Mediterranean habitat, and the Alps experienced seventeen plagues of locusts between 1280 and 1380 (Lamb 1988).

The timing of fluctuations between warming and cooling tended to vary in different regions. Cooling moved westwards from 1200–1500, and warming moved eastwards from 800–1000 and 1700–1900. Thus the period during which the Little Ice Age affected the various Alpine regions differed.

Recent changes

Even during this century, significant changes in weather trends have been recorded, with the benefit of more scientific and quantitative methods and technology. In the Alps above 1,800–2,000 metres, the ground became snow-free three to four weeks earlier between 1920 and 1953 than from 1886–53 or 1954–80 due to warm springs. In the period 1954–80 later melts occurred, as there was an increase of 5 per cent in snowfall, and in 1886–1919 8 per cent more snow fell. Storms may induce lowland melt and thus highland snowfall (Grove 1988). The proportion of precipitation falling as snow at different altitudes has also changed; now only some 50–60 per cent falls as snow at 1,200 metres.

The observation record of the European Alps indicates that a warming occurred of 2°C in summers from 1900–50, and a 2.5°C increase in autumns from 1910–60. The relative amplitude, magnitude and direction of changes were similar between highland and lowland, resulting in similar patterns of change. Regional differences in snow conditions take place in individual valleys – a spatial gradient that is not controlled by altitude (Lauscher 1980). A general retreat of glaciers in the Alps, which occurred between 1900 and 1970, and smaller advances in the 1980s, are attributed to precipitation rather than temperature changes. Temporal changes also occur with altitude – average annual snow cover between May and September at Sonnblick (Böhm 1986) was recorded as eighty-two days in the period 1910–25 and fifty-three days in the period 1955–70, with a mean rise of 0.5°C in summer. There is not a linear incremental change with altitude, however, due to lag effects of melting and summer snowstorms.

The spate of severe avalanches during the winter of 1999 throughout the Alps is thought to indicate events with a return period of 100 years. Thus they affected many areas with serious damage and loss of life in zones which had not experienced avalanche activity in living memory. The combination of an assumption of safety in the light of no recorded events, and the pressures of development of the skiing industry meant that settlements were in the paths of these avalanches. There was a period of heavy snowfall followed by a warm period during which rapid melting occurred, resulting in the failure of slopes.

Elsewhere, studies of the glaciers of Mount Kenya (Rostom and Hastenrath 1994) have shown a thinning of 15 metres between 1963 and 1987, and a further 5 metres from 1987–1993. This arises from three identified phases of climatic conditions: drier years from 1880–1900, warmer years from 1900–1970, and more recently a return to drier conditions. These changes are attributed to variations in the speed of the westerly airflows over the equatorial zone and the Indian Ocean. In addition it is argued that the more recent changes are a response to the greenhouse effect.

Potential future climatic change

The evidence of past changes demonstrates ecological zones moving up- and down-slope in response to temperature and precipitation changes. This is reflected in adjustments of the vegetation cover to different precipitation conditions. However, many mountain landscapes are now largely cultivated, anthropogenically altered landscapes. This means that we won't necessarily see the same adjustment of vegetation on slopes or the reworking of sediments as would have been visible in the past. Agricultural crops and cultivation regimes can be adjusted in response to climate changes, but there may be no natural vegetation to reflect such changes, or space for individual species to migrate, creating new associations. This is particularly true of densely populated mountains such as those in south-east Asia and the Himalayan states. Where forests still persist we may still expect to see species composition changes but on much longer timescales than that on which farmers traditionally operate.

Hulme *et al.* (1995) comment that given the paucity of African climate data and the high natural climatic variability then there is considerable uncertainty when predicting future conditions. They suggest that the vulnerability of African societies is relatively insensitive to this issue, given its marginal impact on an already very variable system. They also comment on the traditional coping strategies of many societies and that vulnerability is more an issue of population growth and politico-economic issues than one of future climatic change.

The elements of climate change that are most important are precipitation and temperature. Changes in the distribution and intensity of precipitation, and in temperature ranges, are the most significant elements of climate change. Extreme events are likely to become more common and thus represent greater danger to livelihoods than net changes. Such events include greater frost risk, summer storms and drought, and floods.

GEOMORPHOLOGICAL CHANGE

Geomorphological processes are closely related to climate as the main factors causing change in mountain environments. It is often difficult to separate climatic causes from human activity, as they are too closely interlinked and produce similar geomorphological outcomes. However, there are a number of barriers to geomorphological change. Brunsden (1993) highlighted the uncertainties in the response of processes to events, and in identifying the nature of triggers and thresholds in the geosystem. The sensitivity of landscapes to change is a function of resistance versus disturbance in the system. Resistance lies in geological strength and structure, landscape morphology, and in the way in which the system works. There are filtering effects that modify the magnitude of response to events. Such effects mean that the impact of an event, such as a large slope failure in the upper valley, becomes filtered down valley. This is reflected in the diminishing volumes of sediment and size of particles as the energy associated with the event is dissipated in the valley on the way down. Thresholds operate whereby the system absorbs a proportion of the effects of change, until the limits of stability are exceeded when the system fails, and the glacier lake bursts, a slope fails or an avalanche occurs (Schumm 1979). GIS-based studies of the spatial organisation and distribution of processes, such as that in the Pyrenees (González *et al.* 1995) can help to identify the key factors or variables responsible for geomorphic change – for example, lithology. The concept of equifinality, whereby different triggers and variables produce similar geomorphological responses, should always be borne in mind. Climate determines processes, but lithology and related factors dictate the form of operation and resultant morphology within the inherently high-energy landscape.

In the western Canadian Rockies five distinctive mountain areas were subject to an analysis of the effect of climate change on geomorphological processes (Slaymaker 1990). As noted earlier, all the models predicted warming and most an increase in precipitation. The effects on geomorphological activity were, however, varied. In the mountains of Vancouver Island and the Queen Charlotte range, there is virtually no periglaciation, permafrost or glaciation. Here the processes are dominated by mass wasting and fluvial activity, and consequently the models suggested an increase in torrential sediment movement. In the coastal mountains, where glacier ice is widespread, warmer and wetter conditions may increase the frequency of glacier surges and summer warming of slopes.

In the dry, interior mountain ranges where subarctic conditions prevail, permafrost activity occurs near the timberline. The greatest changes are anticipated here with respect to the disruption of surfaces and the constructions on them. It is expected that the wetter conditions will increase thermokarst activity. In addition there may be some expansion of glaciers as a result of the wetter conditions. In the higher interior Alpine mountains it is expected that an increase in pluvial conditions will lead to greater snowfalls, which will feed icecaps and also increase avalanche hazards.

There remains considerable uncertainty about the impact of climatic change on permafrost, periglacial activity and the timberline. The time lag between cause and effect may exceed the time-span of the prediction period (to 2050), and thus the impact of temperature and precipitation changes may be considerably delayed.

An important aspect of geomorphological change is the effect of human activities. Removal of vegetation cover leaves the surface open to erosion, and deforestation and other clearance activities such as slash and burn agriculture may be associated with increased runoff and soil erosion. In the Mediterranean during historical times the deposition of Vita-Finzi's (1969) 'Younger Fill' is attributed to the collapse of terrace agricultural systems and abandonment arising from both deterioration of climate and the collapse of political and economic empires. Sediment yields from different land uses is a frequent subject of experimental field plot studies. In the Pyrenees García-Ruiz *et al.* (1995) compared sediment yields from traditional agricultural systems and modern systems. Erosion from land under traditional cereal cultivation varied according to slope character and soil conservation strategies. Shrub cover was shown to be important in reducing runoff. The much greater issues of deforestation and degradation are discussed further in Chapter 8.

IMPACTS ON VEGETATION

A number of studies of the impacts of climatic change have concentrated on the response of vegetation to changing conditions. The growing conditions required by plants, and their physiological responses in terms of productivity and viability, can be experimentally evaluated. These results are then applied to past and potential future scenarios to give an impression of the changes to natural vegetation and any changes that would need to be introduced to crop varieties. Fischlin and Gyalistras (1997) made projections of the possible impacts of climate change at high temporal (annual) and spatial (valley) resolution in the Alps, by linking GCM and local climate scenarios. Results demonstrated different responses in different areas under the same projected doubling of CO_2 regime – from complete disappearance of species and communities to minor change. They propose that human

190

resources can be applied to help vegetation communities change, as well as causing some of the change in the first place.

The different results were in part due to the differing models applied. No response was observed at Berne (Site A at 570 metres), where temperature was used as the limiting factor for the current mixed deciduous forest. The area is currently not drought stressed and winter warming had little effect on vegetation. Site B was in the sub-alpine zone of Bever (1,712 metres) where major species changes in the present coniferous forest were observed, producing a new community.

The contributing factors were identified as changes in seasonal variation, increased winter precipitation and greater summer warming. It is possible that the model may have overestimated forest response and the ability of forest species to adapt. A new equilibrium transition community may itself induce unpredicted changes – for example, the slow migration of species at the end of the Holocene means some are not represented now. Overall, the model at this site suggests greater water availability and cooler summers compared with the present. At the third site (Site C) – St Gotthard (2,300 metres) – tree growth is suppressed under the current climate (it forms part of the north–south transition zone and is above the current timberline), but predicted warming allows canopy formation. A warming of 1°C, together with abundant precipitation, is predicted. Optimal conditions for tree growth (more than 30 days of over 10°C and fewer than 8 months at 0°C) are not currently met but under a doubled CO_2 regime they will be. The existing soils become the primary limitation. Additional influences such as that of local winds at the treeline, precipitation as snow or ice have not been evaluated. Site D – Sion (542 metres in the central Alps) – shows the opposite; the current climate supports mixed deciduous forest but warmer conditions won't support forest due to soil moisture deficits arising from a 28 per cent reduction in summer precipitation.

Mountain environments demonstrate great sensitivities and therefore there are great uncertainties in predictions. Sensitivity to human activity is partly dependent upon the underlying sensitivity to global and regional climate. Possible overestimation of response may arise from estimates of the ability of species to adapt or migrate. There are many unknowns in the complex system – for example the edaphic response to changing climatic and vegetation conditions. The high inherent genetic diversity of mountains may increase resistance and tolerance and reduce sensitivity – the greater diversity of species permits a wider gene pool to be drawn upon in the event of change. In the case of the Berne site, where the composition of forests was maintained, some changes may induce a critical environmental threshold to be crossed. Although models are not forecasts, at the valley scale they have produced integrated, reproducible scenarios of possible change. However, such models need to be spatially flexible and to cope with the wide range of response and the high sensitivity to water and temperature at treelines.

Brzeziechi *et al.* (1994) performed a GIS simulation of vegetation response to climate change in Switzerland. The results showed spatial vegetation changes and a difference between mapped and simulated vegetation. In the Moroccan High Atlas, Parish and Funnell (1999) have explored the effect of temperature changes on the distribution of fruit trees and noted that while some upslope movement might take place it was likely that economic issues would be of greater significance.

The problem of errors and poor quality of data used for models gives rise to these 'starting differences' and increases the uncertainty of predictions. The low resolution of available environmental data is a key problem. Accuracy in measurements is best achieved for elevation, whereas factors such as slope and aspect are less accurately represented. The inherent, continual variation of temperature and precipitation in space and time means that accuracy becomes less important in the prediction of details. Only the mean values are significant here, although it is frequently stated that average values are meaningless in the understanding of real climatic conditions in mountains. The important factor of climatic change is the increase in extreme events. Therefore, to be useful for planning livelihood changes and adaptations, the resolution and determination of processes and their operation in time and space needs to be sufficient to identify at least the one in ten year events on a valley scale. Brzeziechi *et al.*'s study was based on a 1:200,000 soil map, showing strong altitude–climate gradients, and that a temperature shift translated into a zonal shift of communities, although these were not specified.

The relationship between climate change and vegetation cover has been the subject of numerous studies. Kupfer and Cairns (1996) debated the use of forest ecotones (the transitional zone between ecosystems, here the timberline) as indicators of climatic change, calling into question much of the work done to date. They argue that the slow response of ecotones to climate changes means that monitoring such change is not necessarily a good indicator of rates and nature of climatic change. Time lags operate in the adaptation of forests to new equilibrium conditions and to new limiting factors, as well as the human response changing microenvironment and vegetation stresses. Shifts tend to be slow. There is also a non-linear annual climatic change response; that is, fluctuating climate over different timescales means for annual plants or farmers just a series of unusual years, not necessarily that a discernible trend to responses can be planned. More damage is likely to be caused by extremes than year on year small adjustments. It is only after several years of net unidirectional change that a cumulative response will be sufficient to tip the ecosystem or human community over a threshold and respond to consistent changes in agricultural performance. Economic change is much faster than this and, in much less time, can change the landscape of economy more than cumulative decades of climatic change (see Chapter 7).

Climatic change is sensitive to both altitude and latitude. Halpin (1994) simulated changes in altitudinal zone shifts for different mountains in different latitudes using GIS, which illustrates the variability with the nature and extent of changes in mountain ranges across the world. This study arose from a concern with conservation issues, of landscape fragmentation and species migration, and how to address these problems within the framework of establishing reserves. The 'Theory of Island Biogeography' is applied whereby mountain summits and upper altitudinal zones reflect similar problems of species migration to oceanic islands. The distances between isolated peaks, or the disjunction of habitats across mountain ranges, prevent species migration, thus constraining species to the area of reserve. This is particularly true where these reserves are surrounded by intensively cultivated landscapes that form formidable obstacles to species movement. The issue of climate change is critical here, as with warming and the expected upward shift in altitudinal zones species would be more constrained in space. In addition, the lower zones of the reserve would not be supplied by species migrating upwards, as beyond the reserve is usually cultivated land. Species occur at their limits in marginal environments and are therefore more sensitive to relatively small or short-lived climatic changes (Peters and Darling 1985). The rates of climate change, environmental response, species migration, competition, and the nature of the surrounding physical obstacles need to be identified in terms of potential boundaries and reserve limits.

In Halpin's simulation, changes in altitudinal gradients such as a local temperature shift upslope were embedded in a conceptual model using values of 3°C change and a 500-metre shift (Peters and Darling 1985; MacArthur 1972). Expected impacts included a loss of the coolest zones off the summit and movement of the rest upslope. Assumptions applied in planning are based on the largest ranges of altitude and topographic relief giving the biggest range habitats and therefore needing to allow the greatest amount of shift. An altitudinal range of over 1,000 metres allows for a temperature shift of 6°C. However, the extrapolation is flawed as climate change and response are not spatially equivalent; neither is habitat tolerance. A global change of 0.5°C was applied to the GIS (Leemans and Cramer 1990), based on monthly climate data.

New equilibrium ecological climatic zones were modelled. Regional models with a resolution of 55 km² between the 100 metre and 5,000 metre contours were constructed. Reserves with an altitudinal range of over 3,000 metres in different latitudes were compared. The study concentrated on 60–100 per cent natural vegetation assemblages under a +3.5°C and +10 per cent precipitation hypothetical scenario at each location.

The highest percentage change occurred in the northern high latitudes (see Figure 5.1), where the greatest climatic forcing is predicted to occur. A high proportion of the world's nature reserves are located in the northern high latitudes. Common to all the regions, the greatest change occurred in

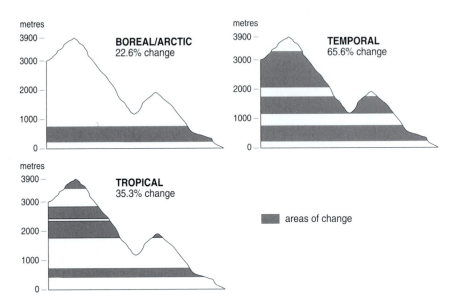

Figure 5.1 Altitudinal vegetation zones, climate change and latitude: a possible scenario (adapted from Halpin 1994)

the lowlands – 10 per cent more for all latitudes. In the wet tropical region, there was a loss of climatic zones with warming, including a widening of the lower alpine frost bands and loss of the cool alpine *paramo*. In the dry temperate latitudes, there was the loss of two zones and expansion of low- and mid-altitude belts. The nival belt was restricted but not lost. The cold arctic example showed an expansion of the basal sub-alpine moist forest. The alpine tundra zones were compressed between the forest and nival zone which was little changed in altitude.

IMPACTS ON LIVELIHOODS

The key impacts are likely to be the increased frequency of extreme conditions – unusually late frosts, frequency of flooding and summer drought, differences in timing and intensity of monsoon, changes in snow cover and duration. Changes in melting times and the movements of glaciers are also likely to be significant. In regions where winter tourist activity is reliant on snow cover, the quantity and duration of such cover will be important, as will the effects of heavier snowfall on avalanche frequency.

The main effects on agricultural livelihoods arising from such changes are likely to include, for example, losses of tree crops due to frosts, impact of summer drought on plant growth, reduction in snowfall and therefore meltwater supply, thus causing water shortages. Increases in flooding events from intensive rainstorms will be likely to cause damage to terraces, valley floors and settlement and communications. Snow cover may be heavier in some areas and lie longer, affecting summer grazing patterns, although this may be offset by tourist activities where snow-related tourism seasons could be extended. Increases in disease (plant, livestock and human) may occur due to higher temperatures. A rise in soil temperatures increases the rate of nutrient cycling and organic matter breakdown but also causes reductions in soil moisture content and thus water stress to plants. Increases in evaporation will enhance this stress, particularly where higher temperatures are combined with more active local wind systems. Increased UV damage to high-altitude vegetation and life is also possible.

Climate change also affects human activities in the non-agricultural sectors. The duration, timing, quantity and type of snow are important. Snow melt supplies irrigation water, water for hydroelectric power (HEP), and the snow itself is the principal resource for the winter sports industry. With respect to HEP, whilst climate change will affect supply, so too will the level of demand for energy respond to temperature and precipitation changes. The seasonal distribution of snow storage, melt and runoff may alter and it is possible that there will be an increased mismatch between an increasing demand and falling supply. However, in New Zealand one prediction suggests a convergence between supply and demand, but other climatic scenarios indicate a reduction in generation potential in spring and summer (Garr and Fitzharris 1994). The net result of modelling suggests that generally supply is likely to increase and demand to fall and that electricity generation is less vulnerable to climate change than other activities.

In the Australian Alps snow cover occurs over 1,700 metres. Any rise in the snowline will effectively eliminate winter sports (Slayter *et al.* 1984). These have already been affected by the El Niño phenomenon, which gives rise to subtropical high pressure over the south-east Australian Alps and thus reducing snowfall.

In the Canadian Rockies, a doubled CO_2 regime may reduce the ski season by between 31 and 44 per cent, resulting in $2 million losses. In the European Alps a delay in snowfall beyond December–January would mean the popular Christmas ski season would be missed. In parts of the Swiss Alps it is suggested that aspect difference will assume even greater significance because north-facing slopes will retain greater depths of snow and alter the spatial distribution of sports activities (Witmear *et al.* 1986).

Anthropogenic impact on vegetation and on landscape modifies the capacity of the natural environment to absorb changes, and thus the response to change and the feedback which may mitigate the effects of change. Hazards may increase and infrastructure and tourist-related structures may be affected by more frequent landslides in heavy snow and rainstorms. This is in part a result of the growing number of physical constructions rather than necessarily the increase in the hazard frequency itself. The occurrence of floods and avalanches and their devastating effects in the last twenty years in the European Alps alone bear witness to this. Conservation of water is likely to become an increasingly critical limiting factor in agricultural activity and a point of dispute between households and villages. On a global scale, issues of biodiversity protection and conservation increase and require more land to incorporate buffer zones to account for future climatic change induced shifts.

6

DEMOGRAPHIC
TRANSFORMATION

INTRODUCTION

In this chapter we examine the mechanics of population change in mountain regions and explore its impact on communities and environments. Growth and decline in population, with its concomitant effects on settlement, land use and livelihood patterns, is a highly significant factor in the transformation of mountain communities. One of the key questions in demography concerns the extent to which population variables can be considered autonomous or whether they are dynamically related to environmental or socio-economic factors which would provide a distinctive 'mountain' element about the process. Unfortunately, with few exceptions, only very aggregate data are available in mountain regions and even this is usually very short term. Therefore much of the work remains not only rather descriptive but also somewhat speculative about the precise nature of the links between population change, the environment and socio-economic development. In the first part of the chapter we explore some of the basic dimensions of the demography of mountains, then consider migration. Finally, the concept of 'carrying capacity' is discussed.

MODELS OF MOUNTAIN DEMOGRAPHY

Whilst it is widely known that Thomas Malthus (1766–1834) established some of the basic principles of demography in his seminal book *An Essay on the Principle of Population* ([1803] 1986), it is not often noted that he made extensive use of material from the Swiss Alps. Many of his conclusions were derived from findings in the Swiss parish of Leysin in the Pays de Vaud. There is a very large literature discussing the general principles of Malthusian demography (Coleman and Schofield 1986), but it is useful to outline some key elements to serve as a basis for later discussion.

According to Malthus, the factors determining the population size of these Alpine communities were very clear and closely related to the *Alpwirtschaft*, the agricultural system that has been described in Part 2. His model of demographic control is based upon the notion of a 'closed community' where there is a 'dependence of the births on the deaths', so that overall the demographic structure resembles a steady state. In this sense, demography in mountain regions has often been described as being particularly close to his 'limits to growth model', shown in Figure 6.1A. This diagram establishes the main variables influencing population size and illustrates the principle that the adult mortality rate will influence population growth. A fall in mortality would cause population to increase, but this will have the effect of decreasing nuptiality (the marriage rate) because social rules insist that a man must have the wherewithal to support a family before marrying. Under certain inheritance arrangements and a fixed supply of resources (land and animals), fewer deaths mean fewer openings for the new generation. If adult mortality rises, the reverse happens and nuptiality also rises, which in turn leads to an increase in fertility and an overall rise in the population.

The model pinpoints the significance of controls on fertility, given a particular rate of mortality. In order to achieve a self-adjusting state (homeostasis) some social mechanism is needed that will provide a control over fertility. In Malthus's work, nuptiality provides this mechanism, so that delayed marriage, and consequently lower overall marital fertility, was seen to be the most important constraint. The process embodied in Figure 6.1A also only works on the assumption that the community is 'closed'; in other words, is autarkic, in which the main determinants of the livelihood of the population are contained within itself. The question of autarky has become something of a contested issue among historians of mountain areas, as we shall see later, but the point here is that the successful operation of Model A depends upon this assumption (see below). Once the local resource base no longer constrains the livelihoods of a community then the dynamics of the demographic change and the factors influencing population growth must be sought on a wider canvas. At this point it is the level of economic activity outside the community, and its effect on the real wage, that provides the stimulus for changes in fertility and nuptiality. The alternative source of income could be provided through seasonal migration or the development of industry ('cottage industry' in the time of Malthus) within the mountains themselves. In some instances, of course, permanent emigration would result, a factor of considerable importance today. This model is shown in Figure 6.1B.

However, neither of these classic models really illustrates the range of factors involved in both historical and contemporary demographic change. Figure 6.1C builds upon these mechanisms to show how each of the main variables is itself a function of a string of other factors around which there is considerable discussion.

Figure 6.1C indicates that there are important controls on both mortality and fertility. Mortality rates reflect the specific age structure of the popula-

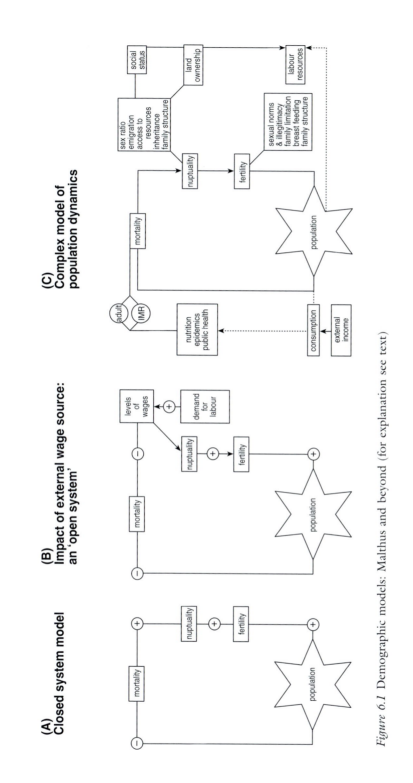

**(A)
Closed system model**

**(B)
Impact of external wage source:
an 'open system'**

**(C)
Complex model of
population dynamics**

Figure 6.1 Demographic models: Malthus and beyond (for explanation see text)

tion and, in the mountains, the occurrence of life threatening events such as avalanches, epidemics and the more pernicious fluctuations in the supply of food. Infant mortality, a well-known indicator of social welfare, is sensitive in mountains not only to the hardships and vacillations of food supply but also to the level of public hygiene. As noted above, changes in mortality first influence nuptiality, but Figure 6.1C highlights the extensive series of factors that impinge on the level of nuptiality. Fertility itself can be influenced not only by nuptiality but also by the social attitudes to illegitimacy, by breast-feeding practices and by family limitation, all of which may apply at some time or another in the demographic history of a community.

More recent demographic work, particularly that which explores the nature of population growth and economic development, has been built around the notion of Demographic Transition, in which the onset of 'modernisation' is associated with a move from high fertility and high mortality to lower fertility and mortality. This has been subject to considerable criticism as it maintains that there is a uniform experience among all societies. Caldwell (1976) and others have illustrated the variety of ways in which different communities accommodate socio-economic and environmental changes through adjustments to their demographic behaviour. Later writers such as Cain (1982) further indicate the complexity of fertility decisions that are not considered in the basic account of Demographic Transition Theory. Of particular significance for the study of mountain communities has been the increasing use of anthropological analysis connecting demographic factors with wider cultural and economic variables. Some of these studies, mainly in the Alps and Himalayas, are discussed below and provide a community scale focus which is often more appropriate for investigating the dynamic nature of population change in the highly variable landscape of mountains.

An Alpine demographic model

In seeking to understand the dynamics of population it is inevitable that we must have recourse to historical enquiry. Whilst we have not set out to produce a historical account of mountain areas (see for example, Braudel 1972 and McNeill 1992) we would argue that a discussion of historical data is fundamental to our knowledge about the transformation of mountain populations. Owing to the data problems noted earlier we are confined to a few regions, particularly the European Alps.

The attempt to construct an acceptable model of demographic transformation depends upon an adequate specification of the *ancien régime*, and this work finds its strongest expression in the writing about the European Alps. This owes much to the fact that historical materials (at least in the form of documentary evidence) are available and that scholarship has been drawn from the several nationalities that today populate this mountain range.

However, the findings from this work, whilst by no means automatically applicable elsewhere, do provide guides to some of the key processes at work in mountain communities, and the academic debates highlight areas of ongoing uncertainty which are useful in the wider arena. Historically, researchers from Alpine countries, such as the geographer Arbos (1922) and also those in the Anglo-American tradition, have been drawn to the Alpine communities, in part because Ratzel (1882) and Semple (1923) used models of these communities to illustrate their arguments for what became known as environmental determinism. Although the underlying assumptions of this approach have been largely rejected, later writers, including historians such as Braudel (1972), drew upon the conventional wisdom that described Alpine communities in term of their marginality and conservatism in the face of socio-economic change. In general, textbooks about mountain areas, for example Peattie (1936) and Price (1981), conveyed this same image, but there exists now a wealth of scholarship, mainly by writers from the region itself, that calls this general finding into contention. Recent work by Viazzo (1989) provides a good summary of many of the debates.

For our purposes, the main issue hinges on the treatment of the pre-Second World War mountain society, the date being chosen to correspond with the point many authors like to signify as the end of the *ancien régime* in the Alps. Much earlier, as noted above, Malthus had developed his model of population (in part, from Swiss materials) and he highlighted the capacity of the local production base to support the population. Winter fodder demands were crucial and the hay yield depended upon the size of herds, as these provided the source of vital manure. In turn, the success of the herd was closely associated with crop production and hence largely controlled the availability of food resources, particularly protein, and draught power. According to the Malthusian model, the distinctive features of these Alpine communities were their relative low levels of mortality and consequent processes which kept fertility in check to provide a homeostatic, low pressure demographic regime. Any tendency to move 'out of line', for example through a sudden increase in population, would be rapidly corrected through the hardships induced through a diminished per capita resource.

The earlier discussion of the *Alpwirtschaft* model noted that it was based upon the operation of a 'closed system' such that the local community was effectively autarkic and thereby solely dependent upon its own resource base. Much of the detailed ethnographic, ecological or specifically demographic work in the Alps explores the nature of this closure, particularly the role of emigration and 'cottage industry'. The problem of 'closure', however, is that it has achieved the status of a myth, in so far as many writers, particularly but not exclusively those with journalistic leanings, have created the image of a 'golden age' of highly self-sufficient mountain communities. In terms of the models set out in Figure 6.1 it is a question of the balance between Model A and Model B. Those who emphasise Model B claim that the existence of external linkages through migration and trade is a sign that the community is 'open'. However, another school of thought, originally led by Grenoble-based writers such as Veyret-Verner (1949), believes that even when there is emigration and local industry, most mountain communities remain dependent on the produce of the local resource base and so it is essentially 'closed' (Viazzo 1989). Their argument is related to the fact that domestic industry serves the local population and provides a source of income during the winter season when agricultural tasks are few. Therefore this activity is a structural complement to the agricultural economy rather than a distinct and separate component. The distinction is partly conceptual, but also empirical in that any measurement of the exact contribution of each element is fraught with difficulty, certainly for past epochs.

One of the most important dimensions of Alpine demography that is not shown in the models presented earlier concerns the precise trajectory of population change. For many years, historians and others using an environmentalist framework proposed a general schema in which the population of the mountains expanded from the middle of the eighteenth century, but the precise details remained controversial. In recent years, employing the techniques associated with family reconstitution and seeking detailed historical documentation, studies have rebuilt the patterns of crucial demographic variables such as mortality, nuptiality and fertility from existing records. Much of the work has been undertaken by anthropologists working in specific communities, for instance Netting (1981) and Viazzo (1989), and they not only provide important insights into the demographic process but also address classical questions about the evolution of the Alpine communities and their relationship with the Alpine environment.

Netting's research was based in the community of Törbel, in Valais canton, Switzerland. The settlement lies in a valley with its lands rising from 800 metres to 2,500 metres and is populated by people of Walser descent. His task was to provide an understanding of the processes that have led to transformation in the population of the settlement over the last two centuries. Figure 6.2 shows the pattern of population change during this period. Until about 1770, the village population was reasonably stable, of between 200–300 persons, but from then onwards, with the exception of a few sharp recessions, the growth has continued until 1950 with a peak of around 700 persons. The overall rate of growth was maintained at around 1 per cent per annum, somewhat lower than many mountain areas elsewhere today. Mortality fell from $30/1,000$ in the eighteenth century to $20/1,000$ in the nineteenth century. For the period following 1800, it is possible to provide details on the main demographic indicators that are shown in Figure 6.3. With the exception of a few years in the 1830s, births exceeded deaths but the margin varies considerably. In the 1830s, the late 1860s, and again at the end of the 1890s, births fell dramatically and there is an increase in

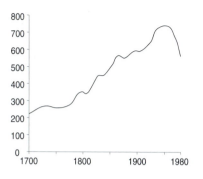

Figure 6.2 Population growth in Törbel
(adapted from Netting 1981)

deaths. Also marked is the decline in deaths in the period immediately before the years noted above.

These figures have been the basis for several conjectures about the principal controls on mortality and fertility, which, when assembled, have been used to examine the original Malthusian position in greater depth and also to propose various mechanisms controlling the fluctuations. In the case of mortality, Netting draws upon circumstantial evidence to associate the marked fluctuations with the onset of harvest failures and epidemics. It is known that in the fifteenth century even high-altitude settlements were on occasion affected by the plague, though the worst effects were avoided. There remains much speculation about the impact of harvest failure in these earlier periods. As we have already noted in Chapter 5, some writers (Messerli *et al.* 1978; Pfister 1978, 1983; Grove 1996) have argued that the accumulation of information from the Holocene and accounts of more recent glacial fluctuations in the Alps provide strong evidence of periods of

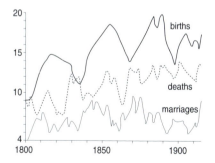

Figure 6.3 Trends in principal demographic variables, Törbel
(adapted from Netting 1981)

relatively cool and wet summers, with a consequent impact on the harvest. In addition, Le Roy Ladurie (1972) has argued that failure of the harvest in mountain areas operated in two ways: first, through the shortfall in food grains; second, through a poor hay harvest. Consequently both grains and fodder had to be purchased to survive the winter, or, more likely, overall consumption was reduced so that the population of both animals and humans became prone to disease and further unable to contribute to the next production round. Outbreaks of smallpox and influenza were noted, and it is interesting that although much greater detail becomes available by the 1850s the precise pathology of diseases is not always clear from the records. One indirect suggestion of an increasing population may have been evidence of severe outbreaks of intestinal infections, probably resulting from water that was becoming polluted with higher densities. Until the 1950s most villages in this area of the Alps still drew their water from unprotected springs.

Whereas mortality was the occasion for ceremony, the same cannot be said for births. Consequently, in many Alpine villages fertility details are sometimes more obscure. This is unfortunate because it is to the management of fertility that commentators have looked when trying to understand the dynamics of mountain populations. As is shown in Figure 6.3 there are quite distinct periods of increase and decrease of births, with peaks and troughs slightly later than those for mortality until the 1890s when the link seems to have broken. The evidence suggests a distinct increase in births per family, from an average of 3.84 between 1700 and 1850 to 5.66 at the turn of the twentieth century. Netting argues that the prime reason for this shift lies in nuptiality associated with a fall in the age of marriage. In earlier periods it was an accepted custom that marriage did not take place until the husband had sufficient resources to provide for family support, and, at least until the 1850s, this would mean access to land/livestock. However, land would only become available through inheritance, thus Malthus's contention that births depended upon deaths. Although migration was another avenue (see below), at least in the Törbel case it appears not to have been more than a supplementary source of family livelihood until the mid-nineteenth century. The resultant pattern of nuptiality was therefore dependent upon the inheritance rules, which have often been ascribed a major role in the demographic story. In Törbel, the partible inheritance system operated, which meant that all siblings would be offered a share in the family property. Elsewhere, especially in parts of the Austrian Alps, the impartible system has operated in which only one heir, usually the eldest son, inherits the property. However, as Netting demonstrates, the contention that the different inheritance rules determine levels of nuptiality may be misleading as Törbel shows signs of fertility control regardless of the formal inheritance system, perhaps because properties were not broken up despite the nature of partible inheritance.

What has been described so far would not be out of place in any agrarian community operating under these specific biological and customary constraints. However, for students of mountain demography the interest lies in further enquiries concerning the timing of births and their 'success'. Within these mountain communities a considerable work burden tradition-ally fell upon women who were responsible for all but the heaviest of tasks, especially if the men were involved in seasonal migration. The pattern of seasonal emigration varied somewhat between different parts of the Alps, but in Netting's study the typical pattern was for males to leave the village in the New Year, and it was also this period in which conceptions took place. The result was that births were timed to take place in the autumn months, after the harvest. The point being made here is that both successful concep-tion and birth are associated with the nutritional status of the mother. Hence dietary requirements are very important, not only on a seasonal basis but in the longer term, because there is some evidence (Viazzo 1989) that the reproductive system may 'shut down' in certain circumstances. Thus in the event of poor harvests and resultant low nutritional status of women, fertility may be low, and vice versa. According to Netting:

> Between 1750 and 1799 mortality began to decline and marriage duration during the women's fertile years went up. An increase in local food supplies could be expected to make people more resis-tant to disease, especially to respiratory illness that occur more frequently with malnutrition. As fewer marriages were cut short by the death of one of the partners, total fertility rose.
>
> (Netting 1981: 159)

From the argument above, it is alleged that the slow rise in fertility is associated with the improvement in nutritional status of the population of the mountain villages. In part, this might have been associated with improve-ments in incomes from migration, but this certainly fluctuated according to economic conditions in the lowlands. However, Pfister (1983) returns to the theme of the introduction of the potato to explain the apparent increase in the resilience of Alpine production. Introduced from the Andes into Europe towards the end of the sixteenth century, it was found in many Alpine areas by the mid-eighteenth century. This crop provided a more reli-able food source during those summers that were cool and wet and therefore has been seen as a key element behind some of the demographic changes noted. However, these arguments are all built around a form of 'determin-istic' thinking associated with the notion of carrying capacity, a concept that in various ways has been the subject of much critical debate and is discussed in more detail on pp. 000–00. This concept implies that the basic behav-ioural patterns of a mountain community can be 'read off' from a model of the physical environment, or that, in the last resort, all patterns of social behaviour lead back to the limitations imposed by this physical environment. Consequently, in mountain areas, given the particularities of the environment and its links to the production process, a distinct pattern or 'culture' can be identified. This neat relationship does not hold beyond very general patterns, as can be seen from other examples of demography in the Alps, by the importance of social institutions such as inheritance, and the role of migration. Viazzo (1989) discusses these issues at some length.

A Himalayan example

Although we are unable, as yet, to provide the historical detail for many other mountain areas it is useful to examine the nature of demographic factors where suitable data exist. A particularly interesting case is that provided by Fricke (1994) on the Tamang village of Timling in the Nepalese Himalaya. The collection of demographic data for this small village faced all the problems of surveying peasant communities but, for the early 1980s, illustrated that crude population growth was a direct function of fertility control in a society where contraception was not used. The age–sex structure (Figure 6.4) indicates neither the bell-shaped distribution of industrialised societies nor the broad-based model that characterises many areas experiencing rapid economic development. Fricke's reconstruction of reproduction histories for the last thirty years found no evidence of a change in the number of births per female. This has remained about 5.25 per female. He notes that this experience corresponds with other data from mountain Nepal but not with the settlements in the lower Terai, where new economic relationships have been absorbed further. The life model of Timling women had up to that date followed a very consistent pattern with marriage and at least one child by twenty-five, and the remainder in relatively quick succession before the age of forty. Fricke argues that whilst the fertility figure is higher than in industrialised societies it is well below that of notional natural fertility, the difference being accounted for by three factors. In the first instance, age and rate of cohabitation sets broad limits to actual fertility and then, second, various reasons for separation further reduce the chances of pregnancy. Finally, he argues that lengthy breast-feeding may diminish the probability of conception, though this is a contentious point.

Mortality information is not widely available and the lack of trained medical staff means that the causes of death are rarely specific. One interesting feature recorded by Fricke is the relatively high incidence of death associated with accidents and related infection. The lack of public hygiene arrangements is an important issue, but conversely the region is too high for malaria. The low levels of hygiene are usually associated with the high levels of infant mortality. Fricke found that about 20 per cent of the children born in his survey had died before the age of one. Overall, about 34 per cent of children had died before reaching twenty years old. This is a

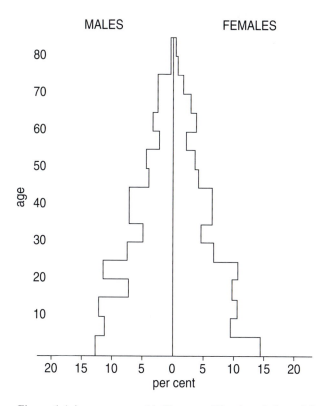

Figure 6.4 Age–sex pyramid, Tamang, Himalaya (adapted from Fricke 1994)

significantly higher infant mortality rate than some other areas within Nepal, where the rate has been on the decline since the 1960s. Linking the fertility and mortality patterns it is possible to see how overall population growth has remained at 1.2 per cent annually, whereas for Nepal as a whole the rate is almost double that figure.

Clearly the data from Timling does not provide us with the time-series that is so illuminating for the communities of the Alps. However, Fricke returned to the village some years after the initial survey to carry out more detailed studies, and in the late 1980s was able to observe some social changes that are likely to have a direct bearing on the determinants of fertility and mortality. The principal changes are associated with the increasing linkages with Kathmandu, and other areas, for work, consumer goods and general cultural values. It was noted, for example, that the number of women who now participated in some type of migratory labour (once the preserve of men) had increased considerably. Overall wage labour, either locally or elsewhere in Nepal or the subcontinent, had almost doubled in the intervening years. Of course, Fricke was only able to document these changes and speculate about their influence on the demography of this mountain village, but it is evident that the issue of migration plays a major role in mountain demography. To this we now turn.

MIGRATION AS A COPING STRATEGY IN MOUNTAIN AREAS

In previous sections we have alluded to the role of migration as a process intimately associated with settlement, population growth and economic change in mountains. Here, a more systematic treatment of the topic is provided. Not all writers have given prominence to this process. In one of the classic mountain texts, Peattie (1936) discusses migration only as a function of transhumance, and notes the existence of depopulation, especially in the French Alps, but makes no further reference to the matter. Since 1945, however, almost all socio-economic studies of mountain areas provide details of the role of migration in the evolution of mountain communities. In particular, labour mobility is recognised as a key factor in the process of integrating highland and lowland areas (Grötzbach 1984; Skeldon 1985)

There are three basic conditions under which population movements take place in mountain regions:

1   In relation to colonisation and settlement development as mountain areas are populated. Examples of this have already been noted in the European Alps, the Andes and the High Atlas.
2   In relation to the farming system; that is, as part of a nomadic or transhumance model in which ecological variations between different altitudinal zones are exploited as part of an ongoing agro-ecosystem. This system used to exist in many Alpine zones and in the Pyrenees, and remains part of the current pattern of farming in the Himalaya, the Atlas and many other mountain regions.
3   Migration as a response to differentials in economic opportunity, security or other socio-economic circumstances. This is the familiar concept of migration and is associated with many theoretical explanations of migrant behaviour.

In fact there are several levels of migrant activity that are important in mountain areas:

1   *Daily.* This might be regarded as commuting, and an example of this has already been provided in the case of men in high-altitude valleys travelling each day to factory employment in the valleys.

208

2   *Seasonal.* This is of major importance, and may be associated with agri-
     culture (see above) or with the availability of non-agricultural employ-
     ment. The length of the period away is very variable and the distance
     travelled may be relatively local or overseas.
3   *Emigration.* In this instance the migrants leave the mountain area,
     usually permanently, although the extent to which all contact is broken
     is very variable. This process is associated with depopulation.

It is also necessary to include the categories of migrants entering the mountains. Some of the cases have already been noted. In the North American
Rockies, the flow of settlers into these mountains included fur trappers,
miners, foresters and the vast array of officials, speculators, traders, and char-
latans of every description who made up the 'frontier' population at that
time. Elsewhere, miners and traders also played a major role in the demographic evolution of mountain settlements, as Viazzo (1989) noted. In this
category also we need to remember that from an early date some settlers
came into mountain areas not just for resources but for religious and amenity
reasons, seeking solace in the landscape. This is 'amenity migration' which
some commentators argue is one of the most important characteristics of
population change in the latter half of the twentieth century (Price *et al.*
1997).

Amenity migration is an area of extremely active research, but studies by
Moss (1993) in northern Thailand and Beck (1995) in Canada indicate that
amenity migration into mountains is a process which can bring new
economic vitality to selected locations. Amenity migrants may be motivated
by changes in social values which come to emphasise (or re-emphasise) both
the quality of the natural environment and the recognition of the importance of cultural character. This, in turn, may reflect the rising valuation of
leisure, education and search for spiritual values.

The process has been assisted by improved transportation and telecommunications, more leisure time and greater general prosperity. The old
tension between an attractive landscape in which to live and proximity to
employment has, in an increasing number of cases, been broken by these
developments to the point where employers now seek mountain locations
for their offices in order to attract high-level staff. Alternatively, a rising
number of people are able to use the Internet to work from a home
located in a mountain environment. However, amenity migration is not
confined to rich countries, as the work by Moss (1993) and others has
shown. Both seasonal and permanent migrants seek these locations and, in
some cases, the patterns of development are closely associated with some
types of tourism. This is most notable where the interrelationship between
facilities and tourism also acts as a means of employment for many unskilled
workers who service the tastes of the more affluent. In addition, Moss also
draws attention to the problems surrounding this type of migration resulting

from the impact on the local populace, most particularly the impact of inflation on the prices of many basic assets and services. He reports that in a number of Asian countries land prices have soared and once relatively cheap local services have been replaced by expensive retail operations.

More general theoretical constructs lying behind the migration process have been extensively discussed (Cohen 1996) and often divided into push and pull factors. To explain in-migration, writers have usually emphasised the availability of resources creating employment or opportunities for settlement. For out-migration from mountains, the key pull factors have tended to be opportunities for employment (and/or enhanced wages) and social attractions. In specific circumstances, migration may be associated with political or religious upheavals, but the predominant factors are usually assumed to be economic. In the case of the traditional 'push' factors, many studies of mountain areas concentrate on the question of population pressure arising from limits set by the ecological resources of the community. Thus questions of 'carrying capacity' and environmental degradation are often cited as underlying causes of long-term out-migration (McNeill 1992).

The dynamics of population change and the concept of carrying capacity

Although the term 'carrying capacity' has a long history in geography and cognate subjects, it has returned to popularity within the framework of discussions examining sustainable development. In the most general sense it is used to suggest that there is an effective limit on the utilisation of resources before those resources are damaged in some way. A particular application has been in the analysis of livestock grazing. Despite many early criticisms, the FAO have been employing the concept in their attempts to provide a framework for the management of natural resources, especially soils (Higgins *et al.* 1982). The concept appears to offer a neat calculus of the relationship between demographic growth and some variable such as soil quality. Furthermore, once the ecological characteristics of the resource are known, then it becomes possible to 'read off' the capacity of this resource to sustain an activity or a particular level of population. Clearly this model has great appeal in the context of the sustainability debate, but it has been the subject of considerable critical appraisal (Jolly 1994; Zimmerer 1994; Scoones 1998). One of these most powerful criticisms focuses on the rather weak recognition of dynamic adaptation in the typical use of the model. In particular, using the carrying capacity argument as a policy guide may have seriously misleading effects as most long-term evidence suggests that, at least in some societies, this 'carrying capacity' is continually being changed by technological input or institutional change. These points are explored in a recent analysis by Marcoux (1999).

The impact of population change on resource management was also a central theme of the work of Boserup (1965). This was subjected to enormous criticism at the time as her thesis ran counter to fundamental tenets in agricultural development thinking. However, some thirty years later, the work of Tiffen *et al.* (1994) once more raised the debate through its provocative title *More People, Less Erosion*, set in the Kenya highlands. Since that time a systematic review of the problem has been produced by Templeton and Scherr (1997, 1999). This work examined seventy case studies in the highlands of South America, Asia and eastern Africa to see if local population growth can be shown to have any deleterious impact on forest production, agriculture, watershed stability and livestock. The centre-point of their work is the specification of a U-shaped relationship between land productivity and relative land–labour costs. It appears that there is a distinct sequence of events following an increase in population: initially productivity falls but the growth of population also increases land values. Land becomes more expensive, which itself may stimulate investment in land improvement. Similarly, relative labour costs fall and together the evidence examined suggests a gradual increase in land productivity as labour-intensive and capital-intensive methods are employed. Good examples of this process include work from Rwanda (Ford 1993) and the Usambara mountains in Tanzania (Feierman 1993). At the very least, it is the dynamics of the relationship that need to be evaluated, rather than employ a static conception of carrying capacity.

EXAMPLES OF POPULATION CHANGE IN MOUNTAIN AREAS

Having established that our understanding of the relationship between population change, migration and resources remains unclear, we can examine a number of case studies that explore the specific impacts of migration in mountain areas.

The European Alps

It will be recalled that the studies of the settlement and demography of the Alps over the last 500 years have debated at length the role of migration. The importance of the debate returns us to a central characteristic of the traditional mountain community: namely, the concept of autarky that was discussed earlier. It has been noted that the traditional Alpine community was 'closed', in which circumstances the local resource base set limits to population growth. Until the advent of the work in historical demography most writers considered migration to have played only a minor role in these Alpine communities until the nineteenth century. The late Robert Netting, writing in 1981, minimised the impact of migration in his classic study of Törbel, and only later did he recognise its importance for the dynamics of the community (Netting 1984).

Allowing for the considerable variation within the Alpine range, the work of Viazzo (1989) summarises some of the key points that have been revealed by historical demography. In his study of Alagna, in the Italian Alps, he argues that although colonisation of the area by Walser communities had been complete by the late Middle Ages, there is evidence to suggest that the population continued to grow at a substantial rate. There is also evidence, however, that migration was taking place, albeit initially on a small scale. Skilled craftsmen, particularly stonemasons from Alagna were reported in several European cities. This migration pattern involved both permanent emigration, with craftsmen taking their families, and seasonal work. The seasonal rhythm, in which men departed during Lent and returned the following November, is interesting as it has been related to the surge of births the following autumn. Whilst birth patterns in such communities have been linked to nuptiality and hence to the timing of weddings at a point when migrants are at home, the evidence discussed by Viazzo suggests that subsequent births are also related in this way. Therefore migration patterns become embedded in the demographic behaviour of the community and are one facet of the mechanisms that control the level of population. It might be noted, however, that this is a characteristic of many migrant communities, whether in the mountains or the lowlands.

The historical pattern of Alpine migration is often linked to the question of the 'carrying capacity' of the resources available to the mountain community. Guichonnet (1948), among others, argues that this can be clearly observed in the late sixteenth century and the seventeenth, when the deterioration in climate (the Little Ice Age) led to an increased flow of migrants to lowland occupations. In this sense, it is presumably lower crop and hay yields that present serious problems for household survival. From the end of the seventeenth century, seasonal migration appears to be an established feature of Alagna and its economy into the present century, despite the fact that, from the early eighteenth century, population grew steadily until the late nineteenth century. Not only does this call into doubt the notion of a closed community, it also suggests that the 'carrying capacity' model has limitations. Is seasonal migration an indication of poverty or of an income maximising strategy? This question is raised as a result of findings that the majority of the migrants were from relatively affluent families, it being the really poor that remained at home. Similar findings have occurred in more recent studies of migration suggesting, of course, that it is only relatively rich families that can support the costs of migration and have the necessary support networks to become involved. Unfortunately, often there is little information about the relative wealth of the people themselves and this makes it difficult to explore this distinction. Certainly there is evidence to suggest that, with the decline in lowland employment in the early twentieth century, real poverty hit some Alpine villages.

Another common explanation of out-migration in mountains is associated with accessibility, suggesting that it is the lower villages that suffer most from migration. Great care is needed with this argument because it depends very much on the historical period, the precise geographical location and the different categories of migration. In the European Alps until the nineteenth century, higher villages, because of their isolation, were more dependent upon seasonal migration, and later suffered permanent out-migration. Conversely, the lower valleys, with improved accessibility and perhaps a better resource endowment, have been able to partake in shorter-term migration and, more recently, commuting, which has helped maintain their populations.

Thus this historical evidence, whilst not denying the significance of population pressure operating through reduced land capability, suggests that environmental factors provide a less powerful explanation of migration from mountain areas than is commonly argued. Over a very long period, temporary migration has been a key element in many mountain communities, equally evident at times of population growth as well as decline.

The Nepal Himalaya

Migration as a strategy for survival is also well illustrated by work in Nepal (Bishop 1998). Nepalis were recruited into the British army in 1816 in India and often the soldiers took their families with them. Some stayed, and in the later years of the Raj more migrants moved to work on the plantations and railways. In those days, it must be remembered that colonial employers had very fixed ideas about the relative strengths of different ethnic groups, and in the Indian subcontinent mountain peoples were often regarded as stronger and more reliable workers than the lowlanders. It was in the 1950s, however, that migration, especially circular migration, became an established feature of households such as those in Melemchi. Bishop argues that labour migration is a typical extension of the previous pattern of life where some family members commonly spent a considerable period of the year living at grazing areas. Today this is considered to be a harder and less socially acceptable occupation than working away from home in wage employment.

Since the 1950s, Nepal has also experienced an internal flow of permanent migrants, as the state encouraged migrants out of the hill areas to valley lands once malaria had been eradicated. This resettlement migration is interesting because it indicates the process by which hill dwellers have sought lower sites that provide the potential for better access to social infrastructure (Shrestha 1990). In Shrestha's detailed case study of the Chack Khola valley, the sequence of resettlement is described, beginning with the period in the late 1950s. At that time, although families had their farms in the valley, almost all ridge dwellers still had their main dwellings on the ridge that was cooler and relatively free of disease. From the early 1960s, as a result of the malaria eradication programme, resettlement began and new settlements were created in the valley bottom. In fact, the resettlement process involved several stages. First a daily trek to the new site, followed by the construction of dwellings at the new fields and occupation only during the agricultural season, to complete resettlement of the family in the valley. Although the main motive for the move to lower areas was the convenience of proximity to farmland there were also other factors. One of the most important was the increasing pressures on the availability of drinking water and firewood on the ridgelands, which had been absorbing an increasing proportion of the daily labour, especially for women. More-over, with the spread of clinics and educational facilities into the valley, access to these facilities had become an important issue.

The ecological impact of this movement has yet to be assessed. Whilst it has been claimed that the reduced availability of water and wood on the ridges was a function of increasing exploitation, the move to lower sites has also brought with it further forest clearance in the river valley. In these valley sites, the forest has been cleared but new agricultural land has been terraced. Recent work in Nepal (Gerrard, personal communication) suggests that on properly terraced land soil erosion may well be within manageable rates, and even many landslides zones are rapidly redeveloped.

However, whilst migration in Nepal may be seen by some as an extension of normal agricultural activities and by others as a process associated with resettlement, the last few decades have seen a new dimension to this process, namely the role of female migrants. Previous models of migration in mountain areas have been based upon the fact that most of the women remain at home taking care of domestic responsibilities, including children, and also performing agricultural duties. In many instances this increase in workload has itself been cause for concern. In Nepal, Thacker (1991) has documented the problems posed by the large numbers of female migrants to the carpet industry in Kathmandu. It is estimated that at the beginning of the 1990s about 20,000 women and children were employed in the carpet industry directly and many more indirectly through outworking. The carpet industry is one of Nepal's top sources of foreign exchange and these migrants come from resource-poor, often female-headed households. How-ever, Thacker argues that the labour intensive practices available, whilst offering some relief from poverty, make the workers open to abuse. There are also gender barriers, as access to new skills, especially mechanisation with its enhanced incomes, is confined to male workers.

Morocco

It is evident that mountain communities have rarely existed without some degree of migratory links. However, most writers argue that two forces – the increase in population and the penetration of capitalism – serve to increase the extent to which mountain communities became highly dependent upon economic opportunities, in the lowlands or often overseas. In Morocco, for example, labour migration has been a structural feature of communities in the mountain areas for many years, to the point where today there are EU-funded initiatives to facilitate the use of remittances for communities to develop local infrastructure (Project Migrations et Développement 1994). In this project, situated in the Cercle de Taliouine, several settlements in the High and Anti-Atlas have been provided with electricity. For instance the village of Taiefst is located at 2,000 metres in the Anti-Atlas and has seventy-five house-holds, many containing migrant workers. A group of migrants in Perpignan, France are supporting the electrification of their village with the assistance of the Migrations et Développement Agency.

Two issues emerge from the Moroccan experience. First, that migration from the mountains, despite its hardships, has also brought with it financial success; second, the fact that migration and the use of remittances have a very variable effect on the community, often dependent upon the resource base of the mountain zones. In the Anti-Atlas, for example, the Ammeln valley, a few kilometres from Tafroute, is renowned for the stark contrast between a largely defunct agrarian resource base and the existence of many extremely stylish dwellings. Many migrants left this area in the early days of French colonialism, and after many decades of hard work have established a premier position amongst Morocco's grocery and related trades. Today these prosperous houses represent the symbols of success. Equally signifi-cant, some few kilometres away the village of Taghout stands on a dry plateau, the site having been established only since the early years of this century. The original site is a ruined *agadir*, a fortified village that over-looks the track up to the plateau from the desert to the south. Large areas of unused terraces surround the new site and many of the eighty or so houses are empty for at least part of the year. This is a typical migrant village, occu-pied at best by the very young and elderly of families who are working elsewhere. Some of the land is planted each year if the rains are satisfactory, but over the years fewer and fewer terraces have been farmed.

The houses, however, are in the modern style, and there are clearly con-siderable investments in property and other features of conspicuous consumption. Most houses now have generators for electricity and water pumps, whilst a telephone exchange has been installed. Moreover, one of the most notable features of this and many similar villages is the fact that the mosques have often been rebuilt or are undergoing refurbishment. In this sense, despite the migration, these villagers are still investing in the cultural values of the community, even though there is little economic support from local resources. It is interesting that recent work in the Rif Mountains (Lazaar 1997), which examines the recent rapid increase in migrant flows to Spain (they had traditionally been directed to France, Germany and the Netherlands), points to the low proportion of investments in productive as against consumption use. The priorities are clear: after satisfying immediate needs it is property improvements, household goods, vehicles and clothing that account for the main use of migrant earnings.

Colorado, USA

The final case study is taken from work by Riebsame *et al.* (1996), noted in Chapter 3, and provides a good example of the impact of population growth in mountain areas which are embedded in a highly industrialised society. The Rockies have been identified as a classic case of the amenity migration discussed above. According to work by Williams and Jobes (1990), quality of life considerations are now a major issue in migration flows which, in the Rockies, have led to population growth accompanied by expansion in tertiary and leisure activities. Between 1950 and 1993, whilst population growth rates declined in the already urbanised zone of the Front Range, the mountain areas have shown a strong resilience to the vacillations of economic fortunes. Thus after a growth spurt in the 1970s, a decline occurred which reflected US economic recession, but the mountain areas have once again 'bounced back' as these areas have proved to be increasingly attractive. Both permanent and part-time residents are attracted to the area, not only because of its inherent attraction but because of the fact that transport and communication facilities are now available at a price that makes residence in these areas affordable by more and more people. A combination of the four-wheel drive and the Internet, along with the changing land management regimes which cater for the 'ranchette' development mentioned earlier, means that residents can live in an idyllic landscape, often operate their own business via telecommunication, and have fast access to settlements. These in turn have expanded to cater for the needs of the relatively affluent. Land prices in some of the areas are now extremely high.

However, it should be noted that the arrival of the 'jet set' in these areas and the establishment of supporting service industries does not mean that the latter are all menial tasks. Far from it. One of the characteristics of the new developments in mountain areas has been the parallel growth of highly paid professional occupations, such as in the medical and related sector, catering for the needs of the affluent. Thus a typical dynamic growth spiral has been established in these mountain settlements, fuelled by the attractions of the locality.

DEPOPULATION

This brings us to consider the fact that migration from mountain areas is often part of a long-term process leading to rural depopulation. For example, in many European mountain areas, including those within the UK, depopulation has been the subject of both academic and official enquiry. It is traditionally viewed as the inevitable consequence of the process of industrialisation and urbanisation. However, it is rather more complex than this simple explanation implies. To understand this we must return briefly to the work of Viazzo (1989) and others in the Alps. He points to various studies which highlight the differential between those settlements which have undergone both total population decline and also demographic stagnation and others which have escaped this process. An ageing population is characteristic of many of these mountain villages, and has produced a fall in marriage and birth rates. Those over sixty may outnumber the under-fifteens by three or four times. In these cases, migration and an ageing population are associated with a loss of key social services – for instance, in the Swiss case the closure of the all-important post office. This process is beautifully described in the novels of Jean Giono, set in the mountains of south-east France. For instance, in *Second Harvest* (Giono 1999) he tells of the decline of a small hamlet in the mountains, the emotional repercussions of this on the remaining inhabitants and then its almost miraculous rebirth through new arrivals. On the other hand, other settlements, such as Davos in the Grisons, have grown considerably. Davos was a village of around 1,600 people in the 1860s, but by the 1930s it had reached around 11,000 as probably the most important tourist centre in the Alps (Bernard 1978). Similarly, the population of Viazzo's research village of Alagna has not experienced a dramatic loss in population and, interestingly, the age structure of the population shows little change between 1935 and 1980. In this instance immigration has been an important factor, not just in the last century but at periods in the distant past associated with local mines. In the second half of the current century, tourism has attracted new settlers to the village.

These local studies suggest that we need also to evaluate the overall impact of a changing demography on the Alpine region. Although many demographers have examined this problem in various parts of the Alps, a recent study by Bätzing *et al.* (1996) provides a geographical overview of the population of the Alpine region based upon commune level statistics. Obviously there is much that can be said about the difficulties involved in creating an acceptably uniform database, but nevertheless the findings of this work are illuminating.

Overall the Alpine population rose from 6.7 million in 1870 to around 11 million in 1990, less than 1 per cent per annum, which is well below the overall rate of increase for Europe's population. This growth masks the fact that in the same period the number of people living in the mountains as a proportion of the total population fell by some 22 per cent, most markedly in Italy, Switzerland and Slovenia. Perhaps the most interesting information to emerge from the study is the fact that 43 per cent of the communes experienced net population loss, but 47 per cent showed an increase. Geographically, the growth was most noticeable in the Austrian Alps between the Rhine and Salzburg resulting from the economic boom in the period from the 1950s. The declines were, for the most part, concentrated in the south-west of the range, including the Aosta area of Italy and substantial parts of the French Alps. Some 600 communes in this region have lost more than two-thirds of their population, some as much as 90 per cent. These figures highlight the fact that any discussion of population change must recognise the variety of trends and its spatial expression, and whilst it is understandable that the problems associated with decline are often highlighted in the political debate so too should those relating to the impact of growth. It is the geographical variability that is so difficult to comprehend and incorporate in strategic planning.

THE DEVELOPMENT OF JAPANESE MOUNTAIN VILLAGES

In the previous paragraphs discussing mountain depopulation we have focused heavily on the experience of the European Alps. However, the issue is important worldwide, especially in areas of dense, relatively affluent populations such as Japan. The depopulation of Japanese mountain settlements has been a major issue in Japanese society (Okahashi 1996). Since the Second World War, a number of studies have explored the process of depopulation, both at the village scale and from a perspective of the restructuring of the Japanese economy. The early work highlighted a distinction between the mountains in the south-western part of the country, where migration involved the movement of entire families from villages, and the north-eastern region, in which depopulation has been associated with workers living away from their home base. However, this simple dualism has been found inadequate because it failed to grasp great variation in the migration process. Ajiki (1993) examining settlements in the Kitakami mountains in the northeast, illustrated the considerable variation between households even in areas prone to high migration, but points out that the process of ageing is likely to lead to the final extinction of certain villages.

More recently, the emphasis has been focused upon the depopulation process as part of the peripheralisation of large areas of Japan as the postwar industrial boom of 1950s and 1970s proceeded. In this situation, some mountain villages have been drawn into the process of growth through the location of industrial activity, usually small or medium scale plants for subcontractors of the main industrial giants. At the same time, these plants have used local female labour; the process of mountain village development, therefore, is intimately linked with the restructuring of labour markets.

The peripheral nature of this development was aptly illustrated once the economy began to go into recession, in the first instance as the result of international oil crises in the 1970s. During recessions there has been a tendency for many of the smaller companies to collapse or relocate in the main centres. The vulnerability of the mountain settlements to this situation has been reinforced even more in the recent Japanese economic crisis. At the same time new forms of employment, mainly in tourism and construction, have developed in some of the mountain settlements as these activities have expanded. However, both are subject to considerable seasonal fluctuations and there has been an increasing awareness that generating more employment, especially of an industrial nature, runs counter to the pressures for more environmentally sensitive use of mountain areas.

This conflict has become apparent as the Japanese have put greater stress on the policy of *Muraokoshi*, or village renaissance. Signs of this process were evident in the 1950s, but the movement has gained strength in the 1980s and involves a network of cultural, economic, political and social institutions, often with conflicting objectives. In the first instance, the Japanese government regards this process as a commitment to preserving the key elements of rural society in the face of large-scale urbanisation. The programme involves the administrative amalgamation of small villages into towns and the creation of new municipalities that could perform more successfully tasks that once were the preserve of the village. As Knight (1994) has argued, this process contains considerable political stress and has required a substantial reorientation of family obligations. Traditionally, Japanese families had close emotional links with their home and village, or *ie*. This link to the *ie* is based on the idea that the household is a corporate entity which continues over time, with its land and resources usually inherited by the eldest son. Today this notion still has some powerful support within religious circles but has been marginalised by the new social institutions associated with municipal development. However, it is worth noting that one of the key elements of the municipal development process is that of civic education which strives to recreate a new social entity based now around the township rather than the village. Nonetheless, even with the considerable out-migration, many migrants remain emotionally, and in some cases economically, attached to their *ie* and it is normal for those migrants to return home for a short period each year to assist in keeping their natal homes and land in order. In particular the symbolic maintenance of grave sites serves to cement this link.

SUMMARY

This chapter has examined the contribution of demographic change to the transformation of mountain communities. Evidence from household, village and broader regional scale has been used to show how the transformation of these communities is an ongoing process, not easily captured by short-term enquiry. Whilst the coverage of available data is very uneven, the basic trends are clear. In the first instance, over the long term, exogenous factors such as the availability of off-farm income or environmental change play an important part in moulding the community. Both population growth and decline are evident in mountain areas and each historical period and locality has its own ways of responding to the problems posed by changes in resource availability. This can have a direct, if delayed, impact on key demographic variables such as fertility and mortality. However, the way in which these variables resolve into particular patterns of growth and decline are dependent upon cultural norms associated with kinship and inheritance. The long-term evolution of these factors is a key issue in many mountain communities and is particularly important when considering policies for development of these areas. Not only is this evident from the Alpine example where regional differences are marked, but also in the Japanese case where rapid economic and social transformation has juxtaposed different cultural norms in a way that poses important issues at the political level.

7

ECONOMIC AND POLITICAL
DEVELOPMENT

This chapter examines what is perhaps the normal brief of most development initiatives in mountain areas: the introduction of new techniques, new
institutions and above all the development of the cash economy. This meshes
very closely with the population and environmental factors examined in
previous chapters but is often considered in isolation as part of the 'modernisation' of mountain regions. This process is often connected with the
penetration of capitalism but it has equal force when considered under the
umbrella of state socialism, for instance in the mountains of the former
USSR. The account that follows is somewhat arbitrarily divided into sections
that look at agriculture, industry and tourism, but it begins with a discussion of the question of the political framework of mountain areas. Much of
the theoretical discussion can be examined in the literature on political geography (Taylor 1993; Agnew 1997).

POLITICS AND POLITICAL ECOLOGY OF
MOUNTAIN REGIONS

The study of politics in mountain areas is very diffuse. In some cases, the
total focus of attention can be located at the level of the mountain region,
where for instance a particular social group may be found. In other examples, it is the relationship between mountain and lowland that provides the
framework, once again raising the question of different scales of analysis.
Equally, the increasing attention to ecological issues is itself a political matter
and points in the direction of the rapidly rising field of political ecology.
Bryant and Bailey (1997) ably discuss this field of research, and Peet and
Watts (1996) provide examples of the varied approaches to this theme.
Political ecology is a deliberate attempt to move away from the apolitical
discussions that have dominated so much of the work on the environment
and places particular attention on the detailed interplay of the various
stakeholders in environmental discourse. Drawing upon several strands
of work which are also common in mountains, a political ecology approach

to mountains would emphasise that all social relations are based upon a structure of power and that the particular geographical or ecological outcomes are the result of this power brokerage. Operating at a different scale, this approach fits neatly with the household, village, valley model and provides also for the dynamics of the decision-making which is bound up with complexity and the evolution of particular systems.

For many mountain areas this process is expressed most aptly through the concept of political marginalisation. It is worth while noting the distinction, developed by Bernbaum (1997) and used in Part 1, between the relatively heavily populated and the thinly populated mountain areas, most of the latter being found in higher latitudes. Notwithstanding the fact that mountain populations have changed dramatically, as we have already discussed, it is inevitable that many of these thinly populated mountain areas are considered politically peripheral. The evolution of settlement, the emergence of nation-states and their accompanying political institutions have incorporated mountain areas from a power base located usually within the lowlands. A good example is the USA, where westward expansion involved absorbing the Rocky Mountains region into the new nation. However, where considerable populations have been settled for some time the process is more complex. The areas have had their own cultural histories which include political organisation, and, whilst perhaps not always identified as 'states', have had sufficient power to remain distinct from other groups. Good examples include the tribal groups in the Yemen mountains, the Baltis, Afghanis and others in the Karakoram/Hindu Kush, and the Berber communities in the Atlas mountains.

In these situations, two processes need to be explored. First it is important to understand the political structure of these originally autonomous mountain states. In Part 2 we have already given this some consideration, for example in the description of the loose but nonetheless powerful confederations of tribes in the High Atlas. This is an excellent example of the way in which local political institutions such as the village council or *j'maa* serve to link households with the wider valley-based political organisations. These combine economic functions such as water management, social linkages (through marriage) and political authority (in watershed scale confederations that are activated to defend common interests). As Hart (1981) has noted, different political factions may at one and the same time have fought each other over a dispute at a local scale, but allied in their efforts to defend their wider territorial integrity.

Second, an understanding of political history is necessary to account for the process by which many of these former autonomous territories have been integrated into larger states. It is in this situation that political marginalisation becomes a critical issue as, having once been relatively autonomous, these territories find themselves subject to extractive economic activity and often suffocating legislation. In particular, attacks on the cultural values of the mountain people, or, perversely, using their values instrumentally as part of political bargaining, underlie much of the history of these territories. For example, the French adroitly used the mountain Berber groups in Morocco in their dealings with the Sultan, after having in the first instance extinguished Berber opposition forcibly. An equally fascinating story concerns the evolution of the Himalayan states and their struggle to deflect the pressures from China, the British Raj, Russia, and post-1947 India; in this latter case the international repercussions are still generating serious conflict.

THE PROCESS OF MARGINALISATION

There are three principal strands to this process which, in any given example, are usually tightly interwoven. The first process is that of internal colonialism (which has already been discussed) and is highly contentious. This theme builds upon the notion of the uneven development that dominated much materialist geography in the 1970s and stresses the mechanisms by which resources are removed from an area under 'unequal exchange'. For mountain areas this usually includes forest products, minerals and also water, the latter rarely considered in the calculus of economic exchanges within countries but very much an issue internationally, for instance in the Middle East and parts of south-east Asia. Additionally there is the question of labour migration, which we have already shown is an important component of the demographic dynamics of many mountain regions. The argument maintains that most labour migrants are poorly paid and hence serve to subsidise the lowland economies, which enables greater accumulation to take place there. These direct processes are further facilitated by the manipulation of official transfers; that is, the payment of taxes and the provision of state benefits to mountain areas. The claim is that net flows are negative and serve to depress mountain economies even further.

Mountain areas have been included in the wide range of studies arguing that this process is at the root of both poverty and environmental degradation. The analysis of the deforestation of the Himalayan foothills, for example in the work of Guha (1989) and Shiva (1989), contains elements of this approach as does work in the Andes, for example the studies of agrarian change (Lehmann 1982). However, such analyses are fraught with both conceptual and empirical difficulty. The geography of mountains, especially their inaccessibility, often makes them very expensive places in so far as any imported commodities are concerned. By the same token, any exports have additional costs. The migration flows occur as a result of the lack of economic opportunities within the mountains, either because of the expansion of population or, more likely, because of the paucity of jobs with the levels of security and paying wages matching those in the lowlands. A good example of the latter is the pensionable employment in military forces, often highly attractive to mountain peoples. Attempts at analysing transfer payments are confounded by the lack of statistical material disaggregated to account for geographical regions, unless these also correspond to some administrative division.

The second strand of the marginalisation process concerns various social or ethnic divisions. Once a mountain area has been incorporated into a larger entity it is often the case that its population will become a minority ethnic or religious group. The precise form will be a matter of the particular historical experience but there is no doubt that this process has played a major role in establishing the marginality of many mountain areas. This has been a powerful factor in Thailand, in Andean states and in Morocco. It is useful to note that, in areas where the mountains are the core zone of a country, for instance Ethiopia, the reverse can be found in that lowlanders find themselves as relative minorities or marginal groups contesting the domain of the dominant group.

The final element of marginalisation is inextricably linked with the previous two and concerns the role of the state, both in its political institutions and economic policies. The political representation and power of mountain regions within the national polity depends upon the precise nature of the political institutions and their operational rules. Two examples can be given. In Switzerland, the evolution of the federal constitution provides for wide ranging powers to reside with the cantons, so that the mountain areas retain considerable powers for local taxation and development strategies. As we shall see in the examples discussed below, India also has a federal structure but with a highly centralised power base in which its mountain regions do not play a significant role. In many other countries, mountain areas are not well served by the political structure, either because local democratic institutions are very weak or because the mountain areas *per se* are not clearly recognisable entities within the political establishment. The EU, in its continued discussions on marginal areas, has encountered this problem.

To illustrate the debate about political marginalisation it is helpful to consider some examples in greater depth. One area where the issues above have been discussed is that of Uttarakhand in northern India. A series of studies by Mawdsley (1997, 1998, 1999) and work by Rangan (1996) explore the background to a recent upsurge in claims by this area for separation from the state of Uttar Pradesh (UP). All these writers are at pains to point out that this is not a case of secession from the federal state but what is felt to be the unsatisfactory governance from Lucknow, the capital of Uttar Pradesh. Uttarakhand is primarily a mountainous area, with peaks ranging from 1,000 to 3,000 metres and steep valleys, and where most of the population is rural, poor and occupying relatively small plots of land. Only in the Terai zone are there significant signs of agricultural development. By contrast, Uttar Pradesh is predominantly a lowland state where agriculture output is relatively high as a result of the considerable investment in 'Green Revolution' technology. It is claimed that the population of the area is culturally distinct from the lowlands and known as Paharis, who have their own languages but also speak Hindi. In addition, their religious affiliation, whilst mainly towards Hinduism, also has animist elements, giving it much in common with neighbouring Nepal. Furthermore, there is a popular belief that the lowlanders consider the Paharis 'backward' and culturally inferior. As these writers emphasise, perhaps the most distinctive social characteristic of the mountain zone is the fact that it has an unusually high proportion of high caste Brahmins and Rajputs, estimated at around 80–85 per cent, whereas the state average is about 11 per cent. This has important consequences for the politics of the Uttarakhand movement (the area is also known as Uttaranchal).

The details of the political activities associated with this movement can be found in the references noted above. It is clear that what was a long-standing, but low-level grievance flared in 1994/5 because of an attempt by the UP government to increase the number of 'reserved jobs' and educational facilities. In the Indian context 'reservation' means specific discrimination in favour of certain scheduled castes, tribes and other backward peoples to ensure that they have access to employment and education. Whereas the national level was approximately 22 per cent, UP proposed to increase this to almost 50 per cent. Consequently, given the particularities of the social composition of this poor mountain region, only some 2–4 per cent of the population would qualify for government employment and education. The uproar associated with this shifted rapidly to embrace the wider issue of a movement to create a separate state. The basic arguments supporting this action conform to the marginalisation discussion above emphasising economic underdevelopment, social differences and, above all, alleged discrimination and neglect by the state government. In this respect it is worth noting that the Uttarakhand area has the lowest number of representatives per head of population in the Legislative Assembly in Lucknow, an immediate cause for grievance. In 1998, a bill was passed in the UP parliament which established the framework for a new state. Despite considerable disputes on details, the process looks all set to be completed in 2000.

This example, and the subsequent extensive deconstruction by many writers, also helps reveal some of the key changes in the direction of political analysis of mountain areas. As Mawdsley argues, the Uttarakhand case is based upon the simultaneous use of arguments for self-determination and for ecological or environmental management. All parties to the debate call upon images of environmental destruction, cultural repression and economic neglect to further their political capital, but there is much that remains uncertain. On the ecological front, the region has, of course, been made famous by the Chipko movement (Guha 1989), but there have been many different analyses of the impact that this has had. For example, despite its international acclaim by those advocating 'bottom up' approaches, and the increasing acceptance by the Indian government that greater protection should be given to the forests, the practical results have been not quite what was expected. Rangan (1996) argues that both the federal and district administrations have indeed passed protective legislation, but that this has had the effect of limiting access by locals to the forest resources. Licensing arrangements have benefited large national concerns. But there is another twist to this question. Much of the debate about deforestation in this area overlooks the tricky question of exactly who the various actors are. The support for local-based logging was, in reality, a support for local medium-scale entrepreneurs who were not necessarily creating the level of employment demanded by the typical forest worker. The local forest workers wanted greater involvement of the larger concerns because of more and better remunerated employment!

Mawdsley argues that many of the arguments about internal colonialism used by the proponents of a new state are difficult to demonstrate. However, she goes on to suggest that the important issue is not that the arguments are supported by evidence but simply that large numbers of people believe them to be correct. This has important implications because the major thrust of the discourse is built around the distinctiveness of the mountain environment of Uttarakhand in comparison to the plains of the rest of UP. Thus a mountain–plains imagery provides the framework in which much of the debate is conducted, neatly dichotomising (and simplifying) what is a far more complex question. Opponents of the separation argue from many viewpoints, including the political repercussions on other states as well as the possible consequences for the UP government itself. Administrative and economic issues include the fact that, as a very poor area, it will have a minimal level of revenue and have seriously deficient internal communications between valleys. The new state will also have grave problems in determining an appropriate site for a capital (internal distinctions being forgotten in the movement's propaganda) and has little prospect of economic development other than some niche horticulture and tourism. To these criticisms the riposte is usually twofold: first, that most of the criticisms hold good for many existing states within the Union and, second, what if Uttarakhand began to levy a tax on its use of water? This latter suggestion opens up once more the whole question of mountain areas effectively privatising their resources and selling them to outsiders. In order to achieve this, such areas need the power to determine their own taxation regime; in other words, some degree of political sovereignty. In the Indian context, as elsewhere (including the UK), this is a highly charged issue.

We have used the Uttarakhand example to explore some of the key features of political marginality and explore regional relationships. Another very prominent example from south-east Asia also emphasises the ethnic dimensions of marginality. This is the case of the various hill communities of northern Thailand, referred to earlier, who occupy an extensive tropical forest landscape between 750 metres and 2,500 metres. Since the 1980s numerous writers have drawn attention to the way in which the hill tribes of northern Thailand have been subject to intense political pressure under the guise of development (Cooper 1984; McKinnon 1989; Tapp 1989; Kampe 1992; Forsyth 1995; Ganjanapan 1996). The highlands provide an arena for the agriculture of several ethnic minorities such as the Hmong, Akha and Lahu. These groups have been migrating into the Thai mountains over the last century, and traditionally practised a form of shifting cultivation, producing rice, maize and opium. At a slightly lower altitude, but important for our understanding of the problem, are to be found the Karen and Lua groups who have been present in Thailand for several centuries and are skilled at mixing irrigated wet-rice cultivation and shifting cultivation. In the lowlands, ethnic Thais predominate; they originally concentrated on wet-rice cultivation but are now involved in a variety of cash crop ventures. Despite the rapid development of the Thai economy from the 1960s, the north-western highlands remain relatively poor and the hill peoples some of the poorest in the country.

Kampe (1992) describes the state policy towards the approximately 750,000 people who occupy this area as reflecting a generally negative view of the hill populations. They are a problem, either because of their predilection for opium cultivation or because, as minorities with links to groups in neighbouring countries, they represent a security risk. Finally, their use of shifting cultivation (swidden) techniques is considered backward and to be the principal cause of rapid deforestation. McKinnon (1989) suggests that the root of the problem lies in the issue of citizenship rights and all the implications this has for the ownership and control of the land. Until 1965, these hill peoples were not considered as citizens of Thailand, and even after new legislation to grant citizenship some 30 per cent still remain unregistered (Kampe 1992). However, the last thirty years have seen a welter of development initiatives with the ostensible objective of incorporating the hill people into Thai society. There has been a 'Masterplan for Development of the Highland Communities and Environment and Eradication of Narcotic Crops'. According to Kampe, in 1985–6 there were 168 government agencies operating from thirty-one departments along with forty-nine international, official and NGO agencies, all working in this area. Official agencies have focused on the substitution of cash crops for opium and, where successful, this has had the effect of integrating highland farmers into the national economy. Other agencies have sought to develop alternative strategies building on the existing knowledge base of farmers and with increasing emphasis on local self-reliance and sustainability. The conflicting perspectives are obvious, but the real problem emerges when it is recognised that the simple set of dichotomous relationships – i.e. Thai versus non-Thai, highland versus lowland, cash crop versus subsistence – is inadequate to examine the situation. This is particularly apparent on closer examination of the impact of the development initiatives.

As noted above, the reduction in opium production required the introduction of other relatively high-valued commodities of which cabbages and orchard crops have been particularly popular among the highlanders. Despite the economic advantages offered by these crops, downstream water users have noted the increase in siltation that has resulted from the clear planting. At the same time, the authorities have greatly expanded the area protected through National Park status, and in a number of cases forcibly removed occupants, despite claims that the land was held under customary rights. Finally, whilst claiming that swidden damaged the forest the various state agencies have been licensing private companies to develop the land for commercial operations. This has involved logging, but also other activities (including road building) which have led to forest removal. The differential impact of these activities is both spatially and socially significant. Some ethnic minorities in highland villages have seen their production and incomes increase as they have been successfully incorporated into the market economy. Others have witnessed a boom associated with employment in new commercial enterprises, including a rapid expansion of tourism partly based on 'ethno tourism'. Certainly many lowlanders who have moved into the highlands have profited by the relatively easy access assisted by state initiatives. In contrast, there are large numbers who have seen their resource base radically reduced. Most of these are ethnic minorities, particularly the Akha, who have been unwilling to relinquish their traditional practices and remain convinced of the sustainability of their forest management practices. In turn this can be related to disagreements within agencies – some government departments and NGOs seeking sustainability through the development of a commercial economy and the exploitation of the forests and water of the mountain areas, others through promoting the 'bottom up' approach based upon encouraging self-sufficiency. A good example of this is the work carried out by the Hill Area Development Foundation operating from Chang Rai. The various stakeholders therefore do not fall simply into the categories listed above but are constantly reinventing the defence of their positions and manoeuvring to gain advantage with respect to mountain resource use.

In this example from the highlands of Thailand we have seen how marginalisation can be part of the development process itself. It illustrates clearly the fact that, in the final analysis, it is the drive for industrialisation and urbanisation that has been the principal direction of Thai development. Issues of forest protection and the sustainability of livelihoods in marginal lands such as the highlands play only a secondary role. Furthermore, in deciding where the benefits of development accrue, it is ethnic divisions that are highlighted and the fact that most minorities are concentrated in the mountains. In this situation, policies for forest protection become a cover for deeply held prejudice. This situation is maintained by a power structure in which, partly because of the problem of citizenship, these ethnic minorities are unable to participate directly. Nevertheless, these communities are far from acquiescent and they have resorted to protests and petitions to Bangkok.

As a final example of the interaction between politics and the transformation of mountain areas it is most appropriate that we briefly examine Switzerland. Sixty per cent of the country is alpine and 10 per cent is hill land with altitudes ranging from 200 to 4,600 metres. Industrialisation occurred in the nineteenth century but, unlike other European nations, it was relatively decentralised, utilising the water resources of the mountain streams, first directly and then indirectly with the use of hydroelectric power. In the twentieth century the sectoral employment pattern has changed radically with agriculture falling from around 30 per cent in 1900 to around 6 per cent in 1990, with a consequent expansion of the service sector. Today, 60 per cent of the population is urban but there are no very large cities. Instead, the geography of Switzerland is characterised by a closely interconnected network of urban centres, many to be found in the mountain valleys. There are many examples of the way in which the Swiss rural economy has been reshaped to face the decline in agricultural prosperity, the booms and slumps of the manufacturing sector, and similar changes in the tourist sector (Messerli 1989). However, this account stresses the nature of the political institutions that have encouraged this continual transformation.

Linder (1994) describes the political history and current structure of the Swiss Federation. He stresses that the combined effect of resource limitations, external political pressures, religious and ethnic minorities and multi-lingualism have produced the particular structure of the institutions that characterise Switzerland today. For our purposes the most important characteristics are the constitutional powers that give cantons, and through them communes, a powerful role in both policy formulation and implementation, and, second, the operation of direct democracy whereby referenda are used to sanction important changes in national policy. Although many countries have evolved federal systems, including India and the USA, the Swiss model is rather different. Owing to its historical evolution, the federal authorities are only able to exercise powers that have been explicitly allocated to them by the cantons. Since 1848 the evolution of the Swiss regional economy has not worked like that of the USA, for example, where the variability among the states is seen as a key element of individual freedom and choice. In Switzerland, despite the considerable structural changes, the political ethos strives to ensure near similarity of basic provision between the cantons whilst allowing for ethno-linguistic differences to remain. Moreover, on some crucial issues the political power of the communes and cantons has been of considerable benefit to mountain areas.

For instance, the utilisation of hydropower essentially for the lowlands and valleys occurred with, for the most part, the express agreement of the communes from which the water flowed. Thus these source areas received payments for the resource, unlike many other mountain areas. In the 1970s, a large federal investment programme attempted to ensure that mountain areas were assisted in the regeneration of tourist and selected agricultural activities. Linder (1994) points out that this programme was implemented in different ways in each canton and was universally approved. Urban-based factions argued that the preservation of mountain settlements and a rural economy was economically inappropriate in the latter half of the twentieth century. However, whereas it is likely that such a faction would be dominant in many other countries, in Switzerland two factors played a crucial part in ensuring the survival of the programme. First, the political institutions allowed a significant number of cantons with mountain areas to defend the programme, and second, the supporters of the programme were able to use the symbolic role that mountains play in the Swiss culture to argue for the maintenance of population in these areas. The programme has had the effect of ensuring that mountain communities have been able to maintain acceptable living standards and to some extent reduce the flow of migrants to urban areas. Thus whilst critics have questioned its economic rationale, there is little disagreement on its socio-political success.

The Swiss example provides an indication of the way in which suitable political institutions have in the past at least protected mountain regions from the worst effects of marginalisation associated with either ethnic minorities or economic modernisation. It may be worth noting that the canton system, flourishing under a strong sense of subsidiarity and direct democracy, maintains checks on the centralisation forces that are so obvious elsewhere. The system is also particularly sensitive to one of the principal characters of mountain areas, namely the diversity of cultural and economic structures linked often to the valley based societies. Whereas in some countries this diversity is treated as a dangerous disease almost treasonable in the eyes of the state, in Switzerland what could have been a recipe for disaster has become the fulcrum around which political institutions operate.

However, we can use the Swiss example to introduce a more general issue in our discussion of the political framework of mountain areas, namely the impact of globalisation. This trend towards the growth of economic and social institutions that now operate transnationally has become a centrepiece of many debates in political ecology. The discussions emphasise the fact that the economic and political structures determining the course of development in mountains are increasingly functioning at a global scale, whether through tourism or direct economic investments. Moreover, key economic policies such as Structural Adjustment, are, in reality, generated by alliances of the main economic powers in an attempt to maintain their hegemony of a global economic system. Macro policies on trade, investment and competition are quickly reflected in the national and subnational policies of other countries through trade and aid. This process is also closely linked to the debates on environmental policy; for example, on the use of mountain resources such as water and trees. The debates on large dams or on Himalayan forest policy are often now mixed with the debate on globalisation (Shiva 1989). Demands for greater commercialisation, the pressures for exploitation rather than conservation and the decline in local economic activities are often attributed to global pressures. However, this does not mean that no local decision-making capacity remains. For instance, Hirst and Thompson (1996) argue that the all-pervasive nature of globalisation is much over-emphasised and that nation-states have yet to be shown to be surrendering to its hegemony any more than has been the case in previous eras. From the opposite end of the spectrum, as noted in numerous examples used already, micro-political action has been emphasised by many writers, particularly in the complex interrelationships between environment, economy and politics. Work on mountain areas remains essentially eclectic in that discussions emphasising both scales of influence can be found in the literature.

We can observe this by returning to the Swiss case where the standard of living is one of the highest in the world. All past experience and policy indicates a striving to maintain relative equality between regions, but some commentators argue that this can no longer be supported. Whilst the Swiss economy remained buoyant the old consensus remained strong, but with the various economic crises since the 1970s some of these fundamental characteristics are questioned. The lengthy processes associated with direct democracy are said to be inadequate to face a fast changing, highly competitive world and the economic support of the mountain areas is claimed to be achieved at the cost of the economic resilience of the more dynamic lowland urban centres. From the globalisation perspective, the highly protective economic and social structures enjoyed by the Swiss, which include some of the highest standards of environmental management in the world, are increasingly at odds with the tenets of the World Trade Organisation. Finally, within the European context, there remains the unresolved division over relations with the EU following the rejection by referendum in 1992 of a move to begin negotiations for possible entry. For the mountains, new initiatives are attempting to co-ordinate policy among all Alpine countries and to establish a transnational body under the Alpine Convention, which will be discussed in more detail in Chapter 10. Both instances represent moves to reconstruct the political institutions appropriate for governance. At the same time, however, many local initiatives have emerged with support at the commune level suggesting considerable resilience at this scale, and currently there is no clear indication as to how the existing Swiss political structure will handle these issues into the twenty-first century.

THE ECONOMIC TRANSFORMATION OF AGRICULTURE

In earlier chapters we have described how agricultural activities have played a central role in the livelihoods of mountain people and examined the many forms in which communities use their natural habitat to provide livelihoods through animals, crops and trees. The approach so far has assumed a static, highly autonomous set of agricultural practices in which the particular agro-ecosystem is treated as the outcome of local adaptations to the specific characteristics of mountain environment. Many commentators regard these traditional systems as being characterised by low intensity resource use and operating on a small-scale, locally oriented basis using traditional skills and functioning within a local institutional framework. In this section we examine some of the main themes in the transformation of the agricultural sector.

In a wide-ranging review of the problems of developing new agro-ecosystems in mountains Jodha (1997) argues that there are three main pressures for change. These are:

- *Demographic change.* An increase or decrease in the number of people in a mountain locality places demands on the natural resources for their survival. In addition to population change *per se* the changing expectations of people have also led to new pressures on agriculture.
- *Market forces.* Increased participation within a wider (regional and global) economic system means that the critical factors controlling agriculture are established increasingly from outside the mountains.
- *Public intervention.* Overall there has been an increased intervention into all aspects of livelihoods. This may take the form of new institutions, as well as the introduction of technology. The state may justify this increased intervention by claiming the need for more efficient public administration, the need to foster development initiatives both locally and nationally, and the fact that welfare issues play an increasing role in state activities.

The net result of these processes, which 'open up' mountain communities to powerful external agents, has been to weaken the impact of community control and sanctions over the key social institutions that previously handled land use management. This is particularly apparent where traditional cultures would rely upon demand management, something that is a very unpopular strategy today. Whilst it is understandable that communities with poor dietary intake would seek to increase appropriate foods and minimise the risks of crop failure, most of the thinking behind current agricultural development strategies is based upon urban models of demand. Equally, as was discussed earlier, migration is not a new phenomenon, being a coping strategy much used by mountain communities facing difficulties with their livelihoods.

If we follow the critique of Jodha then there are a number of reasons why developments since the Second World War have altered the circumstances in which mountain communities handle agricultural change. First and foremost is the fact that the pace of change is too rapid to leave enough time for the communities themselves, and indeed their agricultural resource base, to adjust to new pressures. This is particularly apparent where market-led changes, such as the introduction of new crops, can emerge and decline within the space of a few years. Second, it is asserted that the magnitude of the post-Second World War changes is such as to be well beyond the capabilities of mountain communities alone to grasp and manage. Third, this arises mainly because the principal source of transformations is located not in the mountains but in the lowlands and often in other parts of the world. Thus there is little possibility for mountain peoples to influence the direction of these changes. Rarely can locally based policies overcome political and economic imperatives that may be moving in the opposite direction, a point which links closely with the debate about political structures earlier.

A number of commentators take up these themes. In the Andes, Bebbington (1997) points out that much of the literature on the development of new economic initiatives is either totally pessimistic, assuming that there is no long-term future for High Andean cultivators and graziers, or, conversely, maintains that development strategies must ensure that all localities receive equal attention. Such a view ignores the geographical variability that constitutes the Andean range and is unable to recognise the strength of global and national pressures towards restructuring local economies. Bebbington uses an example from Bolivia to show how some localities, in part due to their local resource base, have been able to adjust to the changing market pressures which elsewhere have led to long-term decline and out-migration. His 'islands of sustainability' suggests that, as always, each historical era leads to the establishment of new geographic patterns of production where the population has been able to 'innovate' and to develop the 'niches' suggested by Jodha. In this process Bebbington suggests that it is a focus on social capital, in other words the local political and economic institutions, etc., which is vital for negotiating the relationship between the mountain community and the wider world. Although developed in an Andean context Bebbington's work has wide applicability and is used here to set the framework for an analysis of the various 'development initiatives' that have become associated with mountain areas.

As has been discussed in Chapter 4, animal husbandry plays a critical role in the livelihoods of many mountain areas, whether in the form of nomadic pastoralism alone or in systems where cultivation and pastoralism are mixed.

Yaks, llama, cattle, sheep, goats and many lesser breeds provide a full range of resources supporting mountain communities and are also the centre of considerable research attention in the debate about securing sustainable mountain livelihoods. From the point of view of interventionalist policy, two themes have dominated: the increase in animal productivity and the management of grazing resources, particularly the question of over-grazing. In this respect it is important to note that most development agencies have until recently made little differentiation between lowland and highland pastoral systems, and the tendency has been to establish policy that emphasises both breed improvement and rangeland management initiatives designed for lowland areas.

One element of this rather undifferentiated approach has been the use of the carrying capacity concept, previously discussed in Chapter 6. For livestock management the core of the argument concerns grazing capacity. Until recently, most formal livestock development initiatives employed a rangeland management approach which uses carrying capacity as a key analytical tool. Livestock carrying capacity assumes the existence of a process of plant succession which establishes an equilibrium (or locally optimum) condition to which carrying capacity can be addressed. However, as work at ICIMOD and elsewhere has shown (Behnke et al. 1993; Miller 1996), this concept fails to capture the non-equilibrium conditions of mountain environments brought about by climatic variability.

In the case of rangelands in the Hindu Kush–Karakorum–Himalaya, research has yet to establish clearly whether specific, local grazing areas can be said to be non-equilibrium systems. In many parts of the Himalaya and eastern Tibet rainfall exceeds 400 mm, a figure considered by Behnke et al. to be the threshold below which non-equilibria can be observed. However, Miller (1996) suggests that specifically local conditions, such as the mixture of drought and precipitation occurring as violent blizzards, may well provide equally non-equilibrium conditions.

At the very least, more research is needed, not only in terms of climatic variability in mountain regions but also on the impact of 'opportunism' in pastoral management. In this approach, there is no 'long-term plan' per se but a grazing strategy based upon the nature of the forage availability each season. This requires a rapid response to grazing opportunity through high mobility and rapid changes in stock levels. In describing such as a strategy as a basis for livestock policy agencies involved would be doing little more than 're-inventing the wheel' as this is precisely what pastoralists in many parts of the Himalaya (and elsewhere) have been doing (Jina 1995; Ehlers and Kreutzmann 2000).

Many pastoralist groups have occupied mountain areas to utilise the diverse ecological conditions provided by 'verticality' to support their livestock (Ehlers and Kreutzmann 2000). A commonly found institutional framework for livestock management was founded originally upon socially defined (clan, tribe, etc.) access to pastures at different seasons of the year with communally determined rules of use. A well-documented case of the transformation of such a system is that of the Ait Arfa in the Middle Atlas of Morocco south of Azrou (Bencherifa and Johnson 1990, 1991). It is an area with rather poor soils except in limited areas of bottomland, a patchy distribution of adequate water supplies and marked seasonal climatic contrasts where heavy snows or serious frost can be damaging to herds and crops alike. In this example, we can observe the influence of external pressures and the evolution of different institutional structures noted above. The principal agropastoral feature of this area was the fact that, until the 1950s, there were large-scale seasonal movements of both people and animals between highlands and neighbouring lowlands as grazing conditions changed. Nomadic pastoralism of this type was not prevalent in other areas of highland Morocco, where for the most part permanent settlement prevailed but animals moved. The Ait Arfa are part of a Berber-speaking tribal confederation known as the Beni Mguild who, probably about 300–400 years ago, moved to this part of the Middle Atlas. At the beginning of this century population densities were low (6–10 persons/km^2), and the natural resource use was organised according to the tribal rules so that each subgroup had recognised access to resources of both the highlands and the neighbouring lowlands. Generally, in October with the onset of wintry conditions, tribal groups with their tents and animals moved to the lower lands in the west. During the period in the lowlands extensive cropping activities took place but, with the onset of warmer conditions, the herds and some of the families moved back into the highlands. Bencherifa and Johnson maintain that this migratory pattern was not a direct outcome of seasonal pasture shortage but of poor technical adaptation to the extremities of climate experienced in the higher areas.

The onset of French colonial rule in the twentieth century began a process that altered the resources available to these mountain pastoralists. To understand this it is necessary to explore the changing institutional framework in which the pastoralists operated, which is closely linked to the 'reconfiguration of space' by the French. The colonial administration gradually imposed a 'cosmogony' based upon an image of settled populations, each group having defined rights with respect to a fixed spatial unit (Bencherifa and Johnson 1990: 188). Tribal lands were formally delimited. In the case of the Ait Arfa this land area was where summer settlement and grazing took place. As this procedure unfolded throughout Morocco, it resulted in very rigid 'ownership' patterns emerging. In the lowlands adjacent to the mountains, the land was allocated to French settlers as part of the programme to improve agriculture. Consequently, the Ait Arfa no longer had access to the lowlands for the winter settlement. In addition, some land areas, especially forests, became 'state lands' and traditional rights of ownership were denied although some usufruct rights remained.

Morocco also experienced a period of rapid population growth in the middle of the twentieth century, and this region saw a doubling over a period of some fifty years. Whereas in other regions migration to urban areas became a critical outlet for a restructured rural society, this was not the case among the Ait Arfa. In 1985, less than 10 per cent of the population enhanced their livelihoods through occupations outside the area. Thus the key response lay with changes in land management. Two developments are evident, both supported in different ways by state agencies. First, there has been a strong growth in agriculture, especially in the Guigou valley where traditional irrigation practices have long been practised. Whilst still notably subsidiary to pastoralism, cultivation of cereals (barley, wheat and maize), and particularly fodder, has increased along with vegetables. This places heavy demands upon local household labour and now outsiders provide most of this.

Second, the animal husbandry regime has had to change, as long-distance movements are no longer possible. Animals, mainly sheep, now have to graze all year within the confines of the recognised communal grazing of the highlands. According to Bencherifa and Johnson the new routine utilises the agricultural residues in the late summer and from October animals move to collective pastures. It will be recalled that in the old regime these had to be abandoned in winter primarily because of low temperatures, but today, to survive the winter, herders have built animal shelters and also rely increasingly (about 15 per cent) on fodder purchased in nearby markets such as Azrou. In the early summer, whilst crops are cultivated, animals are moved to state lands in the south and graze where they can.

One of the results of the restructuring of the geography of this area has been to place the natural vegetation under considerable pressure. Although it would be easy to see this case as typically one of overgrazing on communal land, the circumstances are more complex. Access to grazing, particularly during the winter, is established through individual membership of the tribal group but there are considerable variations in herd size. Also, those with smaller herds also contract out their grazing and herding services to outsiders who own sheep. This is seen as an acceptable way to generate income. However, what is now lacking is socially acceptable and enforceable consensus about the levels of stocking and the dates of access to grazing. Moreover, a range management project in the area dating from the 1970s had the effect of encouraging each herd owner to maximise individual herd size!

This example in the mountains of Morocco serves to illustrate three things. First, that state sponsored land partition is a particularly significant form of institutional change that initiates modifications in agro-ecosystems, not all of them necessarily for the better. Particularly significant also is the increasing inability of locally constituted authorities to establish appropriate rules of land management. Individuals can appeal to outside agencies in their striving for economic success. This is a very difficult problem to resolve. Second, in this instance it was not the direct influence of the 'market' that played a key role, though at a macro level such considerations did apply. For many years, the pastoralists had had links with urban wool buyers who were associated with carpet making. Neither has the market provided necessarily clear signals in the sense of pointing to other agricultural activities that might reasonably expect to generate new forms of income. Third, however, these external pressures have to be mediated through the principal cultural values of the local population to whom animal husbandry is granted much higher status than either crop cultivation or migration. Finally, state intervention in the form of development projects does not always assist. The key to the long-term survival of the local agro-ecosystem would be to increase the local output of fodder so that the livestock becomes closely related to crop cultivation in a manner similar to other mountain areas that also have restricted pasture. However, recent visits suggest that whilst there are signs of some development in this direction, and also crops such as potatoes and vegetables have found greater favour, another avenue has been for the bigger herd owners to truck their herds to pastures far away.

A different experience of the impact of economic change can be found in a case study taken of the evolution of animal herding in the Yolmo valley, Nepal, provided by Bishop (1989, 1998). In the decades before the 1960s, the households in the village of Melemchi, which is situated about 50 kilometres north of Kathmandu at 2,600 metres, practised an agropastoral system based upon transhumant herding. Allowing for difficulties of measurement, the village saw an increasing population during the early part of this period that has now slackened appreciably. In this system, the traditional product was butter and cheese derived from the milk of the zomo, which is a cross between a yak and a cow. The animals and their pastures were owned and managed by individual families rather than through some communal organisation. The zomo herd was moved seasonally in search of suitable pastures, and because of the labour needed to milk the animals several members of the household were required to be in attendance at the temporary shelters called *gode*. In most instances the pastures are not more than a day's walk from the village, but are highly variable due to altitude, aspect and local soils. According to Bishop's account, originally most Melemchi households were purely pastoralists but agricultural activity gradually became important and families adapted their farming system to handle both. Thus animals were brought to the household for a short time in October to manure the land just before ploughing and the planting of wheat, barley and potatoes. Problems now arise in the summer planting of maize because this takes place when the herds are pastured some thousand metres higher and labour therefore is in short supply.

The major change has been a shift from an emphasis on the zomo as a source of milk to herding yak and cows for the purposes of breeding zomo.

This change, recorded since at least the 1970s, has meant that the herds have been restructured and now the key product is the calves, not butter. Most of the persons responsible for the yak/cow herds have been older men, as the task, whilst still tricky because of species incompatibility, meant that the labour intensive job of milking is no longer necessary. Even if the survival rate for calves is low, the resultant profits are higher than for sales of butter. However, a shift in the animal husbandry regime is predicated on a wider structural change in the economy. The number of households operating *godes* has fallen from twenty-four in 1971 to eleven in 1993. Households have become increasingly reliant upon wage labour, with young males often away for long periods. Whilst migration within the subcontinent is by no means a new feature, today migrants are to be found in Taiwan and other distant locations.

There are distinct social reasons for the particular way in which the animal husbandry system has shifted. Despite the significance of migration, many men still try to retain links to Melemchi resources. Fathers feel that it is essential to pass herds and grazing to sons. Consequently, at least for the period under review, older men who find it more difficult to find work elsewhere are able both to generate a cash income and build up capital to pass to the next generation. At the same time, successful migrants often return with relatively large funds and are able to purchase land or animals as they establish new homesteads.

One of the problems with this modification of the animal husbandry system has been the increasing concentration of livestock within the 'middle zones' of Nepal, especially near villages. This is a universal problem for pastoralists in so far as increasing demands on household labour, including the education of the young, have meant that fewer persons are available to use pastures away from the villages in other ecological zones. This is exacerbated in the case of Melemchi as the village is within the confines of the Langtang National Park, which has increasingly acted as an administrative authority with respect to some land use practices.

Like the previous case from Morocco, Melemchi is interesting in so far as it documents the gradual readjustment of animal husbandry practices to changes in the broader economic and political environment of the locality. There has been little direct state intervention in the agricultural system, but nevertheless the changes have been profound. Unlike the previous example, recent history has been built around the role of migration and the currently pervading view that, ultimately, men return to the village to marry and establish a family. Simply projecting this trend forward might well cause considerable concern for the sustainability of the livestock sector because of a shortage of locally available grazing. However, this is not necessarily a useful exercise because the circumstances indicated here have shown how, in response to new conditions, the community modifies its agropastoral activities. We have to say that the form that these might take over the next fifty years or so remains uncertain, and could just as well see a reduction in grazing demand as the more normally predicted increase.

Developments in crop husbandry have also played a major part in the transformation of mountain areas. As Stevens (1993) notes, in mountain areas changes in cropping are often closely linked to livestock management due to the production of fodder, especially for the winter. In an example drawn from the Khumbu valley in Nepal, he describes how crops such as buckwheat, barley, the potato and radish, selected over many seasons for their hardiness in respect to altitude and disease resistance, have all been important components of the agro-ecosystem. However, over the last fifty years, this area has seen a distinct shift from the growing of buckwheat, especially at high altitudes, to a predominance of the potato. This has been part of a slow evolutionary process that is often ignored by many development agencies. Potatoes were probably introduced into the area in the mid-nineteenth century, but it is only in the period since the 1950s that potato has come to dominate the cropping pattern. In addition, the significance of this change varies from village to village. Stevens is also rather sceptical about the association between the introduction of the potato and population growth, a linkage that has been discussed in Chapter 6 and extensively debated in the context of the Alps (Netting 1981; Viazzo 1989) and proposed for this region by Fürer-Haimendorf (1964).

Much of the discussion of this process revolves around the different varieties of potato that have been introduced because each has distinctive growing, storage and taste characteristics. Some varieties are known for their ability to withstand drought or to be resistant to blight infections and others can give high yields under certain conditions. The length of the growing season is also significant for survival at high altitude sites. Also, from a taste point of view the preference is for a variety that does not have an excessively 'watery' taste. The introduction of the potato was a major innovation in the region, not only because it led to an overall increase in food production but also because it is associated with agricultural intensification. The amount of land required to meet household requirements is thought to have fallen as a result of the productivity of potatoes versus other crops. It is further claimed that many abandoned terraces and, in some cases, abandoned houses result from the increase in potato cultivation (Stevens 1993). This was because any fields that were relatively marginal for successful potato cultivation were withdrawn from agriculture, despite an increase in population.

The significance of the potato crop to the regional agriculture can be judged by the fact that regulations were in place in the 1950s to prevent the spread of blight, a serious disease for potatoes. According to Stevens, the varieties introduced in the 1970s (such as the 'yellow' potato) were brought into the area by individuals who had 'discovered' them whilst working elsewhere. Once again, it is clear that a combination of significant increases in yield and disease resistance made these varieties attractive. In the 1980s the so-called 'development' potato (*riki bikas*) appeared, again apparently the result of an individual obtaining some from an agricultural extension worker in a neighbouring area. It is also interesting to note that Stevens examines the increasing importance of the potato in relation to the declining significance of buckwheat. He argues that the conventional explanation of increasing population and the relative productivity of the potato do not explain the fact that the geography of crop change does not always match up with the available evidence on population growth. There are some areas where land shortage appears to be present but buckwheat is still grown! Another possible explanation lies in the fact that a strong market for potatoes has existed for several decades. Potatoes have been exported from the area and also the growth in tourism can be associated with a marked increase in local demand. However, during his investigations in the late 1980s Stevens was not able to find direct evidence that the decline in buckwheat was a result of increased opportunities for cash sales of potatoes.

As noted above, this example of potatoes from the heart of the Himalayas placed great emphasis on the initiatives of individuals rather than agency activity. Stevens notes that most farmers seem happy to experiment with new varieties rather than distinctly different techniques, although he cites a few cases of the introduction of micro-irrigation through plastic pipes.

As an alternative viewpoint on crop cultivation we can use the example provided by Negi (1994) who examined the effectiveness of new grain high-yielding varieties (HYVs) in a mountain environment in the Kumaun Himalaya, northern India. Despite a long-established history of terraced agriculture, the area produced only about half of its grain needs, the remainder being purchased from the lowlands. Under the successive Indian development plans HYVs have been introduced to the lowlands and, more recently, attempts have been made to spread these varieties to mountain farmers. The experiment suggested, as in the lowlands, that the HYVs achieved good results under irrigation and with appropriate inputs but in ordinary rainfed areas there was little difference in yields. Moreover, the poor production of straw under the new varieties was a significant drawback as crop production is closely integrated with animal husbandry. All in all, the mountain farmers were not impressed by the varieties being promoted by the Department of Agriculture!

Another area of rapid development in some mountain zones is the introduction of new tree crops. A good example is the introduction of fruit tree cultivation that has become a significant feature of many mountain communities, including the Atlas, the Himalaya (Plate 7.1) and Indonesian Highlands (Kreutzmann 1988; Suryanata 1994). Of particular interest is the way in which tree crops link with other activities within a mountain economy. Tree crops have long played a significant role in the agro-ecosystem in parts of the High Atlas. Along with pastoral activities and the cultivation of grains

Plate 7.1 Grafting fruit trees, northern Pakistan

and vegetables, many households possessed perhaps a plum, cherry or apricot tree. At the same time stands of walnut trees (*Juglans regia*) provide an important source of cash income. These trees can be found above about 1,200 metres, alongside the irrigation channels (*seguias*), or on steeper slopes that are unsuitable for cultivation. The walnuts, harvested in December, can be stored and sold to visiting buyers or at the local market. In addition, once they are no longer productive, the trees may be felled for timber, which is in high demand for woodcarving. This has been boosted in recent years by the interest shown by tourists in wood products.

The French introduced commercial production of fruit trees including cherries, plums, apricots and apples, although originally most of this was located in the cooler northern areas. In addition, tree planting has been accepted as a useful adjunct to programmes aimed at reducing soil erosion. For instance, in the Rif Mountains, Boujrouf (1996) writes of the promotion of fruit tree cultivation as part of the DERRO project. With government assistance in the task of selecting suitable cultivars, tree fruit production, particularly apples, has expanded dramatically. Orchards have become a common sight on the *dir* (the transition zone between the plains and high mountains), especially in the Middle Atlas. However, since the 1980s there has been a rapid spread of apple tree cultivation into some High Atlas valleys, particularly where relatively good transport exists for marketing. A

Plate 7.2 A small orchard, Reraya valley, Morocco

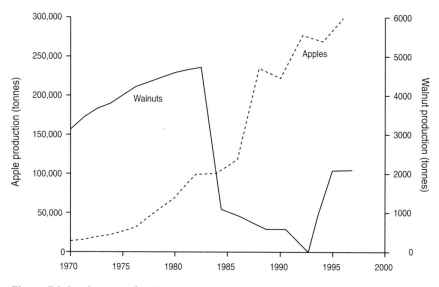

Figure 7.1 Production of walnuts and apples, Morocco
(from Parish and Funnell 1999)

particularly noticeable example of this is the Reraya valley, south of Marrakesh (Plate 7.2). In a detailed study of the village of Imlil, Miller (1984) notes a small number of fruit trees such as plum, etc., but a decade later Dougherty (1994) wrote that apple trees far outnumbered walnuts, a trend that is confirmed nationally in Figure 7.1. For the farmers, this represents a distinct shift in terms of cash crop income but not an entirely new direction of production. They have benefited from the work on varieties carried out at the national level although no specific extension programme has been established for this area. Currently, the 'red' and 'yellow delicious' varieties are in favour due to their hardiness in the face of spring frosts and their desirability on the market.

There is no doubt that apples do offer some distinct advantages over walnuts which have been the main source of cash income derived from agriculture for many households. Currently the financial returns from growing apples are almost twice the income from walnuts. Also the fruit trees begin to provide an economic return more quickly than walnuts, usually after three or four years. In the high valleys this is a significant factor because of losses due to extra severe spring frosts. When walnut trees die apples now replace them. Another advantage, discussed by local farmers, is that although apples need to be planted on better soils than walnuts they can be planted successfully on the downslope edges of terraces, thus helping to stabilise soils. However, an interesting evolution can be observed over the 1990s in that, from households planting a few trees along the edge of terraces, there are many more instances of little orchards now being planted, although these are usually undersown with barley (Parish and Funnell 1999).

The commercial cultivation of fruit in mountain regions represents an interesting example of the 'niche' production mentioned by Jodha (1997). In tropical or subtropical areas, the highlands represent areas of relative coolness and so are suitable for more temperate crops which find a demand in the increasingly 'westernised' urban centres, and, in some cases, a high value export market. Whereas most of the development of mainstream cereal or livestock farming has meant that mountain farmers cannot compete in national or international markets, these 'niches' offer significant opportunities. The High Atlas example also indicates how mountain communities have absorbed new cultivars into their agro-ecosystem, for the most part with little direct assistance from the state. Unlike the potato example, the bulk of the produce is sold for cash and contributes to household income. Moreover, the cultivation methods initially demanded little more than extending existing skills, but once trees are planted in orchards then new procedures, including the question of spraying, become important. In the High Atlas case, these are only just being considered.

In many plans for the development of mountain agriculture, agencies place considerable emphasis on encouraging the production of high value crops. In Thailand, for example, many forested upland areas are being cleared, not for swidden but for crops such as cut flowers and cabbages. Some of this development involves locals, but more often than not it is initiated by lowland capitalist interests encouraged by the Thai state. Similar projects exist in the Andes, Karakoram and the East African highlands where flowers, beans and other horticultural products are now widely cultivated. However, whilst many of these investments indicate the onward expansion of the market for foodstuffs, there is another area of production which tends to receive less publicity in the textbooks, though its impact on local incomes and the environment is enormous. This is the production of drugs. Farmers in south-east Asia, in Pakistan and Afghanistan, the Andes and the Rif Mountains collectively produce most of the world's main narcotics. Opium and heroin provide significant incomes for their producers in Asia, coca for farmers in Bolivia, Peru and Colombia, and cannabis (*kif*) has become a major product in the Moroccan Rif. In all these areas, the production is actively discouraged and punishment can be extreme. On the other hand, it is clear that factions of the elite in these countries are closely involved in the trade. The Federal Bureau for International Narcotics and Law Enforcement of the USA generates much of the pressure for control and eradication and is supported by the governments of other 'rich' countries. The result is that behind many of the attempts to introduce 'legitimate' cash crops to mountain areas lies a bigger issue – that of drugs control. This is particularly true in the Andes and in Thailand.

Moore *et al.* (1999), in a paper entitled 'Kif in the Rif', neatly pose the problem of drug production, and explore some of the background and impact of the cultivation of cannabis (*kif*). As in many other narcotic-producing areas this crop is part of the traditional cultivation system which, through the global linkages now possible, has become extraordinarily valuable. They report that in 1996 some 74,000 hectares of the Rif were under cannabis and the most rapid expansion has been on marginal land (and, in part, also the most inaccessible for obvious reasons). According to a study by Laouina (1995) it is estimated that cannabis provides an income for about 200,000 persons, and it is claimed that its cultivation and trade now represents the biggest single source of foreign exchange for Morocco, roughly some 60 per cent of legal exports. In addition, the almost insatiable demand, coupled with the relatively limited alternative sources of income, means that unsustainable cultivation methods are often employed. Above all, however, it is the economics of production that are so telling. Moore *et al.* provide figures which show that the net profit (Dirhams (Dh)/hectare) for cannabis is in the order of 27,700 Dh for leaves and 46,500 Dh for resin, whereas olives could produce 10,000 Dh and cereals 2,600 Dh. Only tomatoes potentially produce a near alternative at 20,000 Dh. However, the other issue concerns the size of the initial investment. For a hectare of tomatoes an input of 40,000 Dh is required whereas for cannabis the equivalent sum would be 6,000 Dh. Put frankly, in economic terms there is no choice!

Clearly this raises complex issues that closely link environmental management, rural development and social behaviour and the influence of international agencies. It is a case where the market does not give the right signals, but, as experience in Thailand illustrates, attempts at eradicating production have met with violent resistance. Although programmes have been able to reduce narcotic production, not only does this offend certain traditional values but it is also a significant loss of income to relatively poor producers.

This section has examined a small sample of some of the transformations taking place in the agriculture of mountain areas. Many more examples could be found but the emphasis has been on those initiatives that have been the result of farmer response rather than 'clean slate' development programmes. Mountain areas remain relatively poor in many countries and this reflects in part the preponderance of agricultural activity in a world in which the most remunerative income generating activities are to be found in the service sector or industry. Thus it is important to examine the extent to which these activities have a direct impact in mountain areas.

THE DEVELOPMENT OF FOREST RESOURCES

The traditional uses to which forest resources are put were outlined in Part 2. Here we are concerned with the changes to these traditional uses as a result of exploitation by the state and other agencies. In addition, we introduce the conflicts that arise as a result of these changes in usage patterns. Foremost among the questions for policy-makers and developers are decisions concerning where, how and why intervention is needed (Chauvin *et al.* 1997). The issues of deforestation, and the protection of forest resources, are highly emotive and political, and a focus of policy initiatives will be discussed in Chapters 8 and 9. Other recent overviews of the development of mountain forests include Hamilton *et al.* (1997) and Price and Butt (2000).

The commercial exploitation of forests by agencies other than the immediate community has a long and colourful history. Thirgood (1981) and Meiggs (1982) have both examined the nature of timber resources and deforestation in the Mediterranean Basin, which has occurred throughout later prehistory and historical times. Exploitation on a large scale in this region had already begun by the Bronze Age when metal production had been instigated, for example in south-east Spain. During the Classical period, historians refer to the clearing of forest and development of different agricultural systems. In addition, the more intensive mining, and most importantly the development and maintenance of extensive fleets by both Greeks and Romans, required constant supplies of a variety of timber. Different species were used for different purposes, for example suppleness versus strength. Although writers such as Plato and others refer to intensive clearance, there is little indication, according to Meiggs, for concern over degradation until somewhat later. Supplies certainly do not seem to have been exhausted, although reference is made to the strategic importance of timber stands on islands (Cyprus, Corsica) and in the eastern Mediterranean (Turkey, Lebanon, Syria).

Pliny refers to various checks on the wholesale decimation of forest. The economic value of other products such as resin, pitch, and cork meant that some conservative strategies were in operation. Religion (protection of sacred groves) and distance (the uneconomic transport of low value goods over long distances) also served to protect some areas. The collapse of the Roman Empire was associated with forest regeneration as cultivated land was abandoned, so that by the fifteenth century the Byzantine Empire was able to use resources very near to Rome to construct and repair major basilicas. By the Renaissance, however, the growth of shipping and naval based power, based in the Italian city states of Genoa, Venice and Pisa, again required vast resources of timber for shipbuilding. Regions with substantial timber resources were therefore exploited. Larger ships needed stronger and larger timber.

A number of travellers' records refer to the substantial clearance of large areas of forest. Papal lands were sold off in Spain, for example, in order to generate revenue, and these lands were stripped of their timber. In addition, land tenure laws permitting the claim of ownership of land to those responsible for clearing it were also in place. By the nineteenth century, many parts of the Mediterranean had been cleared. In Turkey, the southern Taurus mountains were largely cleared but excellent forests survive to this day along the Black Sea coast, protected by their relative inaccessibility and poor transport networks. The effects of the two world wars, which saw major clearance of forests for strategic reasons and in order to fulfil the needs of local and invading populations, led to the realisation of the extent of clearance and the beginnings of a change in attitude. Certainly the replacement of charcoal and wood with coal relieved the pressure on forests, but growing populations, the need for more agricultural land and increases in numbers of grazing animals, especially goats, restricted natural regeneration.

The intervention of the state has in most cases caused conflict with the local population. This is due to the misunderstandings on both sides of the different perception of forest values, and the different priorities and timescales to which national governments and local farming communities operate. Conflict commonly arises from the exclusion of local populations from the forest resource, which as we have seen is an integral and critical part of a balanced sustainable livelihood. All too often, local farmers have been chased off lands to which they held traditional usufruct rights, although these were not upheld by law. They then watched as state agencies cleared the forest and replaced it with plantations of commercial species. The farmers then observed the forest being decimated by logging companies holding licences granted by the same government which had excluded the farmers on the basis of the irretrievable damage they were perceived to be causing. Consequently those areas retained by farmers were placed under greater land pressures and became severely degraded, and farmers encroached into state reserves due to land hunger and the need for forest products. Such has been the experience of Nepal and other areas of the Himalayas, where colonial expansion by the British caused widespread, substantial felling of mature forest to provide timber for construction, particularly for railways.

It is equally true of many south-east Asian mountain regions at the present time, where conflicts arise from the different needs and priorities of the state and hill tribes. This becomes increasingly politicised by the intervention of military and business interests, national security issues, and international concerns over biodiversity and other ecological issues relating to forests and their uses. The introduction of a ban on logging in Thailand in 1988 in response to severe flooding in southern Thailand was based on an immediate reaction to misinterpreted data (see the discussion on the causal linkages between deforestation and lowland flooding in Chapter 8). This, despite apparently bowing to international demands for logging controls for ecological reasons, actually played into the hands of business and military elites who had interests in importing timber supplies from neighbouring countries – Burma, Laos and Cambodia. By 1989 there were eighteen concessions for teak cutting in Burma, and by 1993 forty Thai companies were involved (McKinnon 1997). Illegal logging continued in the border regions under cover of the greater number of convoys travelling on newly constructed roads.

Meanwhile, the local population was increasingly at the 'wrong end' of the law by continuing to try to sustain a living by traditional swidden agriculture in areas from which they were now excluded. The increased presence of the military also made life more difficult for them. In addition to this, the state licensed a number of companies to clear land for plantations and permitted logging on a rotational basis over a wide area of Thailand's forests. The emphasis on eucalyptus plantations for export of wood chips, together with the permitted logging, meant that Thailand took a relatively ambivalent stance at the Rio Summit in 1992 concerning the conservation of the world's forests (England 1996). Their own record was sufficiently poor for them to wish not to draw undue attention to it.

The most significant element in the modern development of mountain forests is the concentration on single, high value commercial products, particularly timber. Government agencies and private companies have developed timber stands in both temperate and tropical areas, so that today timber as a key resource is concentrated in mountainous regions. For example, according to Tuchy (1982) 50 per cent of Austria's productive forests are to be found in areas of steep slopes. This also indicates that forest plantations are important, not just for timber *per se* but also for the stabilisation of slopes, avalanche protection and the provision of biodiversity (Hamilton *et al.* 1997). In the USA, many forest lands occur on federal or state-owned land, so the costs of these other 'services' are borne by the taxpayer. For example, in Colorado, USA, most of the forests are administered by the federal Forest Service which has developed a number of policies aimed at providing a sustainable timber yield at the same time as providing water catchment protection and recreational amenity. For this purpose, protection from forest fires is a major endeavour. Elsewhere, private entrepreneurs are encouraged to manage their forest holdings in such a way as to ensure that these extra functions are not ignored, but, of course, the private landowner does not benefit directly, and without appropriate laws or transfer payments there is little incentive to manage the forest accordingly.

Clearances for plantations tended to introduce large tracts of exotic species, particularly eucalyptus, which is fast growing and commercially viable for wood chip and pulp production. In some areas, such as Thailand, development initiatives and farmers' own actions have resulted in the increase of planting on private land. Lychee orchards are grown as a cash crop in the lower hill regions of Thailand, whilst in the upper regions tea plantations have been set up by companies for export, but also by local farmers to supply the local market. The additional planting and nurturing of fodder and fuelwood species is discussed further in Chapter 8.

There is also increasing pressure to develop non-timber resources (Price 1990; Edwards 1996). In traditional usage patterns a wide range of products was normally extracted from forests (see Chapter 4), but today international markets exist for nuts, syrup, ratan, resin and, of course, specific medicinal products. According to Hamilton *et al.* (1997) such developments are found in the highlands of Thailand, India and Indonesia whilst on the lower slopes of mountains in North America the production of maple syrup is common. In Italy, in the Val di Taro, Parma, mushroom collection is now organised to provide not only local demand but international sales. This has involved not just technical issues for preservation by drying but also complex legal arrangements, for instance the registration of a recognised label. Today, the dried mushrooms are worth at least ten times more than the fresh product. This demand has raised local incomes and created employment (Zingari 1999).

In the European Alps, the problems facing the effective use of forests are primarily economic. It was mentioned earlier that the added value to the farmer of forests as 'services' was negligible. In order effectively to maintain a healthy forest, a certain amount of tending, thinning and rotational clearance is needed. Naturally this is costly, given the labour intensive nature of the work, the impracticality of using much machinery, and the inaccessibility and the need to minimise damage. Where there is little hope of recouping the costs, a more extensive management regime develops, or they are abandoned (Meyer 1984; Price 1987). Both Meyer and Ott (1984) refer to the possibility of a long-term investment that may be recouped on the global market in the future when other timber supplies are scarce. However, this requires effective management now, and there is no guarantee that such markets will develop, or will be economically viable.

However, it is certainly true that the need for forests is increasingly in the service sector, and increasing proportions are being accorded conservation or preservation status. This is difficult in many areas – in central Europe for example, large tracts of forest are being damaged by long-range pollution (especially acid deposition); in the western Carpathians, it is estimated that 85 per cent of trees, especially the evergreens, are affected (Jansky 1999). Although the national policy in this region is ecologically biased, it is difficult to implement such protection measures where pollution is caused by other nations and cannot be diverted.

Although many developed countries have designated much of their mountain lands, where the densest forests are generally located, as public and protected, such areas cannot be treated in isolation. Forests cannot be left to themselves, but need considerable technological, ecological and economic nurturing to remain healthy, even when the main pressures for cutting for fuel, etc. have been lifted. Attitudes to management have changed from an isolated, single product, mono-species, and exclusive approach to a more sensitive, community based approach. This tries to incorporate not just the government-defined national needs, but also those of its marginalised peoples, and utilises their knowledge of the diversity, potential uses and sustainability of mountain forest resources (Hamilton *et al.* 1997).

MINING ACTIVITIES IN MOUNTAINS

As a result of their often complex geological history, mountains provide particularly important sites of mineral deposits. The topography of steep valleys and eroded slopes produces exposures that have been very important in the process of early resource discovery. Consequently, mining has been and remains one of the principal sources of economic activity in mountain areas. For example, much of the mineral wealth of the countries of the former USSR is to be found in mountain areas, whilst the Andes provide some of the principal global sources of copper, lead and zinc. Many small developing countries are highly dependent on the minerals from their mountains, for example, Papua New Guinea. This is not a recent development, as we explored briefly in Chapter 4. Whilst such investment is often hailed as a critical element in the process of economic development, the scale and impact of the modern activities in mountain areas has become a particular source of concern as society is more aware of the negative consequences. At the same time mining companies themselves are slowly becoming more sensitive to the questions of environmental damage.

Mining is one of the most powerful anthropomorphic agents of landscape destruction. Denniston (1995: 34) states that 'Of all the economic activities in the world's mountains, nothing rivals the destructive power of mining.' Few areas are without signs of former exploitation and today many areas continue to undergo severe damage. Fox (1997) provides a broad overview of the process in mountainous regions and notes that the impact on mountain areas depends upon the stage of the process. Generally little destruction of the landscape takes place at the prospecting stage. Many prospectors find that the mountain topography provides good exposures that aid the task of geological mapping and assay work. The inaccessibility of many areas plays a crucial role but even this has operated in the past to the advantage of prospectors who were able to preserve the secrecy of their findings. On the other hand, Hughey (1997) reminds us that not all the damage is physical and suggests, unlike Fox, that the prospecting stage has also socio-economic dangers in that the legislation in many countries fails to ensure that the local populace has a serious opportunity to question and evaluate the proposals. The resources available for them to do this are simply not available and large mineral companies have a notorious record of influencing local opinion with their publicity, much of which has been shown to be inaccurate.

Once the extraction and concentration operations commence, then land-scape damage almost inevitably follows. Vegetation is cleared, overburden removed and the spoil from tunnels and extraction deposited nearby. In addition, most mining operations use large volumes of water, the mountain areas often providing a relatively abundant supply. However, the chemical wastes from ore tailings can be particularly toxic as they contain arsenics, xyenides and other heavy metals, with devastating results for riverine life and vegetation. Soils in the vicinity of the mine become poisoned, and water downstream unusable. This is apparent in the mountain regions of eastern and central Europe where mining extraction policies have favoured the uncontrolled destruction of the landscape. The result is a remodelled land-scape, which may be quite localised in some cases, but in others the impact is huge (for example, the damage surrounding some of the large mines in the Andes). Whereas small adits may have been the norm in many earlier workings, the large opencast operations that are favoured today have a major landscape impact. In many instances, these mines are part of a multinational enterprise that, with the support of the domestic government, brings the full force of capital and technology to bear on the local environment. A good example is the Ok Tedi opencast copper mine in the Star Mountains of Papua New Guinea (Denniston 1995). Over a period of eleven years this mine exported 600,000 tonnes of ore concentrate to Japanese smelters, and when the operation is complete, a mountain that previously rose to 2,330 metres will have been demolished. Denniston notes that this mountain was considered sacred by the local people, who, needless to say, had little impact on the decision-making process. The PNG government would argue that without the exploitation of this resource there could be little social and economic advance for its population. The dilemma is obvious, and even if sophisticated environmental impact studies are undertaken, there is no obvious way of weighting the 'cultural value' of this mountain.

In some areas, however, resistance to the exploitation of mountain mineral wealth has had some success. A good example concerns the plans by the Ecuadorian government and the Mitsubishi Corporation to develop a mining enterprise in the Toisin Range in north-west Ecuador (Zorilla 1999). In this instance the mining project aims to exploit an area of cloud forest primarily for copper, but the area has high levels of biodiversity and already neighbouring regions have seen the forest removed for plantations. In 1995 a group of local activists and others from Ecuador formed a body called DECOIN to campaign against this further encroachment into the moun-tain region. Zorilla describes in detail the various moves that brought together local farmers and representatives from the government and mining corporation, offering the first chance of negotiation. DECOIN astutely used the media to publicise the problem and through the increasingly wide network of contacts obtained a copy of an Environmental Impact Report (prepared in Japan) which set out very clearly the extent of the likely damage.

At one point the campaign involved violence as the mining camp was burned down, although all the equipment was removed and sent to the municipal authorities. The subsequent charging of the movement's leaders further exacerbated the public outcry against the project to the extent that further work on the project has been postponed.

Nonetheless, we should be aware that despite the negative effects on the landscape, mining has offered some mountain communities a new and relatively higher source of income than farming. In the case discussed above, one of the principal concerns of the opposition movement was to seek suitable alternative sources of income, largely through improved sustainable agriculture. In many mountain regions mining has been the only reason why settlements have occurred. A glance at the map of the Rocky Mountains in the USA indicates the number of settlements which originated as mines. In many instances, a mine may be opened in such a remote area that all the workers are effectively immigrants. Some may have travelled enormous distances – for example the labour movements associated with the opening of the US West. In other cases, the first sources of labour may be from farm workers in the locality, but this is very variable for several reasons. First, farmers may be unwilling to take part in mining because it is considered both a dangerous and low status job. In some cases, the employers require men with previous skills and this has become more the case as the technological sophistication of mining has increased. In this instance, premium wages are paid to attract labour from elsewhere.

It may be supposed that the development of mining might reduce migration from mountain zones. In fact the position is very varied because it depends upon the nature of alternative employment and the social status attached to the job. It has already been noted that mining did not attract some residents of mountain villages in the Italian Alps, but nonetheless mining communities became established in mountain areas and, in some cases, were absorbed into the local population. Mining may also bring with it the associated infrastructure, particularly roads. In earlier times infrastructural development may have been sparse, but more recently companies investing in mountain locations have built substantial housing, water supplies, schools and medical facilities which have generally become available for the populace of the area.

Therefore the balance of advantages for a mountain community and for the mountain environment as a whole is not straightforward. Taking the example of mineral development in PNG, for those who have been employed, directly or indirectly by the mining enterprise, significant benefits have accrued, at least over the short term; for those who lost land or other resources such as forest in the process, the result has been devastating and it is only in the recent past that mining companies have become mindful of their duty to compensate for this loss. As the examples illustrate, the literature is replete with accounts of state backed plundering of land (Dunaway

1996; McNeill 1992). Mining companies often became a 'law unto themselves' in these remote mountain areas, and providing that the host state receives revenue, little is done to compensate the mountain communities.

However, as Fox points out, in more recent years there has been a move towards a better recognition of both environmental and social responsibility. Subsequent debates about policy for mining in mountains point to the fact that larger companies are gradually introducing appropriate procedures for checking on the environmental impact of their operations. National governments are beginning to recognise their role in implementing appropriate legislation, and international agencies are well aware of the issues. As yet there is no internationally accepted concordat on mining and it remains very difficult to police the activities of smaller companies. Moreover, the biggest question remains the sustainability of mining operations and its long-term impacts in the mountain region. As Dunaway (1996) neatly argues for the Appalachians, not only were the local resources removed and the environment damaged, sometimes irretrievably, but also the mining activity itself was ephemeral. Once a mine is worked out (or the market collapses) the facility is deserted, leaving only the abandoned railway, rusting and dangerous infrastructure and a polluted landscape which few poor governments can afford to attempt to rehabilitate. One source of redevelopment of the mining infrastructure may be investment in tourism to visit the old mining environment.

WATER DEVELOPMENT AND HYDROELECTRICITY

As almost every elementary geography textbook will note, most of the world's major river systems have their headwaters in mountain regions. In many cases, as a result of political geography, water from the same mountain range may supply several different countries with consequent problems both in the headwaters and downstream. The hydrology of these mountain catchments has already been described briefly, but in this section the focus is on the impact of headwater development for the mountain area itself. And that, in a nutshell, is the issue! Whilst mountain communities exploit water for irrigation, power and consumption, it is the demands on this resource by populations outside the mountains that have increasingly determined the nature and pace of its exploitation. In parts of the world where water is in short supply, for instance in the Middle East, the political control of key headwater areas underlies much of the conflict. For the mountain communities themselves, this exploitation has seen the construction of dams and associated hydroelectric schemes and their reservoirs, often leading to the enforced resettlement of population, or at least the loss of valuable agricultural land. Equally important, however, is the fact that once a catchment is

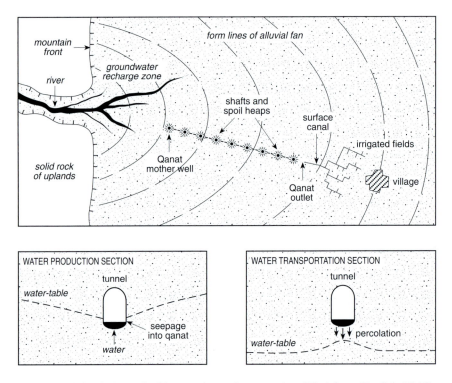

Figure 7.2 The design of a *khettara* (*qanat*) system (modified from English 1968)

linked to a particular water development scheme there is a strong likelihood that it will be the subject of conservation legislation. This usually means that the residents of the area no longer have free use of the land, and forests are protected, or often planted, on what might have been open rangeland. This itself has been the cause of disaffection between mountain and lowland people.

Morocco provides an example of the way water demands of the lowlands impinge upon mountain life. The lowlands are encircled by the Atlas range and the Rif mountains from which most of the main rivers flow. The use of these mountain catchments is not new. In the twelfth century the city of Marrakesh developed with water supplied from the High Atlas some 40 kilometres to the south. This utilised the *khettara* system, which is very similar to the *qanats* of ancient Persia (English 1968). As the diagram shows (Figure 7.2) the *khettara* is an underground canal, dug about 10 metres from the surface. In the foothills of the mountains they intercept the water-table, and a slightly steeper gradient permits a flow far out across an otherwise dry plain. According to Pascon (1977) about fifty were constructed from the

254

mountains to the city in the twelfth century and the water enabled some 15,000 hectares to be irrigated.

Although this system fell into disuse with the fluctuating fortunes of Marrakesh, individual surface canals (*seguias)* were later built. However, it was the French colonial machine that set in motion, after 1912, what became a major programme of water development based upon the mountains. In the first instance, the objective was to use the water for irrigation, as the French *colons* or agricultural settlers were unable to compete without this water. The French played heavily on the image of water development in California to promote new crops and the development of irrigation 'périmètres' (Swearingen 1988). This theme was repackaged after Independence in 1956 and became the 'Million Hectare' target. At the same time urbanisation and industrial development has seen a big increase in the demand for electricity and also for domestic drinking water, and from 1967 a 'Politique des Barrages' has meant that a considerable proportion of the development budget has been invested in these schemes. Altogether, Moroccan economic development has been heavily reliant upon the exploitation of water resources, mostly from the mountain areas. For example, one of the first major dams was constructed in 1929 on the Oum er Rbia, with a volume of 32 million cubic metres. However, the scale increased rapidly, so that by 1987 the Moulay Hassan on the Lakhdar in the High Atlas was constructed with a volume of 9,500 million cubic metres!

To explore the impact on mountain areas it is useful to trace the development of the Lalla Takerkoust dam on the Oued N'fis which flows north from the High Atlas to the Tensift river in the Haouz plain. This project was initially constructed between 1929 and 1935 with a capacity of 50 million cubic metres and then in 1980 this was enlarged further. The electricity from this scheme has been a crucial element in the development of irrigation (pumps, etc.) and industrialisation in Marrakesh and the region. It also plays an important role in flood control. The N'fis river has a regime which is markedly peaked with a spring snow melt and so the reservoir is able to provide a relatively even flow throughout the year. Downstream, the largely privately owned land has been developed for cereals or citrus and olive plantations. Upstream the land consists primarily of mountain slopes which are either grazed or, where possible, terraced for typical mountain agriculture. Apart from the loss of land in the creation of the reservoir, the other issue concerns soil conservation. The principal reason for the increase in the capacity of this dam was the fact that it had lost at least 25 per cent of its capacity through silt accumulation, a very common situation. Consequently, from the time of the French a key policy has been to attempt to limit soil loss in catchment areas (Dressler 1982), not so much for the farmers of those areas but for the sustainability of the water programmes. These policies have been enforced quite severely at times and there is no direct compensation for loss of livelihood in this situation.

As noted earlier, forests have been planted and grazing controls implemented with varying success, but the general tone of the policy has been,
until recently at least, to assume that the problem lies with peasant agricultural practices in the mountains. Interestingly a recent detailed study of the
upper N'fis above the site of the barrage points to the problems of this view
(Maselli 1996). There is no doubt that the landscape in this basin is
degraded, with both vegetation and soil condition indicating a gradual
decline in species richness and soil fertility. Maselli notes that the continuation of grazing and agricultural practices developed long before the present
do not augur well for recovery. However, this apparently typical example
of the environmental damage resulting from peasant practices is misleading.
A reconstruction of the vegetation history of the basin shows that whilst
the environment is 'fragile' it had not suffered severe degradation until the
1960s when there was large scale commercial tree felling in the area. Far
from being the result of local farmers, it was the officially authorised action
of entrepreneurs from Marrakesh desperately seeking timber for building
and firewood. Given the very rapid and extensive clearance, there has been
little chance that this ecosytem could recover what resilience it had, and
local households have little option but to continue their farming system,
albeit with some modification. Consequently, soil loss has increased.

The factors involved in the Moroccan case highlight again the privileging
of the lowland over the highland communities and this can be seen on an
even grander scale in the many examples elsewhere where large dams have
been built in mountains. One well-known case that has received considerable publicity is the Tehri dam in northern India (Bandyopadhyay *et al.*
1997; Singh 1997). This project began in 1976 and has a planned capacity
of 3,539 million cubic metres providing flood control, hydroelectricity, irrigation and domestic water supply for urban areas. Several features of this
example add to the points made above. The resistance of the population
faced with resettlement is one notable feature. By 1999, the waters of the
reservoir were set to flood the town of Tehri and a new settlement has been
constructed. Critics point out that this is inadequate compensation for the
expected loss, not only of the town but of some ninety-two villages. One
of the outcomes of this movement has been to encourage various impact
studies that have revealed further complications. In this case the dam site is
located in a seismically sensitive area where there is a high probability of a
serious earthquake. At the time of writing, the dam has survived a quake of
the order of 6.9 on the Richter scale, but many worries remain. As a number
of mountain regions also have this characteristic then clearly the risks associated with dam building, especially very large dams, can be considerable.

The political consequences of increasing resistance to dam construction,
for both environmental and social reasons, are considerable but the debates
have become global with organisations such as the International Rivers
Network providing a critique of large dam building operations. In a review

of the positive and negative aspects of dam construction, the World Bank accepts that such schemes are controversial. Nevertheless it argues that in almost all the schemes considered the benefits have far outweighed the costs, including the costs of adequate resettlement programmes, environmental safeguards and other mitigating measures (World Bank 1996). It does recognise that the awareness of both environmental and social consequences, particularly those of population displacement, has become much greater than in the 1960s, and the Bank itself issued guidelines in the 1980s. However, as the report notes, many of the schemes in remote areas or involving minority groups (often in mountainous regions) have been managed unsatisfactorily with regards to these issues. In 1995, the Bank withdrew assistance from Nepal's Arun III project, in part for ecological reasons after co-ordinated protests. This has resulted in plans to develop a series of small dams in Nepal, particularly with the energy and water demands of Kathmandu in mind. Nonetheless, these projects still continue, as the example below illustrates.

Probably one of the most controversial highland water projects has been the Lesotho Highlands Water Project which aims to utilise the catchment areas of the Orange River to provide water resources to the main industrial area of South Africa (see Figure 7.3). The scheme involves a network of dams on the tributaries of the Orange River and the diversion of water into the river Vaal, so supplying the principal industrial region of South Africa. It is not a new project, being originally discussed in the 1950s before Lesotho became independent and when the rapid development of South African industries was becoming evident. At the present time one dam has been completed and another is under way, but there have been considerable objections. Both the Lesotho and the South African governments are firmly behind the programme, although recent reports have stressed that the estimates of water demand that favoured the scheme are no longer thought to be accurate. The highland area has been affected in a number of ways. For the first phase of the plan it has been estimated that there will be a direct loss of some 925 hectares of arable land and about 3,000 hectares of grazing land – small figures until it is realised that Lesotho is short of both of these land categories. Compensation remains a problem. In the first instance, those who lost land were to receive fifteen years' annual grain supply and houses were to be built. However, resettlement has been very slow, and with the increasing problems for the rural economy in Lesotho the chances of recreating a new farm based economy are not strong. Consequently, recent attention has been focused upon the possibility of creating rural business enterprises in the highland areas similar to those that have been tried in many other mountain areas. Unfortunately, given the fact that products from the Republic usually swamp the Lesotho economy, there is little optimism about this scheme. Nonetheless, there are benefits to both Lesotho and South Africa, the original agreement providing a royalty split of 56 per cent to Lesotho and

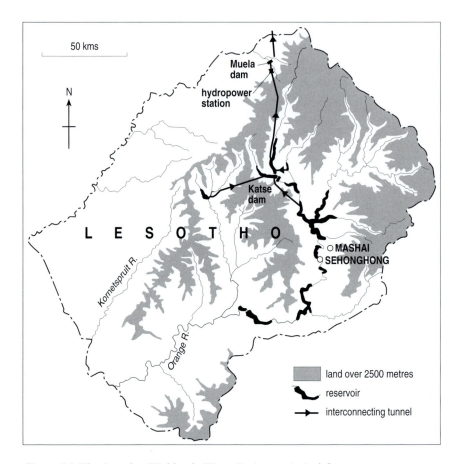

Figure 7.3 The Lesotho Highlands Water Project: principal features

44 per cent to South Africa. Net benefits are reduced significantly after paying interest on the capital borrowed for the scheme.

The debate about water development in Lesotho provides an interesting link with the earlier discussion of the political economy of mountain development. As was noted, the precise way in which mountain resources are exploited, and the balance between gainers and losers, depends crucially on the institutions that govern these resources. We can turn again to examples from Switzerland to note that hydropower development in the mountains has provided one of the key engines of growth for most lowland cities; it has also, to some extent, provided the resources for the support of rural communities within the mountains themselves. The legal authority responsible for the utilisation of mountain water in Switzerland is the canton that receives a royalty payment for the use of the resource. This institutional arrangement does not guarantee that the mountain communities automatically benefit from the use of the resources because there are examples where lowland companies have been able to use the system to secure supplies with very low payments to the canton. However, the institutional arrangements at least guarantee that, in principle, mountain communities receive some, albeit negotiable compensation. It is arrangements such as these, coupled with the strong political power, that are increasing being sought by mountain communities around the world.

Water rights and payments have an impact on more than just the mountain communities themselves. In the Swiss case noted above the financial arrangements for the provision of hydroelectricity by the cantons became a political issue in the wider debate about ratification of the Alpine Convention (Price 1999b). This Convention, which will be discussed in more detail in Chapter 10, was signed in 1991 but Switzerland only ratified it in 1998. This was because the mountain cantons thought that the Convention placed too great an emphasis on environmental protection at the expense of economic development. In 1996 a deal was formulated whereby the mountain cantons received more income from their rents for hydroelectricity as well as higher federal payments to communities who withheld future agreements to develop hydropower. The bargaining was in exchange for acceptance of the Convention. This example could only merge in a situation where the political structure had already provided the mountain communities with suitable withholding powers, not a common circumstance as we have already noted.

The development scenario posed by the Lesotho scheme suggests that the construction of dams for water development and hydroelectricity in mountain areas is facing unprecedented resistance. Donor agencies are showing an increasing reluctance to face the adverse publicity from environmental groups and others (especially NGOs) concerned with the social repercussions of such programmes. Following a conference jointly sponsored by IUCN and the World Bank (IUCN/World Bank 1997), a World Commission on Dams (WCD) has been launched. It is interesting that the current president was, until recently, the South African minister in charge of water affairs, and who before the end of the apartheid regime was bitterly opposed to the Lesotho Highlands scheme. The WCD aims to encourage rethinking about the strategy of building dams, especially in areas such as mountains that are environmentally sensitive. Although this offers the possibility of a new look at many large dam schemes, it cannot really address the political problem of the way in which environmental and social issues of mountain water development receive appropriate representation.

Probably the most well-documented programme of development for a river system in a high plateau or mountainous area concerns the Colorado

River in the USA. Graf (1985) provides a detailed account of the relationship between the basin water regime and measures to manage the water. At the time of writing, numerous disputes remain unsolved concerning the appropriate use regimes on this river. In many years, the water of this river system is literally all 'used up', despite a century of dam construction in which at least twenty major dams have been built. Foremost among these has been the Hoover Dam, commissioned in 1936, and the Glen Canyon Dam in 1964. The debates about the management of this system, built essentially to supply water and hydropower to the burgeoning cities of the south-west USA and also farming communities, continue in the form of an increasingly vociferous critique of big dams. In the last decade or so, in some catchments dams have been decommissioned, which introduces other problems – for example the management of large silt deposits.

Even with all the resources of a prosperous nation available, the institutional framework for managing these large basins remains unsatisfactory. The legislative and management structures have often failed to resolve, or even noticeably reduce, conflicts in a manner likely to ensure long-term 'best practice'. As the US literature reveals, even after the often interminable debates about construction the fluctuations in attitude and changes in the personalities leading the programmes, as well as ecological and economic events, have meant that there is little long-term stability. This is an important consideration for the analysis of mountain regions because it demonstrates clearly that water development has all the elements of a complex system discussed in Chapter 1. Consequently, the framework for analysing the impact of these schemes on mountain areas must have the capacity to handle such circumstances.

TRANSPORT AND COMMUNICATIONS

The topography of mountain areas, and often their associated remoteness from centres of population, means that systems of transport and communications play an important role in the life of their communities. Not only are some locations extremely difficult to reach, requiring the negotiation of very steep gradients, cols and narrow passes, but many routes are closed for part of the year due to snow. Modern engineering skills have made it possible to forge more direct routes and sometimes improved gradients that have allowed the construction of high-speed railways and motorways. However, all these developments remain subject to the hazards of avalanches and landslide, often despite enormous effort and cost to minimise the impact. For example, in the mountains of western China, routeways are regularly cut, and Hewitt (1997) cites the example of an area where, in a two-year period in the 1970s, there were a thousand landslides onto the railway system.

Plate 7.3 Lorries on the Karakorum highway

The development and control of particular routes into and through mountain areas has played an important part in historical events and continues to be crucial to both economic and political development. In the Andes, Karakorum and Alps selected routes were controlled initially by local groups who were able to exact a tax for safe passage. In the High Atlas mountains, these groups became particularly powerful because only they could ensure safe passage for the trade between the Sahara and the Mediterranean on which cities such as Marrakesh depended. Some of this control was associated with 'protection', but it also involved detailed knowledge of the pathways and safe routes at different times of the year. Control over an all year round pass could be a major element in the rise to power of particular groups, as the history of the Atlas tribes indicates.

Of course the engineering feats are tremendous. The tunnels and passes of the Alps and Rocky Mountains, for instance, stand as monuments to engineering skills. In Asia a series of roadways such as the Karakorum highway, between Islamabad (Pakistan) and Kashgar (China) and the route between Lhasa (Tibet) and Kathmandu (Nepal) (Allan 1988) were constructed with military requirements in mind, but often followed the path of pre-existing routes that date back many centuries (see Plate 7.3).

Elsewhere it has been the railways which have been given the greatest publicity for their cross-mountain construction, no more so than in the case of North America (Avery 1989). Whilst many writers emphasise the engineering wonders of the 'opening up' of the West, it should be noted that the mountains also posed enormous financial burdens on railways which were far from obviously going to make large profits from these exploits, at least in the early stages. White (1991) notes the political and financial scandals surrounding the companies in their early days. Whilst companies such as the Union Pacific, the Atlantic and Pacific, and the Southern Pacific (among others) all constructed routes through the Rockies, their principal goal was to link the east and west coasts of the USA. The Union Pacific made it first in 1869. Nonetheless, it is clear that these railroads made possible the development of the large-scale mining industries within the area, along with bringing large numbers of people to settle in the booming towns. Other towns along the railroad routes also grew in importance as a result of the wide variety of trades that grew up, each serving as a railhead for the surrounding area. Today, many of these are important cities within the mountains, though their railway function is now of relatively little significance. As far as the mountain regions were concerned, the main railway companies dominated the carriage of passengers and freight, at least until the 1930s when passenger traffic began to show a decline in favour of road transport. By the 1950s, the movement of passengers to the mountain regions had become dominated by the roads and the airways. However, as the mining industry and other bulk producers continue to develop at some sites, the railways remain a crucial player in the transport of freight.

The wider impact on the mountain areas from improved transport has been discussed by a number of writers (Allan 1988; Kreutzman 1991; Stadel 1992). While the difficulty of transport and communication acted to limit trade between mountain and lowland and through mountain routes, it also served to preserve, in some senses, the political and social integrity of the area. However, as pointed out already, improvement is a dual-edged sword, making it possible for mountain people and products to reach lowland markets but, equally, facilitating the movement into mountains of lowland products. Much of the tourist development discussed below would be impossible without these improvements. This is particularly true in developing countries where mountains have only recently been on the tourist agenda and where access can still be difficult. Whilst this is no problem for the keen mountaineer, it is a serious drawback to any attempt to widen the tourist market. At the same time the increase in traffic through once secluded areas produces noise and air pollution. This has been particularly apparent in the various passes through the Alps, but is also evident elsewhere. Stadel (1992), in his study of Brenner, examines these problems in some detail and, once again, it is a matter of considerable controversy whether these routes should be further upgraded to handle more traffic.

One of the most interesting effects of improved transport concerns the spatial variability of its impact. In mountain topography, where there are principal and tributary valleys, the routeways tend to follow the main valley, as this is usually the most cost effective. For settlements on this route the improved transport may have a beneficial effect, not simply through the improved access but because they become local magnets for population leaving more remote areas. This is particularly evident along the route of the Karakorum highway and also in the growth of settlements in the nineteenth century in the main Swiss valleys. Here, industry was attracted both by improved accessibility and, of course, by the availability of hydropower. Conversely, settlements in the other valleys lose population as part of the out-migration process when people make their first steps to the towns along these main routes. This has occurred in the Pyrenees with the construction of new roads. At the same time transport improvements should not always be viewed as simply highland–lowland linkages. One of the most important factors of roadbuilding programmes has been to link adjacent, but previously relatively remote areas. This has often meant tunnelling through a col, or performing some other complex engineering feat, but facilitates existing exchanges often using 'old' technology. Thus donkeys may be the principal means of transport but they will use sections of a modern road where this saves time. These improved intra-montane linkages, for example in the Atlas Mountains, may well prove to be a major factor in the development of new scales of social and economic co-operation.

Although most of the literature concentrates upon transport it is the developments in communications that perhaps hold the greatest challenge for mountain areas. It is often the case that whilst relatively remote settlements have simple roads or tracks, few have the telephone. This is not a luxury in an age when even small enterprises need to be competitive and the marketing of high value goods requires access to market information. Furthermore, as expectation of improvements in health and welfare rise, it is often important to be able to summon medical or other aid. This is particularly true in areas which are popular with tourists.

Of course, the possibility of radio links has been available for some years but telephone services have, until the last decade or so, been limited by the need for wired links. This is now unnecessary, either because of high-frequency local transmitters or, increasingly, the use of satellite-based systems. They have suddenly become cost effective and the impact could be considerable. In the Anti-Atlas Mountains, for example, a common sight is the old, rusted telegraph pole, now devoid of wires, which once served as the vital link from rural communities to the nearest town and onwards. Today a network of high frequency boosters serves to allow digital transmissions to these areas, and an increasing number of people have access to mobile phones. Whilst the mountain topography produces difficult 'shadows' for radio transmission, technical improvements and lower costs have meant that a suitable network of boosters can provided good coverage even in these difficult areas.

The availability of sophisticated telecommunications and improved transport does mean that the mountains are becoming increasingly attractive to those people who wish to live and work in beautiful landscapes. This is all part of the background to amenity migration that is discussed below.

INDUSTRIAL DEVELOPMENT AND URBANISATION

In Chapter 4, the impact of new industrial activities and urbanisation was introduced through the concept of proto-industrialisation. This section provides a brief commentary on the fact that, contrary to popular imagination, many mountain areas are quite heavily industrialised and the sites of extensive urbanisation. The image problem arises because of the effect of modern transport facilities that now connect these high-altitude settlements. Rail, road and air routes often minimise the fact that access to these sites was once difficult. In addition, many of these cities are themselves in valleys, though at a high altitude and surrounded by towering peaks or extensive plateau. A good example is Grenoble in the French Alps. This city, with its population of about 450,000, lies in the valley of the Isère which has a reputation for occasional serious flooding. Only a minute proportion of the local population is engaged in agriculture, whereas 40 per cent are linked to service occupations. Many foreign companies are to be found in the locality, which is very attractive to skilled employees precisely because it has access to the mountainous terrain. However, probably the best-known, large-scale cities within high mountains are those of the Andes. In Colombia, the capital Bogotá (5.3 million) and a string of other cities, including Cali (1.7 million) and Medellin (1.6 million), lie in mountain areas. In Ecuador, Quito is a city of over one million; in Bolivia, La Paz and Santa Cruz are sizeable urban areas, as is Santiago in Chile. However, for this section it is not just a question of population size. These cities also represent important employment centres where, in addition to the service sector, there is extensive industrial activity. Both foreign and domestic capital has been invested in industrial activity, ranging from very 'high tech' electronics manufacture through to small-scale labour intensive units producing a vast range of commodities. Although Bogotá lies at an altitude of around 2,600 metres, its urban and industrial problems are just like those of many Latin American cities. Thus problems of capital formation, managing large multinationals and facing ever-increasing numbers seeking employment are issues of very general nature (Mohan 1994). Interestingly, given the common geographical situation in Andean America where most west coast lowland development needs resources from the mountain zones, the rapid growth and industrialisation of the mountain cities is providing more competition for basic resources, especially water. According to Anton (1993) this issue

264

has reached crisis proportions in some Andean countries as potable water, as well as irrigation and industrial supplies, faces increasing demand.

However, some of the fastest growing urban centres in mountain areas can be found in the western United States. This has been mentioned earlier but its significance needs to be emphasised as each surge in the US economy sees greater pressure on the attractive landscape of mountain states. In Colorado alone, since 1987 nearly 650,000 hectares of farmland have been lost to 'development', a sizeable proportion of which has been in mountain areas. Planning control has attempted to confine urban expansion to specific small towns and cities, but as in other parts of the USA the ranchette development continues.

MOUNTAIN TOURISM

This section examines one of the most dynamic and controversial aspects of mountain development, namely the tourism industry. Although the archetypal developments of mountain tourism are often related to the Alpine or North American experiences, mountain areas in regions as diverse as Scandinavia, New Zealand, the Andes, the Hindu Kush and Himalayas are now experiencing both the gains and also the stresses from this industry. Moreover, from fragmented, usually upper-class origins in some zones, or pilgrimages in others, the current use of mountains for leisure is now a global industry. Tourism in general has grown at a phenomenal rate since the Second World War as economic conditions and social norms shifted towards a more leisure focused society. In the USA for example, where National Parks in mountain areas were initiated in the last century, growth from 1945–75 was of the order of 500 per cent. In Nepal, in 1964, twenty tourists are said to have visited the Khumbu area in which the Sagarmatha National Park (Mt Everest) is located; by 1972 this figure rose to 3,200 and at the beginning of the 1990s was thought to be between 10–15,000 (Brower 1991). In many instances the early international interest was stimulated by climbing but has now widened to include package tours and other elements of mass tourism (Plate 7.4). The Karakorum/Hindu Kush region is a good example, where this transition has taken place but remains fraught with problems due to the political instability of the area. After many years of closure China now promotes tourism in Tibet, Sichuan and Yunnan. Moreover, tourist development has become inextricably linked with problems of environmental management in many mountain areas (Price 1997).

In all cases there is no doubt that tourism is a mixed blessing. In rich countries, as mountain communities have been absorbed within the urban industrial economy, so traditional livelihoods have failed to provide the levels of living increasingly demanded by the population. Often, this incorporation

PAKISTANI PEAKS

			(WORLD) (RATING)
K.2	28251	2	
NANGA PARBAT	26660	9	
GASHA BRUM.I	26470	11	
BROAD PEAK	26400	12	
GASHA BRUM.II	26360	14	
GASHA BRUM.III	26090	15	
GASHA BRUM.IV	26000	17	
DASTIGHIL SAR	25868	20	
KUNYANG KISH	25760	22	
MASHA BRUM SE	25660	24	
RAKA POSHI	25550	27	
BATURA.I	25540	28	
KANJUT SAR	25460	29	
SALTORO KANGRI	25400	32	
TRIVOR	25329	36	
TIRICHMIR	25290	41	
CHOGOLISA	25110	46	
SHISH PARE	25000	49	

Plate 7.4 Sign indicating major mountain peaks in Pakistan

has been the major factor in the demise of this traditional economy. In other parts of the world, tourism has become a key element of development programmes usually based upon the encouragement of foreign visitors. Positive economic benefits in some areas must be weighed against studies which reveal that not only are the socio-economic effects of tourism sometimes negative, but so too is the environmental damage that may result.

Mountains as sites for tourist activity

Price *et al.* (1997) provide a useful review of the properties of mountains considered attractive to tourists. Today, there are many different types of tourism in mountain areas ranging from the physically strenuous climbing, trekking and skiing, to the more leisurely sightseers and the longer-term visitors, often called 'amenity migrants', which have been examined in Chapter 6. Consequently the attractiveness of mountains lies in a complex of properties perceived differently according to interest. For some it is the ruggedness of the landscape, for others the obvious availability of suitable snow-covered slopes for skiing. Others find that the landscape itself, its scenic qualities and vistas, provides a strong emotional or psychological draw. Increasingly, the 'atmosphere' of the mountains, including the relative silence and clean air, attracts ordinary lowland pursuits such as golf and general sports complexes.

At the same time, accessibility is an important factor in tourism development. Tourists by definition do not reside in the area in question, and many mountain areas are by their very nature marked by difficulties of access. The presence of trails, roads, railways, chair-lifts, etc. provides a major element of tourist infrastructure for a mountain area. However, improvements in transport present a paradox as increased access reduces the attribute of remoteness which itself is a sought-after quality for some tourists. It further serves to pinpoint the 'process' of becoming part of the tourist circuit. For any given area, it is early travellers who provide information of the 'other' to a wider world. As we shall see below, European history is replete with this travelling, which is often closely associated with colonialism. At this point the traveller (climber, etc.) would use whatever means of local transport is available, more often than not spending long days trekking to a destination in some remote mountain valley. Writing about the Swiss Alps Beerli says:

> Up until the nineteenth century, a journey through the Valais was an adventure in itself, not recommended without a sturdy mount and trustworthy guide. The roads were in a wretched condition and the local inhabitants were distrustful of strangers, often hostile. Thus the first English tourists in Zermatt were taken for thieves and sheep rustlers, and had to be rescued by the priest from the aroused villagers.
>
> (Beerli 1967, quoted in Friedl 1974: 96)

The local populace might provide accommodation for a traveller merely as a kindness or social duty, although some gift or small payment was normal. However, once the opportunity to generate an income through the regular provision of facilities was recognised then tourism as such began to develop. At this point, under the influence of the market, the larger the flow of tourists the greater the income, and anyone engaged in this form of business seeks to improve transport to maintain or increase business. Obviously, solitude, an original 'attraction' of the mountains for some, is jeopardised by this process but, at the same time, the expansion of facilities might make them cheaper (or more easily available) for a wider clientele.

There are particular circumstances when the concept of a mountain means more than a set of 'attributes'. As discussed in Part 1, for large numbers of people mountains have symbolic and religious significance and their visits to them become a pilgrimage. Good examples of this are sites in the Pyrenees and Mt Fuji in Japan. However, even for those whose main desire is for recreation, mountain tourism often has to be 'created' by the image construction of those involved in the business. In the Alps, the attractions of clean air, healthy exercise and 'status' were all 'manufactured' – on the one hand by popular medical writers and on the other by society's affluent members. Even the desire to visit ancient cultures and their artefacts, such as temples, arises often from an interest in these matters developed by the media.

The historical origins of tourism in mountains

As was noted above, mountain areas have long been the destination for outsiders seeking solace, spiritual uplift and recreations. In many instances special sites in the mountains provide places where contact can be made with a deity although the mountain itself does not possess sacred properties. Elsewhere, the deity resides in the mountain. One of the most famous is Mount Fuji in Japan which has played an important role in the symbolism of Japanese culture and, despite physical desecration, retains important mystical powers for many. Similarly, Mount Kailas in Tibet remains one of the most sacred places for both Hindus and Buddhists. Pilgrimage to many of these sites has been taking place since Antiquity and today may constitute the principal source of visitors. According to a study by Sreedhar (1995), tourism accounts for about half the GDP of the Indian state of Uttar Pradesh, and 60 per cent of these tourists are pilgrims.

As Kariel (1993) points out, tourism in the Alps dates back to ancient times. There were Roman health resorts with their baths, casinos and well-equipped villas. These symbols of power can be recognised in a more modern era in the hill stations of colonial India such as Simla or the French settlements in the Middle Atlas Mountains such as Ifrane. However, in Europe, for much of the period before at least the eighteenth century, highlands or mountain ranges were considered to be unsafe places. This arose either because of the very real natural hazards and the presence of vagabonds, or equally because of the cultural image of these areas which usually portrayed them as the haunt of demons. This was particularly true of the Alps which from ancient times had represented the boundaries of 'civilisation', namely that of the Mediterranean. The *montes horribiles* were to be avoided at all costs, if they had to be crossed it was to be accomplished quickly. From the late eighteenth century onwards this view began to shift, along with the cultural fashions of Europe, with the development of the Romantic Movement. Paintings by Calame and Diday expressed the strong emotions of nature with sombre forests, soaring peaks and clear light glacier images. One of the key figures of the period was Jean-Jacques Rousseau who stated that:

> il me faut des torrents, des roches, des sapins, des bois noirs, des montagnes, des chemins raboteaux a monter et a descendre, des precipices a mes côtes, qui me faissaent bien peur.
>
> (*Confessions* IV, quoted in Grossjean 1984: 75)

This evocation of the purity of nature and the implicit challenges it presents found strong support among the burgeoning groups of natural scientists and others who began to see the mountains as places to visit and the peaks to conquer. Debarbieux (1993) argues that this period saw the creation of a particular mythology of Alpine peaks, especially Mont Blanc. Whereas many other peaks were already noted by the eighteenth century for their strong spiritual values, Debarbieux argues that Mont Blanc is a modern creation. This has as much to do with the way mountains filled a space in the minds of artists, natural scientists and others of that period, as it has to do with any specific physical quality. It had no definite name until 1742 and knowledge of its height only gradually became more precise between 1685 and 1787, the first ascent being in 1786. The symbolism of man conquering nature became very significant in this period, creating a new geography of mountains as more and more explorers, artists and climbers sought to 'tame' nature. The result has been the subsequent burgeoning of visitors to places such as Chamonix.

In order to examine the way in which the growth of tourism has affected specific localities, three case studies will be examined in some detail. They represent a continuum from the poverty of Nepal, through the emerging capitalist economy in Morocco, to the affluence of the Alpine economies. First, we will look at the tourist impact on the Himalayas, especially in the Everest region; this will be followed by the experiences of the High Atlas Mountains in Morocco; and finally an examination of selected Alpine villages will show how certain problems have emerged in tourist development.

The Himalayas

A number of writers (Pawson *et al.* 1984; Brower 1991; Stevens 1993; Sharma 1998) have examined the development and impact of tourism in Nepal, particularly in the Sagarmatha National Park (Mt Everest). Until the 1950s, Nepal was largely unknown to outsiders except for the capital, Kathmandu, where the British Raj maintained a representative. Climbers had, in fact, been coming to the Himalayan range through Tibet since the nineteenth century. However, after the successful expedition of 1953 when Edmund Hillary and Sherpa Tenzing Norgay climbed Everest, a process was set in train that ultimately led to Nepal becoming one of the most popular tourist destinations of those seeking 'wilderness' experience.

In the first instance, most of the visitors were those in climbing expeditions, often with the patronage of nation-states. Later, trek organisers based in Kathmandu established the infrastructure of guide, porters, etc. so that a wider clientele could visit the area. Finally, considerable numbers of individuals, often seeking 'spiritual' rejuvenation, also made their own way to Nepal. Although climbing permits were instituted, there was no control on the actual flow of visitors, unlike in Bhutan. The figures for the growth of tourists have already been noted, but it is important to recognise that while the largest number of tourists arriving in Nepal itself originate in the Indian subcontinent, the visitors to Khumbu and the Everest region are predominantly from the richer regions of the world, including North America, Europe and Japan.

At the time of the 1950s expedition, the Everest area could only be reached by a fifteen-day trek, usually from Kathmandu along mountain paths, although pre-Second World War expeditions approached from the north. This was later reduced to about five days as new roads were constructed within Nepal. This alone stimulated a considerable increase in visits by mountaineers who had not the time or resources to spend over a month simply gaining access to the peaks. Later an airstrip was built and so today many are able to fly direct to the Khumbu area.

Even before the 1950s, Sherpas from the area had been used to support the early expeditions and so when the focus of activity shifted to Khumbu there were already experienced porters and guides. However, the 1960s onwards saw rapid growth in the employment of local Nepalese in providing trekking support and in the development of accommodation. Whilst some trek groups camp, local Nepalese offer accommodation in 'Sherpa Hotels', which prove popular especially for independent travellers. At the other end of the spectrum foreign capital funded an international hotel in the area which has had mixed fortunes. Today, potential visitors from around the world can arrange their trips to Nepal and the Everest region using computer links to reserve hotels and look at new pictures of Everest nearly every day.

In addition to the provision of labour, demand for fresh vegetables, and the provision of accommodation, tourism has led to the development of other economic activities. Small-scale traders run many small stores in the settlements of the area stocking goods primarily for the tourist, including some remarkably esoteric tinned products such as tinned snails. According to Brower (1991) many of these goods are expedition 'left-overs', but more generally goods are obtained from Kathmandu or, in some cases, from Tibet. One of the most significant developments arising from tourism has been the provision of livestock for trek purposes. Supply of 'pack-stock', usually yaks or zopkios (male crossbreeds, e.g. yak and cattle), is a thriving business where money generated in tourism finds a profitable investment.

Although the Khumbu area is considered to illustrate the more successful economic outcome of participation in the tourist business, there are some clearly negative effects. In particular, whilst some families whose members have been successful at securing positions on expeditions have generated sufficient capital to invest in new enterprises, the majority of the area's population only participate on a very small scale, providing menial labour at very low pay rates. The peak tourist times depend upon precise location and in some cases generate a demand for porterage and similar tasks which conflict

Table 7.1 Leakage and linkage from mountain tourism in the Himalayas

Location	Tourist/year	Linkages	Leakages
Ghandruk (Nepal)	15,000 overseas	12 per cent households involved	Two-thirds expenditure
Northern Gorkha (Nepal)	1,500 overseas	Little, only along trails	Most, because almost all are organised groups
Upper Mustang (Nepal)	Fewer than 1,000	None	All items imported
Badrinath (India)	At least a million, mainly domestic pilgrims	Little	40% trade controlled by outsiders
Hunza valley (Pakistan)	1,600, of which 60% are overseas	Little; mainly guides	Most items imported

Source: Modified from Sharma (1998)

with the needs of agriculture. The harvest of potatoes and barley as well as the period of hay gathering is a time when labour input is particularly high. The result is the excessive labour demands placed upon women, a tendency for older children to be removed from school and a steep increase in agricultural wage rates. For those without suitable labour resources or finance such a process is one of the first steps to a gradual decline in agricultural activity and perhaps impoverishment.

It is useful at this point to present the findings of Banskota and Sharma (1997) and Sharma (1998), who examined the economic impact of tourism on a number of localities throughout the Hindu Kush/Himalayas. This latter study shows that not all localities appear to have the localised benefits suggested for Khumbu. This arises because of the fact that leakage from the local economy occurs in so far as many of the real beneficiaries of tourist activities in the mountain areas are located elsewhere. Table 7.1 provides data from this study and it is notable that even in the case of the pilgrimage circuit of Badrinath, which has had over a million visitors per year, there are only very weak linkages with the local population. A more detailed study of Ghorepani and Ghandruk suggests that about 76 per cent and 68 per cent of food and consumer durables were, in fact, brought into the area.

However, the problems of economic returns are related to environmental management. In the case of Khumbu, this is linked to the formation of the Sagarmatha National Park. This was designed to draw tourists to the area and at the same time provide an umbrella of control over some of the more problematic side effects of tourism. Brower describes in detail some of the factors that were brought into play to ensure the creation of the park, and these provide an interesting insight to a common feature of most extant parks in mountain areas around the world – namely, the low level of consultation with the resident population. Much of the pressure came from international conservation groups and Brower argues that they employed a campaign which somewhat exaggerated the negative effects of tourism on the local environment. Often short visits by conservationists provided the basis for powerfully presented reports in an endeavour to influence the authorities in the kingdom. Whilst there is no doubt that many of the problems highlighted existed to some degree, the exaggerated presentation gave rise to actions which since have proved problematic. This National Park was gazetted in 1976 following a New Zealand commission report on management structures. It has an area of 1148 km^2, with some of the highest peaks in the world including Sagarmatha (Everest). Some 69 per cent of the park is barren above 5,000 metres, about 28 per cent is used for grazing and 3 per cent is forested.

The management rules define acceptable practices in the park, whose overall objectives comply with international standards and attitudes towards environmental management. However, at the outset local Sherpa participation was very limited. Certainly, as has been typical in many similar institutions globally, most of the staff were not locals and there has been a long-running battle concerning activities such as poaching and fuelwood collection. The park itself symbolises many of the environmental problems associated with the growth of tourism. Attempts to control the loss of vegetation conflict with the high demand for fuel needed, both for cooking and, increasingly, for the heating of water for washing in hotels. Although trekkers are encouraged to use bottled gas, local families still rely heavily on fuelwood. However, perhaps one of the most notable features of the Sagarmatha park has been the problem arising from garbage which has reached such horrendous proportions that an NGO exists with the prime aim of keeping the park clean. Between July 1995 and 1996 the Sagarmatha Pollution Control Committee collected 145 tonnes of rubbish that could be disposed of by burning and 45 tonnes of unburnable garbage. In 1995, rubbish from expeditions alone amounted to 2 tonnes disposable and 1.5 tonnes of non-disposable material, including 195 gas cylinders, 524 oxygen cylinders and 603 kilograms of batteries (Nepal 1997). At the same time, there has been concern about overall hygiene in the park from so many persons using makeshift toilets. Overall the environmental impacts of different types of tourism can be summarised in Table 7.2.

Table 7.2 The impact of tourism

Positive impact	Negative impact
Tourism promotes:	Tourism damages:
• ecological awareness	• forests through excessive use of fuelwood
• conservation provision	• fauna, as a result of hunting
• pollution control	• land, if campsites not maintained
Tourism also:	Tourism generates pollution of air, water, noise and waste
• supports maintenance of the landscape	
• encourages research activities	
• provides alternative incomes	Excessive tourist activity can:
	• damage landscape
	• threaten local hygiene
	• reduce the value of natural scenery
	• congest roads and facilities

Source: Derived from Singh and Kaur (1986)

We can sum up our review of Khumbu by saying that it illustrates not only an area which contains some of the most magnificent mountain peaks in the world but also the way in which a once relatively isolated community has been inundated with tourist pressures. Today it is highly integrated into the global tourist business (note the Internet connections), but whilst this has brought many benefits to the locality it has also emphasised social differentiation, posed problems for environmental management and raised questions about the destruction of local cultural behaviour.

The High Atlas

Many of the same issues arise in the Atlas Mountains of Morocco. However, this case is an example of an area which, although less popular than the Himalayas from a strictly mountaineering aspect, is much more accessible to tourists from Europe and also an increasing number of Moroccans seeking recreation and solace from the pressures of urban life.

The French initiated tourism in the Atlas Mountains during the 1940s with the creation of a ski resort at Mischliffen near Azrou in the Middle Atlas and Oukaimedene in the High Atlas. Climbers under the auspices of the Club Alpine explored the peaks, including Toubkal, which at 4,165 metres is the highest peak in North Africa. Climbers were followed by trekkers, initially from France and then from other European countries, who set up refuges, especially along the route through the Ait Mizane valley. This process was associated with the arrival of the road at the end of the Mizane valley in the 1930s and created the present village of Imlil. As a result, a new settlement, the 'Imlil Garage', developed with small stores and a local market (Miller 1984).

By the 1960s tourism in Morocco was beginning to expand generally, and alongside the ancient towns of Fez and Marrakesh, coastal venues such as Agadir built up their clientele, most of whom were foreign. This was part of a major strategy on the part of the Moroccan authorities to expand the tourism potential as a source of both foreign exchange and employment (Berriane 1993). Much of the development was through organised package tours visiting several centres, usually by road. Therefore, an increasing number of tourist buses began to visit the lower regions of the mountains, especially the towns with markets, such as Amizmiz, and also famous kasbahs. This business encouraged traders to develop sites alongside the main trans-Atlas autoroutes, but very few tourists stayed overnight in the mountains themselves. There are few 'tourist hotels' within the mountains and since the 1980s some concerns have been voiced that 'inappropriate' hotel construction would take place.

As part of its overall schema to offer mountain residents new employment opportunities a programme of developing 'soft tourism' was initiated in the Azilal region. This was a co-operative venture between the government and the universities of Grenoble and Marrakesh (Bellaoui 1996). This programme began in 1984 and, in various guises, extended over the subsequent decade. The project aimed to provide the physical and organisational infrastructure for the development of trekking holidays in the High Atlas, specifically in a manner that would involve the local populace and not require the construction of hotels, using foreign capital, which were architecturally out of scale with the mountain environment. At the same time it was intended that local economic activities, including crafts and associated agricultural production, would be stimulated by the creation of a market by the tourists. As part of the programme a 75-kilometre road was built to the Centre de Formation Aux Métiers de Montagne (CFAMM) at Tabant and, in addition, a small hydroelectric plant installed for electrification. Five high-altitude refuges have since been constructed and approximately thirty 'gîtes d'étape chez l'inhabitant' created to provide appropriate accommodation for visitors.

There is no doubt that tourism to the province has increased significantly. Available figures suggest that there were 4,963 visitors in 1988 and that this had increased dramatically to around 24,000 by 1996. However, some of the objectives have not been achieved, especially those which attempted to expand agricultural production and its possible links to tourist activity. Thus whilst there has been an increase in the number of bee-keepers and some development of apple production it is not clear whether this would have taken place anyway given the widespread increase in apple cultivation in the mountains (Parish and Funnell 1999). In fact Bellaoui (1996) suggests that the net effect of the project has been to replace a once highly integrated economy with one based more on specialisation. In addition he reports that, as we noted above in the Himalayas, even with a noted growth in the number of visitors using the facilities it is estimated that less than 10 per cent of the expenditure of the foreign tourists entered local circulation. The rest accrued to other elements of the tourist chain outside the region. In the opening years of the project, most of the *gîtes* were established by the local elite and later also by guides. The guides and muleteers generated most of the earnings, and the strong kinship networks ensured that business was confined to their families. Consequently, relatively few local inhabitants participated in the project. Moreover, despite the attempt to develop artisanal production, most tourists purchased their souvenirs in the big cities, particularly Marrakesh.

The project also raised important questions about the socio-cultural impact of tourism. The structures established to cope with visitors have been explicitly geared for foreigners, especially French. There is often little social contact, mainly through the fact that virtually no visitors speak Berber. Clearly the foreign visitor brings cultural styles and behavioural patterns which do not correspond to local norms. The domestic nature of the accommodation makes the contrast in styles very marked. It might be noted, however, that in many instances the tourists do not eat with the family. Pézelet (1996) suggests that the conditions established by the authorities for the licensing of a *gîte* are designed more with the foreign visitor in mind, rather than as a good representation of the lifestyles of the local populace. It is also often argued that the influx of tourists encourages men to develop a desire to seek work in Europe and to question their marital fidelity. Whilst these claims are undoubtedly true to some extent, it must be said that the precise significance of this impact is not only difficult to measure but also highly differentiated by age, sex, locality and occupation. In many instances the visitors only reinforce stereotypes that are already presented in other ways (for example through radio, television and video), all of which are increasingly finding their way even to some relatively remote areas.

One of the most interesting places to explore this problem is the village of Imlil which has been mentioned earlier. Having been a pioneer location for the early climbers it has become a very popular destination for a wider spectrum of visitors, both from overseas and from the Moroccan towns. In addition to the early accommodation in several small hotels and the Club Alpine hostel, recent years have seen the redevelopment of the former kasbah as a site for educational parties and other travellers. Many families now have links with the large urban centres and some have family members working in Europe. Thus friction within the community appears less a response to 'alien' ideas than it is to the problems posed by sharply increasing economic differentials between those who are able to use their contacts and experience to develop new enterprises and those who perhaps feel marginalised by the direction of events. There is also competition with neighbouring centres (e.g. Arremdt), which have also built up their own clientele. An interesting example of the problems concerns a plan to develop a tourist circuit that would run along the Mizane valley to Imlil and then turn to cross the col into the neighbouring valley at Tacheddirt and thence to the ski resort at Oukaimedene. This proposal evoked widely different reactions both within Imlil and at other villages. Motorable roads exist at both ends of this proposed circuit, but between access is either by mule trail or, at best, jeep or lorry. The proposal has been welcomed by those in the villages who are trying to expand tourism through attracting new types of visitors. Already, mountain biking has proved popular and others see a major advantage if they can capture business from coach-based trippers, as already happens in the lower valleys. On the other hand, the development of this route would mean that the comparative access advantages currently experienced by Imlil (not for buses) would become available to many other villages in the locality. Equally, those who have specialised in the trekking/climbing business recognise that widening the tourist base may well change the overall environment for their clients who look for a 'wilderness' atmosphere. At present there has been no further progress with this proposal.

As in the Nepalese example, the development of tourism in the Atlas raises important environmental problems. At the aesthetic level there is the question of the construction of unsuitable buildings and the effects of road construction. The considerable increase in rubbish and the problem of sewage is also apparent. The detritus of modern society with a considerable volume of non-biodegradable material has now reached serious proportions at some sites, and according to local medical officials is a cause for concern, especially for infections to children playing nearby. Simple cesspit systems have been in use for many years, but with the increase in the larger accommodation facilities these have become problematic. Neither is there a system for the direct supply of potable water, as most supplies are taken from the various irrigation ditches that criss-cross the area. Further discussions to improve the situation are taking place at the time of writing. This is all part of a gradual improvement in the infrastructure of some of the mountain villages. Imlil is one of the first in the area to receive networked electricity, although some villages had their own generators, and a relatively new health dispensary exists.

Where there are motorable roads into the High Atlas, there are also examples of amenity migration. Given the relative proximity of Marrakesh and the good roads onwards to Casablanca, areas such as the Ourika valley are inundated with weekend visitors, often to rented properties in the valleys. There are a number of foreign-owned properties but perhaps the biggest longer-term issue concerns the purchase of land by Moroccans who aren't from the area. This is an issue that has provoked some strong views amongst the local inhabitants. Some argue that it represents a serious problem because it causes inflation in land values and limits access for local people, others pointing out that sometimes the new arrivals are able to use their influence to obtain greater assistance from central government.

The Moroccan example has raised a number of points about the influence of tourism in its mountain areas. The government encourages tourism, both to increase the input of foreign exchange to the national economy and to provide a new source of cash income to the mountain families. This latter is considered important for direct welfare objectives and, indirectly, as a mechanism for reducing the flow of migrants to the rapidly expanding urban centres such as Casablanca. While large-scale tourism now exists in many locations, the mountains have, until recently, remained somewhat peripheral to the main circuits. Many people pay short visits to the mountains but relatively few stay for any length of time. On the other hand, to encourage longer stays would mean building more accommodation which itself raises questions about suitable architectural designs that would blend with the environment. An experiment in 'soft tourism', whilst successful in handling a significant increase in trekkers, appears to have offered little in the way of extended benefits to villagers. As in many other examples, only those who were fortunate to be directly involved in the business appear to have gained, and even in this instance there is a considerable leakage of benefits from the local economy. Nevertheless, at some sites the impact of tourism in formerly agricultural communities has been considerable – especially where improvements in infrastructure have followed. Culturally, whilst tourism does bring new styles of behaviour, etc., its influence may be exaggerated in so far as many mountain communities already possess knowledge of (and for some the desire for) an urban lifestyle. In Morocco this lifestyle is promoted by the local media and demonstrated through the increasing numbers of Moroccan visitors escaping from the urban centres. As affluence increases, it is likely that more and more Moroccans will seek out the mountains for vacations. When environmental disasters occur, such as the flash floods noted in Chapter 4, the state becomes directly and very publicly involved and the political profile of the mountain communities is raised, albeit briefly.

The European Alps

The countries of the Alpine region are all major players in the tourist business. However, by 1995 Alpine tourism might be said to have reached a crisis situation. In Austria for instance, the rate of unemployment in the tourist sector is markedly higher than in any other. The situation in Switzerland is similar; winter sports centres are engaged in destructive competition and tourism in structurally weaker areas is declining (Pils *et al.* 1996). Despite this, the Alps accounted for almost one-fifth of registered tourists worldwide. Whilst it is possible to trace this development back to the nineteenth-century climbers and trekkers, much of the growth has occurred in the twentieth century, especially in the last forty years. For many

Alpine regions the changes became most notable after the Second World War. Until then, the Alpine area was one of Europe's industrial and agricultural zones with particular emphasis placed on industrial development in valleys based upon water power. Agriculture, too, has seen a sharp decline as the economic returns from farming were unable to match those from other sectors. There have been several waves of migration from the mountains – first to lowland areas in Europe and, during this century, overseas. This changing economic structure was accompanied in some areas by a shift to tourist activity (Lichtenberger 1988; Brugger *et al.* 1984). Indeed Lichtenberger argues that the geography of the Alps in the twentieth century fundamentally reflects a transformation of an agricultural society to a leisure society. The governments of Austria, Switzerland and Germany have been particularly active in providing support for tourist development as part of a programme to bring employment to mountain areas. In France, however, much of the investment in ski resorts has depended heavily on private capital.

Although settlements such as Zermatt, Grindelwald, St Moritz, Davos and Chamonix are known for their development as internationally famous tourist sites, the dynamics of change are well illustrated in the remote locality of Kippel (Friedl 1974). Kippel is located in the Latschental valley, a tributary valley to the north of the Rhône, and has been settled at least since the twelfth century. Until the Second World War it was primarily an agricultural settlement. Tourist development began in 1868 when a hotel was opened at another village in the valley. In 1906 a small hotel was opened in Kippel, followed by several small guesthouses, and throughout the period before and after the First World War hotels began to be built in the villages and on the Alps themselves. These facilities catered almost exclusively for summer season tourists who represented the hardy walkers and mountaineers of that period but whose existence had little impact on the 200 or so residents of the village. Before 1913, travel to Kippel involved a long trek on a cart track from the Rhône valley, but with the building of the railway at that time, which crossed the mouth of the valley, relatively easy access from the station at Goppenstein was possible from the main cities. The railway also greatly facilitated the passage of those who, in the pre-Second World War period, ventured into employment in the growing industries of the main valleys. The transition of the economy can be illustrated with the figures shown in Table 7.3. It was a transition that occurred in two stages. First, as the state restructured rural education to develop a workforce with technical skills, younger men from the village travelled to the industries elsewhere. Subsidies for transport and the improvement of roads led to the well-known Swiss Postbus service which made it possible for some villagers to commute on a daily basis, although many remained away for longer periods.

However, perhaps the most notable feature of the period from the 1960s has been a second transition, illustrated in Table 7.3 as a marked rise of jobs

Table 7.3 Changes in the percentage shares of economic activity, Kippel

	1941 (%)	1950 (%)	1960 (%)	1970 (%)
Agriculture	76.0	70.8	36.0	14.2
Industry	4.4	9.7	41.7	31.6
Commerce/tourism	19.3	19.5	21.9	54.0

Source: From Friedl (1974)

in the tourist sector. This has been prompted by the investment in winter tourist facilities, as these became an increasingly important part of the tourist market. A British company built a ski lift and some residential accommodation was offered by a few of the villagers. The pattern has been to build a house suitable for several families and let out all but one section to tourists. These tourists have replaced the animals in the traditional two- or three-storey structure of Alpine houses. At the same time, much more of this new employment has gone to women, who now have become equally important in the wage earning stakes. There has been an increase in the number of shops and other suppliers catering almost exclusively for tourists and more facilities for skiers have been constructed subsequently. One of the main impacts of the development of these facilities by an overseas company has been the fact that some of the villagers have realised the value of their land by selling up to the company. This money has enabled residents to construct their own holiday accommodation, sometimes blocks of flats, and so by the 1970s Kippel was experiencing a building boom, which continued into the 1980s. In a manner very similar to that described in the Obergurgl model in Chapter 1, the community itself, with assistance from the canton, has improved avalanche protection and widened the facilities to include camp-sites and car parks. Today, hotels in Kippel offer the full range of 'Alpine' leisure activities, so in addition to skiing, climbing and trekking it is possible to find activities ranging from snow-mobile use and mountain bike hire, through to making cheese and feeding the goats for the less adventurous and children. The valley is now only three hours from Geneva, one from Berne, and it is only ten minutes to the cable car for a trip up to the alp for skiing or trekking depending upon season.

Nonetheless, tourism has not been able to rekindle the economic viability of the old community. As Table 7.3 shows, the proportion employed in agri-culture declined considerably between 1950 and 1970 and has continued to do so. The transitional process has not been without its traumas however. Intergenerational differences were very marked in the 1970s as an older group attempted to retain some links with an older lifestyle, both culturally and economically. With state assistance now geared not only to maintaining livelihoods but also to the desire to maintain the landscape, which itself is probably the most vital asset of the valley, some have retained an interest in farming. However, Kippel is now part of the leisure society that has increasingly come to dominate some Alpine communities.

Given the different geographical and historical backgrounds within the Alps it is important that we take note of the varied responses to the impact of tourism. In an interesting paper by Kariel (1993) four communities in the Austrian Alps were examined in a series of studies over the period from 1979 to 1990. These ranged from an internationally important snow sports centre to a relatively new community developed as a result of the building of a hydroelectric scheme in the 1940s. Kariel argues that in the space of twelve years it is possible to highlight subtle differences in the attitudes of residents towards tourism. In the earliest interviews, most residents expressed very positive attitudes about their experience of the rapid growth in tourism over the previous two decades. The older members of the community could remember the hard times associated with the lack of employment in the valleys as agriculture increasingly failed to provide an adequate source of income. The arrival of the tourist industry had been the one thing that had helped reverse out-migration and had been a major factor in encouraging improvements in infrastructure. Asked to list any negative impacts, respondents at this time noted that some of the old communal habits had virtually ceased, as families became more insular. This was ascribed to the fact that many families now actively competed with one another to attract tourists. They noticed also that there had been a very large increase in prices for many goods because these were now only available through commercial purchase rather than produced within the domestic system or locally bartered.

The second survey confirmed that by 1986 almost all families had become enmeshed within the leisure economy and clearly appreciated their economic role within the wider economy of Europe. However, it was quite evident that by 1990 there were clear differences emerging between the younger and older generations. This was most marked with respect to two factors: first, attitudes to tourism and the environment; second, attitudes to work and leisure and family life. The 1980s had seen a growth in summer tourism parts of the Austrian Alps as holidaymakers kept away from what they felt were highly polluted beaches in the Mediterranean. The increase in demand raised important questions about further investment in facilities and, unlike the earlier study, the later enquiries found strong views among some residents that there should be strict controls on expansion. It was argued that the environment had become overloaded and that there were signs of significant damage through excessive use of pathways, etc. It was clear that those who had spent their lives building up their tourist businesses did not subscribe to this view, but their offspring were much more 'eco-friendly'. In part this may be because the younger groups were influenced by the wider debate on the environment and felt they ought to act accordingly, whereas their parents saw only the need to ensure that their incomes were maintained or expanded. The conflict became particularly noticeable within the community political arena as strident objections to expansion plans were voiced, splitting the comfortable consensus that had been maintained for several decades. However, the discussions also indicated that the younger generation was not prepared to sacrifice family life in the way they felt their parents had done. As part of the process of building up their various leisure businesses, many parents were accused of neglecting family life and concentrating instead on the welfare of the tourists. This attitude also spread to the community at large, and therefore the younger generation wanted to see the balance corrected.

There are several interesting points that emerge from this study. First, even allowing for the short period involved, the evidence highlights the dynamic nature of attitudes. Second, the apparent changes in attitude to tourism reflect a combination of influences such as age group and national culture. Finally, despite the importance of tourism in a strictly economic sense, those who had experienced only the more successful years questioned the drive for further expansion at the expense of environment balance. Of course, it is not clear whether these attitudes represent a typical development of perspective as a new pattern of social behaviour is introduced, or whether these attitudes are specific to the socio-economic framework of the communities. For instance, if the tourism takes a downward turn, as is suggested, will the environmental interest still remain?

Environmental impact of tourism in the Alps

In order to pave the way for the discussions of policy in the final section of the book, we will end this discussion of Alpine tourism with a brief examination of the environmental effects that so far have been noted only in passing. Krippendorf (1984) makes the point that the real capital of tourism in the Alps is the landscape, not the assembled equipment of cable railways, ski lifts and hotels that adorn the view of many Alpine valleys. Whilst this might be true of the basic physical activities of mountain areas, recent investments have merely extended some of the recreational facilities from the lowlands to the highlands. Indoor swimming pools and, above all, golf courses are perhaps the best examples built in an attempt to secure the widest range of visitor activities. However sensitive the design, the construction will always interfere with the landscape, often in dramatic ways such as the regrading associated with skislopes. Moreover, because facilities are usually highly localised around a particular site, the resultant concentration of tourists itself leads to further damage. Not only is there the direct destruction of the landscape through overused ski-runs, pathways and other features but pollution in watercourses and the atmosphere from waste and automobile fumes. The number of scenic areas in the Alps that have remained free from tourist or transport infrastructure has halved between 1963 and 1993.

It is necessary to be careful about the meaning of landscape in this context. For many writers, the Alpine landscape is either one of snow-covered peaks and slopes, or a rocky landscape surrounded by farmland with green fields and grazing animals. Built-up areas with roads, housing or large leisure facilities are not part of this image. However, the decline in farming as an economic activity, and the 'urbanisation' of the mountains in so far as few people now derive a substantial part of their income from rural activities, has meant that the conservation of the landscape is placed in jeopardy. The fact is that even if there are vacillations in the fortunes of the leisure industry, it is likely to survive, although its focus may change. In the Alps this has already happened in so far as in the last thirty years the area has shifted from being used by an 'affluent' clientele to one where more and more of the tourist trade has shifted to the mass market. The implications for environmental management depend upon two factors: the extent to which development regulations can be enforced, and in the long term the degree to which the public can act to ensure suitable environmental conservation practices.

Tourism in mountain areas is growing apace. The 'market', i.e. the visitors themselves, travels to mountains for a variety of reasons, but it is the various functions of the geographic environment (ruggedness, isolation, snow, clean air, etc.) of the mountains which represent the main resource. Governments see this 'resource' as a major opportunity to increase foreign exchange, and to offer new sources of employment to replace the traditional livelihoods of mountain people that have collapsed as the areas become enmeshed in the urban economy. In both rich and poor countries there are clear social and economic impacts within the communities as well as environmental issues. As the leisure industry becomes ever more global and competitive, the vicissitudes of the international economy come to impinge directly on the mountain communities and, having been part cause of the decline of agriculture, a serious economic depression could once again cause despair in mountain communities. Thus the mountain problem becomes one also of regional and national economic policy. Consequently the long-term sustainability of mountain tourism becomes a critical factor in government plans for the development of these areas, a matter taken up in Part 4.

SUMMARY

We began Part 3 with reference to the 'cog diagram' which highlighted the essential dynamism of mountain environments and communities. A closer examination of the evolution of this diagram (from Figure 1.4b to 1.4c) indicates that it illustrates both changes in the size and relationships between particular cogs (variables) and the appearance of new ones. Thus it serves to focus attention on the fact that once we look at how socio-economic and environmental transformations take place our ordering of key variables, and indeed the way we establish linkages between them, may well change. This is clearly indicated when we look back at the material in Part 3. This has been organised into three sections, each exploring the dynamics of environmental, demographic, and economic and political change. For mountain areas we would argue that these broad categories provide a focus for the identification of the factors which are critical for the transformation of mountain areas. Although these elements were present in Part 2, there is one major difference – namely, the balance of emphasis. In Part 2 the greater emphasis was placed on the direct or indirect impact of the physical environment on mountain livelihoods in the past. In Part 3 we have argued that the transformation of mountain areas today owes more to the impact of increasing state intervention and socio-economic factors.

Yet this conclusion must be treated cautiously. Clearly, in terms of the current focus on mountains, their political incorporation, demographic restructuring and economic development are major causes for concern. So, however, are the environmental changes involved, though these are today increasingly viewed through the spectacles of human involvement. The relative importance attached to any of these factors therefore depends upon the time and spatial scales through which the transformation process is viewed. At a local level, in terms of the community livelihoods, some of the most immediate concerns involve the maintenance of livelihoods in the face of declining competitiveness, or rapidly rising aspirations.

Today, action is needed to tackle problems of poverty in some areas, but in others the issues arise from affluence. At the macro level many mountain regions are considered only crucial resource areas for national economies and so their fortunes are a function of the vacillations of global and national economic demand which may shift over several decades. In this instance, autonomous action by mountain communities is limited unless there is, in most instances, a shift in the recognition of 'asset ownership' so that mountain communities are in a stronger bargaining position. This is the debate about empowerment and governance. Long-term shifts in the climate or other natural phenomena are harder to appreciate and, more importantly, harder to take action on, but nevertheless are an inevitable part of the environmental dynamics of mountain areas. The transformations taking place in this context are interesting. They highlight very different perceptions of the value of important environmental parameters which, although often self-evident to local communities, require seemingly complex accounting methods to justify action on the part of official agencies.

It is important to maintain a strong sense of history in examining the transformation process. Despite the image projected by many authorities, mountains are not and have never been 'static'. Our historical knowledge is unfortunately too limited in many cases to produce a realistic long-term account. However, over the last century or so most mountain areas have

283

witnessed an increased pace of socio-economic change. In the European Alps during the nineteenth century, from the early years of the twentieth century in the Andes and in the last few decades in the Himalayas and other ranges, very startling changes in the socio-economic and political structure have taken place between generations, and today evolve within generations. Much of the stimulus for change has occurred by virtue of the increasing integration between mountain and lowland. Today, with few exceptions, most of the debate on the key problems of mountain areas takes place mainly in lowland cities whose perception of the issues may be quite different from the views of those living in the mountains. Or was, because, as we have seen in our discussion of amenity migration and tourism, in some areas the separation between the two has become increasingly hard to justify!

Part 4

MOUNTAIN POLICY
AND INTERVENTION

Mountain areas have been drawn into the centre of political action for three interrelated reasons. In the first place we have seen that they have become part of a global debate about the destruction of the natural landscape and its implications, both for lowland areas as for the mountains themselves. To counter this situation an increasingly international programme of conservation has included projects in many mountain areas. Second, mountain regions remain centres of political dispute and, in some cases, actual warfare, for example in Kashmir, eastern Turkey or the highlands of Mexico. Finally, closely related to directly political issues are those associated with attempts to transform the economic and social framework of mountain regions involving 'development programmes'.

We can refer to our 'cog' diagram (Figure 1.4d) to note that one of the key features about the construction of formal programmes, whether for conservation or social development, is the fact that they are almost without exception focused on specific locations. Moreover, policies are in the main

sectoral, being the responsibility of specific agencies within either governments or international organisations. Equally, such policies are framed within
closely defined time limits and with established targets. Examining Figure
1.4d this is represented by the rectangle, which artificially cuts across the
dynamics of a system, perhaps including selected elements but for the most
part excluding linked concerns. Of course the reality is much more fluid,
spillovers and fuzziness of project boundaries both spatial and social are the
norm so that the real impact of the project is uncertain. It is interesting to
note that whilst many projects deal with issues involving both risk and uncertainty this is not an element to be found in the discussions of the overall
programmes themselves! The implications of this will be discussed later.

This section begins by re-examining the debate about degradation in mountain areas, specifically addressing the Theory of Himalayan Environmental
Degradation that has dominated much of the discussion about mountains in
the last two decades. This is followed by a review of conservation policies that
many writers believe to be crucial for the survival of mountain ecosystems.
Both of these issues are closely related to the nature of development policies
implemented within mountain areas. The third section examines the conceptual framework of these policies and allows us to re-examine some of the
examples already described earlier. This brings us to the crucial question of
the formulation of policies specifically for mountain areas. In Part 1, we
described the basic history of national and global concern, but in Part 4 the
development of a specific mountain 'project' is described with an examination
not of local policies but of broader-scale programmes. This leads on to a
consideration of the implications of some of the theoretical ideas introduced
in Part 1.

8

DEFORESTATION AND DEGRADATION

INTRODUCTION

The issue of the deforestation and degradation of mountain landscapes is a classic example of the complexity and uncertainty surrounding the state of knowledge and understanding of natural and anthropogenic processes in mountain environments. As a result of the uncertainties in data, processes and interactions of different parts of the system, the understanding of degradation and deforestation issues is hazy but the implications for policy are profound. Changing attitudes are required to promote the introduction of new approaches to policy and methodologies for tackling the wide-ranging effects of mountain degradation.

The debate about mountain degradation has centred largely on the Himalayas, especially Nepal. Many visitors have commented on the state of the forests and the visible effects of erosion since the country was opened to foreigners in 1957, but it remains to be seen to what extent the flourishing of studies and development interventions has really been helpful. Ives and Messerli (1989) produced an exhaustive critique of the 'Theory of Himalayan Environmental Degradation', pointing to its over-simplicity and generalisations, and questioned the validity of the data and the assumed linkages between variables. They have attempted to present a more realistic view of the Himalayan ecological situation than that presented in the media. This 'debunking of the myths' is itself a complex and to some degree uncertain exercise. The very data criticised for giving rise to the myths cannot always be any more reliably employed in debunking them! However, effectively they place into some sort of perspective the realities in time and space of the nature and rates of degradation processes, including deforestation. This goes some way to dispelling the unfocused hysteria that still pervades discussion of the Himalayas. However, the legacy of this hysteria remains, and the new approaches recommended by Ives and Messerli and others are only just coming to fruition. This is reflected in the growing awareness of mountain environments and their importance for wider societies, and the placing of mountains onto the world political and developmental agenda. This chapter explores the theory of degradation, more recent perceptions of degradation and deforestation, and mechanisms available to mitigate the negative effects.

THE THEORY OF HIMALAYAN ENVIRONMENTAL DEGRADATION

The theory which Ives and Messerli discuss represents a synthesis of the situation throughout Nepal in particular, but the critique is relevant to almost any mountain region experiencing growing populations and resource pressures. The following basic points summarise the sequence of degradation that is considered to afflict many mountain regions:

- Population growth arising from improved health care, in particular malaria suppression. (Figures given for Nepal indicate +2.6 per cent per annum (1971–81) – a doubling in 27 years.) Natural growth is augmented by immigration from India into the region.
- The growing population increases demands for fuelwood, construction timber and fodder requirements. In addition, forest and scrubland are cleared to provide more land for cultivation. This results in great pressures on the forest cover. (An estimate is provided suggesting that more than half the forests of Nepal have been destroyed between 1950 and 1980, with the prediction of no accessible forest remaining by 2000.)
- Clearing of cultivation land increasingly encroaches upon steeper, less stable land, resulting in massive soil erosion through increased incidence of landslides and surface erosion.
- The loss of forest and fuelwood supplies results in the increased burning of animal dung, traditionally used as organic fertiliser. This results in loss of soil productivity, soil structure and moisture making it more susceptible to erosion. The loss in productivity has to be compensated for by an increase in the area cleared for cultivation.
- Other hydrological changes resulting from the removal of the forest and increased surface susceptibility to erosion include increased flood hazards, more runoff during monsoon rains, siltation of plains, and the drying up of springs and channels in the dry season. These are manifest in silting reservoirs, changes in river channels and deposition of sediment on the lowlands.
- Transboundary issues arise from the downstream effects of flooding, and siltation occurs on the Ganges and Brahmaputra delta and adjacent plains. Plumes of sediment and the extension of islands in the Bay of Bengal, together with increased flood damage, are claimed.

The increasing global concern, and awareness of issues such as climatic change and the role that forests play in the carbon cycle, has reinforced this perception of an 'eco-crisis' on a massive scale. A strong lobby from the West has emerged which campaigns against the felling of trees in such areas. The visible erosion, siltation and flooding problems are assumed to arise directly from the inefficient, uncaring, reckless hill farmer who becomes the scapegoat for this wholesale destruction. Through their critique, Ives and Messerli have done much to establish a more realistic view of the problem and its perpetrators, and their work has since been supported by more recent research examined below.

Much of the debate is founded in the definitions of terms such as 'deforestation', and the perceptions of the nature of forest resources and environmental capabilities. Deforestation by a commercial logging company may mean clear felling or cutting of selected species, or the replacement of natural forest with plantations of selected, often alien, commercially viable species. Deforestation to the hill farmer may mean the occasional selection of trees for construction purposes, the lopping of boughs for leaf fodder, or the permanent or temporary clearance of land for agriculture. Both institutions may have strategies for forest management, but on different scales and for different ends; the former is aimed at distant destinations and economic gains, whereas the latter represents adaptations of livelihood strategies for survival.

With respect to biodiversity and carbon reservoirs, there is a huge cultural gap between the value and importance of forests advocated by western scientists and environmental campaigners and the perceptions of forests as a renewable resource serving many local users. In this latter view the forest is essentially a convertible resource whereby the clearance of trees makes way for more productive cultivated land (see Chapter 1). The nature of degradation is also questioned. Casual observers of the Himalayas lament the occurrence of landslides and erosion, but these are a fundamental part of an actively uplifting tectonically active mountain region, and the local farmers have become adept at managing slope stability and failures. Smadja (1992) describes the conservative strategies of hill farmers in the Nepal Middle Mountains and concludes that losses as a result of small-scale slope failure pose no real threat to food supplies. It has also been demonstrated that commercial logging, road construction and mechanical felling is much more damaging to the landscape in terms of erosion than the more restricted activities of the indigenous farmer (Ives and Messerli 1989). In addition, the intervention of the state into traditional forest management practices has not only excluded the locals from a source of livelihood, and hence induced protest (for example the Chipko movement), but also increased pressure on non-reserved areas and opened the way for commercial exploitation. The interaction between economics and politics has often been at the root of failure to implement effective policies, as in Thailand (see pp. 305–306).

Finally, the linkages between deforestation and flooding, landslides and sedimentation are assumed, as is the perception of widespread dung burning as fuel. Many of these linkages have been demonstrated to be at best tenuous, and at worst downright misleading. In order to evaluate the nature of the 'degradation problem' in the Himalayas, the following discussion will consider first the questions of the assumed linkages, the quality and reliability of the data on which they are based, and finally the historical context of deforestation.

ASSUMPTIONS AND LINKAGES IN THE DEGRADATION CYCLE

Erosion rates

As shown in Chapter 2, the sediment transport system of mountain environments is conducted in a series of stages. Movement of sediments is concentrated during periods of high flow, in between which they are held in temporary storage in channels, alluvial fans, etc.; there is therefore a significant lag time before sediment eroded off the slope surfaces reaches the plains. Hamilton (1987) indicates that for a watershed of about 10,000 hectares, off-site sedimentation rates will not respond to changes of use, and thus sediment yield, for decades. This implies that sediment deposited on the Ganges plains is not related to recent erosion of slopes but to a long history of such processes. There are some 5,000 metres of sediments on the Ganges plain (Sharma 1983), undoubtedly originating from the Himalayan mountain range, but the volume represents a long period of rapid erosion of actively uplifted mountain masses.

The natural rate of erosion of the Himalayas is relatively high, and the sediment load of the rivers draining the mountain watersheds significantly higher than other river basins. Ives and Messerli (1989) summarise data for river sediment loads both within India and globally. There are generally much lower figures for non-Himalayan rivers. The Kosi river (drainage basin = 62,000 km^2) yields an estimated 55,480 t/km^2 annual soil erosion from fields, and the Ganges (1,076,000 km^2) yields 27,040 t/km^2, whereas for the Nile the figure is 2,978,000 km^2 which gives only 740 t/km^2, and for the Congo (4,018,000 km^2) 320 t/km^2. The major dams along the Nile are an important, but not sole, cause of the considerably lower figures. These figures, however, even within India, are highly variable according to location, seasons and measurements undertaken, giving rise to considerable uncertainty of the true nature of the sediment yields.

Ives and Messerli (1989) provide comparisons for rates of denudation within the Himalayas of between 0.51 and 5.14 mm/yr, with exceptionally high estimates in Darjeeling of 10–20 millimetres for catastrophic storms, demonstrating the considerable variability within the region. The aridity of parts of the Himalayan catchment gives rise to higher erosion rates due to storm rainfall on much less vegetated coverage. There is also the difference in relative relief – not only are the Himalayas the highest region of the world but they are also being continually uplifted, and so the potential erosion from the supply of material, tectonic disturbance and the hydraulic energy of the rivers is much greater. Finally, the nature of the rainfall is significant; in the Amazon the tropical rainfall tends to be more aseasonal and gentler. Much of the Himalayas is dominated by the monsoon climate giving rise to dry seasons where vegetation cover may wither or be seasonally degraded, and wet seasons of high precipitation falling in heavy storms. Such storms provide much greater potential for surface erosion than gentler rain onto more continually saturated soils. In summary, therefore, the Himalayas have a naturally high rate of erosion associated with tectonic instability, high relative relief and seasonal rainfall patterns.

Forests: landslides and erosion

One of the assumptions surrounding the issue of forest clearance is not only that it will increase erosion but also that it will increase landslide activity. However, a number of studies have demonstrated that the removal of forests is not necessarily linked directly to increased landslides. Indeed, it is not always the case that forest clearance leads to increased erosion. In the case of erosion, it is the removal of vegetation cover itself that is crucial, rather than the removal of specifically forests. Studies have shown that the replacement of forests by grassland or comprehensive shrub vegetation may actually be greater protection for the soil than the original forest cover, particularly if this was open or degraded. It is also the case that the effect of raindrop impact on soils and displacement of particles may be enhanced by drops falling from leaf tips (Hamilton 1987). This is demonstrated by the appearance of compacted rings on bare soils around the trunks, corresponding to the edges of the canopy where concentrations of raindrops may fall. Where drops fall on grass, turf or shrubs, the turf binds and protects the soil, and absorbs the energy of impact. Shrubs have the effect of catching the drops and thus reducing the distance of fall, and hence the compacting or dislodging effect. Once the surface is exposed, however, similar surface erosion can occur. In the case of forests it is the growth of understorey plants, or the protection afforded by a litter layer, which is critical to the protection of the surface from erosion and the increase in infiltration capacity of the soil. The collection of litter for animal use, therefore, represents a more significant cause of increased surface erosion than the selective lopping of branches.

The second element of this debate, the relationship between forests and landslides, has also been questioned. The study of landslide frequency, magnitude and distribution in relation to forest cover has shown that deforestation actually has very little impact on the frequency of landslides in only a few locations. These areas are subject to shallow landslides and may be geologically prone to small, frequent failures. The removal of trees in these areas may cause the rotting of roots and the reduction in binding of slope materials and consequent failure. In most other areas, however, tectonic activity, particularly earthquakes, will cause landslides whether trees are present or not (Carson 1985; Hamilton 1987). Haigh *et al.* (1995) analysed the association between deforestation and new roadcuts, and slumping and sliding in the Kumaun Himalaya. They concluded that there was no correlation between these factors.

On the other hand, it is claimed that commercial timber exploitation is a much greater cause of increased landslide and erosion activity. The construction of forest roads on steep slopes contributes significantly to their instability by changing the angle of slope and angle of rest of surface debris. Disturbance on a major scale occurs with the use of large machinery, and the opening of roads and tracks concentrates runoff in these areas through clearance and compaction, thus opening the way to gully development. Laban (1979) indicates that 5 per cent of Nepal's landslides are related to road and train construction, and that the lowest frequencies occur on terrace slopes. This is supported in the Hazards Mapping project (see p. 133). In addition some landslides may be deliberately triggered as a precautionary measure. Therefore the evidence suggests that the impact of forest clearance on erosion processes depends heavily on the method of clearance itself, and the nature of the land use that follows.

Forests: flooding

A number of hydrological changes have been attributed to the clearance of forests. Whilst it is undoubtedly true that deforestation does cause changes in the hydrological regime of mountain regions, the effects are not necessarily those to which general and media concern point – the drying up of springs, increases in flood frequency and magnitude. Any changes in forest cover will affect the hydrology of a catchment.

Trees are great storers of water, and release vast quantities of water back into the atmosphere by evaporation or transpiration. As indicated above, any ground cover, be it forest litter or grass turf, will reduce erosion, increase infiltration and help prevent gullying. The loss of forest removes what is not so much a sponge as a pump, and generally results in the rise of local groundwater levels. Deforestation, therefore, is not the cause of springs drying up, and the removal of trees allows more water to be retained in ground storage. This is also extended to the dry season low flows of rivers – in a forested catchment, dry season flows are likely to be lower under other vegetation cover, other variables being equal, as the losses by transpiration are much reduced. It is therefore not deforestation that would appear to be the cause of the loss of water availability during dry seasons (Hamilton 1987). This has also been demonstrated in the case of reservoir construction, where afforestation of the catchment slopes has been introduced in order to reduce erosion and has resulted in a reduction of water availability for the reservoir and possibly not greatly reduced erosion rates.

Finally, there is the question of flood frequency and magnitude. In monsoon climates, flooding is an inherent part of the stormy nature of the rainfall. Gilmour *et al.* (1987) indicate that there is little overland flow even during the monsoon in the Middle Hills of Nepal, except during the heaviest rainfall events. This is not the case in the semi-arid regions where sparser vegetation cover and lower infiltration rates give rise to more frequent and spectacular overland flows. Flooding also results from the melting of snow and ice from higher in the catchment that is effectively beyond the influence of most forest stands. The presence of trees cannot therefore change the nature of the precipitation or its distribution in time and space, or the rates of melting of snow and ice during the spring and summer. Trees can only absorb and transpire water that has infiltrated into the soil, unless they are situated adjacent to a water body. Flooding is therefore an inherent characteristic of mountain catchments and the presence or absence of trees may have little relevance to the occurrence of floods. The natural seasonal variability of flows, and the poor data relating to flow rates, sediment yields and flooding, mean that it is not possible to demonstrate that there has been any change in flood frequency or hazard, or whether the mountains are the direct cause (Hofer 1993). There is growing evidence (Hofer 1997, 1998; Grossjean *et al.* 1995; Messerli and Hofer 1995) that local monsoon rains on the Ganges plains and adjacent sub-Himalaya, rather than precipitation on the Greater Himalaya, have the most significant correlation with flooding events. For example, the Megahalaya Hills are important due to their proximity to the plain, high rainfall and rapid runoff. By contrast, peak flow events generated in more distant mountains have often been levelled to baseflow by the time they have reached the plains.

Forests: farmers and fuelwood

The assumption of the theory given above casts the indigenous hill farmers in the role of careless exploiters. The perception of wilful, wanton destruction of forest has been perpetrated by observations of the landscape taken out of context of the wider subsistence and political and economic spheres. Some observers interpret forest clearance for cultivation ever further upslope and continual pressures from grazing and fuelwood collection as unmitigated destruction. However, as mentioned above, there has been a growing appreciation of the indigenous knowledge and management strategies of hill farmers that puts such clearance activities into a different light.

The farmers view forests as either renewable or convertible resources. These attitudes determine the pattern of use, and will change over time according to the greatest pressures facing the community at a particular time. Such changing perceptions of resources have been more formally expounded elsewhere (see Chapter 2). In the case of renewable resources, the forest is managed in order continually to provide a range of products that are essential elements in the subsistence economy. These include fuel-wood, construction timber, various foods and medicinal plants, grazing and leaf fodder, and litter. Gathering of these may be controlled by use rights determined by membership of a tribe, family, and village or by consensus of a community. Equally, a landlord may control rights of access to resources, as was the case with many Indian princes, or where monasteries own the land. Management controls may exist or evolve over time in response to shortages experienced by the user group. Some of these management systems have been shown to be sustainable and effective over long periods of time, but under the current conditions of growing population and changing economic bases such systems cannot cope. The account by Maselli (1996) of the N'fis valley discussed in Chapter 7 is a good example. Much of the problem lies with the intervention of external systems of manage-ment associated with international and national development agencies or former colonial administrations. These institutions have different priorities and attitudes to resources, and their past influence highlights the impor-tance of a historical perspective when examining these areas. Indications that the indigenous system is beginning to fail may be manifest in the lowering of the density and quality of forest cover, poor regeneration of trees and replacement of trees by scrub.

Viewing the forest as a convertible resource permits the clearance of trees in order to make way for alternative activities – cultivated terraces or grazing for example – or to select and promote particular species mixes more appro-priate to the needs of the community. The result of conversion of forest is, of course, its removal. However, although this is often decried by ecologists as a loss of biodiversity, wildlife habitat and carbon reservoir, in the eyes of the subsistence farmer it is a logical step. Cultivated terraces of maize, wheat or rice are much more productive and sustain significantly greater numbers of population than does forest. This does not mean that they cease to need forest products, but that the need for food production is more immediate and pressing. Alternative strategies for providing fodder and fuelwood are put into place – planting on terrace edges for example, which is discussed in more detail on pp. 302–304.

An important aspect of the conversion of forest land is that it often increases the surface stability of slopes. Well-maintained terraces are more effective at controlling runoff and erosion than degraded forest. Good grass turf is more effective at reducing raindrop impact and surface erosion than bare ground under open canopies. Soils are also improved significantly as a result of manuring and tilling over a period of time, increasing infiltration, soil moisture storage and structure as well as fertility. These effects are secondary benefits, rather than the initial reason for forest conversion, but they do illustrate effectively the benefits of conversion and the logical practicality of farmers making optimal use of marginal conditions.

A final misconception about the consequences of fuelwood shortages involves the burning of dung. It was observed in the theory above that as a result of fuelwood supplies being in short supply, or too far distant to be collected (especially given shortages of labour), greater quantities of animal dung were dried and burnt as fuel. Observers reported this to be the case, and indeed dung is burnt. However, Bajracharya (1983) and others explain that this is generally a last resort and that farmers recognise the critical importance of manure for crop productivity. Burning of dung was observed in the Kathmandu valley, but this was common during the winter season when otherwise it would go to waste, it not being the right season for application to the fields (Ives and Messerli 1989). Far from wasting a precious resource, therefore, villages were actually adapting it for a different use according to the finely tuned seasonal needs of their livelihood.

Uncertainties in data

A number of the linkages concerning the effect of forests upon hydrology, erosion, landslide and flood incidence, and the attitudes of farmers to forests, can be demonstrated as misunderstood or plain false. With the breaking up of these linkages, however, there still remains the reality that erosion and forest degradation and clearance do occur in the Himalayas, and often in highly destructive and destabilising ways. The farmers are also poor – not just in terms of material wealth but in choices of how to operate their livelihoods. Growing populations impose real threats to the productivity of the environment, not least because of the pressure they place on the resources and on the institutions that manage them. However, the extent of resource use and degradation is subject to great confusion as a result of the poor quality and quantity of data describing them. Ives and Messerli (1989) and others (Thompson and Warburton 1985a) highlight the difficulties arising from the data quality in assessing rates of fuelwood consumption and forest clearance in particular. The disparities arising from different methods of measurement and definitions create a very high degree of uncertainty in knowledge of the true situation.

There are two areas of particular significance – the accuracy of measurements (this refers not only to the determination of what is actually measured, but also how it is measured) and the extrapolation of this data in time and space that not only creates but also compounds these uncertainties. Both elements are fundamental to the assessment of the realities and the nature of the degradation in mountain environments.

First is the quality of data. It has already been mentioned in the context of climate that the scientific data of many densely populated, developing world mountains in particular are lacking in both time and space. Records tend to be short in duration and widely spaced. Different standards and systems of measurement are employed, making compatibility difficult or meaningless. In the same way, important disparities have been shown relating to data for fuelwood consumption, forest cover change, landslide incidence, flood hazards and even population growth. As these represent all the critical variables in the degradation debate, the compounded uncertainties indicate no real basis for asserting that there either is or is not a crisis at all.

The first point at issue is the actual rate of forest clearance. An alarmist estimate published by the World Bank (1979) indicated that half of the forested area in the Himalayas was lost between 1950 and 1980, culminating in the complete removal of accessible forest by the year 2000. Writing in 2000 it is obvious that this situation is far from having been realised. This is not to deny that forest clearance has occurred, but questions the way this is assessed and interpreted in terms of land use changes. Fürer-Haimendorf (1975) compared forest cover in 1957 with 1973, and noted changes in use (fires were no longer kept burning all day) and severe depletion of the Khumbu region. This area has become a centre of concern following its incorporation into the Sagarmatha National Park (1979) and the burgeoning tourist pressures. Houston (1982), however, suggests that forest cover in the Khumbu Himalaya has not changed, and has possibly increased. It is this type of disparity between observers that has given rise to the reassessment of the degradation debate. The use of time-series air photographs and, more recently, satellite imagery has enabled estimates of cover change to be made. Though very helpful the data raises its own problems and all imagery must be subject to careful ground truth enquiries. In many cases the actual coverage of crowns may not seem to be significantly reduced in the imagery, but on the ground the density of tree cover is often much reduced. What in the imagery seems fairly continuous cover is in reality considerably degraded in quality. Byers (1987a, 1987b) demonstrates that in the Khumbu Himalaya there is no evidence for large-scale deforestation, and indeed there has been less clearance and degradation since the 1950s than previously thought. Human activity is more likely to have been spread over several centuries rather than be concentrated in the last few decades as the theory stipulates.

The case of fuelwood consumption is an oft-cited example of the uncertainties in the data used to assess resource use. Recent estimates of the supply balance for the Hindu Kush–Karakoram–Himalaya region made by the International Centre for Integrated Mountain Development (1998) suggest that in Nepal only about two-thirds of fuelwood demand at the national level is matched by sustainable production. However, at the district level the situation is varied, with some areas (e.g. western Nepal) having a surplus and others (e.g. the central hills) being in marked deficit. In the northern mountains area of Pakistan, the same study suggests that fuelwood supply exceeds demand by a factor of 1.6.

However, this attempt to provide some estimates of a fuelwood energy balance and hence an indicator of deforestation should be viewed with some caution. Earlier estimations of the consumption per capita of fuelwood in Nepal vary by a factor of twenty-six, and the figure is sixty-seven if extreme estimates are included (Thompson and Warburton 1985a). This disparity arises, amongst other reasons, from uncertainty in the unit of measurement. In some cases, dry weight of firewood is used; elsewhere it may be unspecified, or contain unseasoned green wood. In addition, in some areas wood substitutes may be used, on a regular seasonal basis or as a result of shortages in supplies. Other studies use volume – a measurement of 'load' – which may relate to distance carried and who is carrying it, a donkey being able to carry more than a person and further. For example Macfarlane (1976), cited in Fricke (1984), refers to a load as being 3 feet long by 1.5 feet round, and Stevens (1993) takes a load to be 30 kilograms. The distance travelled to collect firewood is also repeatedly cited as a significant factor in the quantity collected, according to labour availability, and the recession of forests to distances far away from the village is a causal factor in the increase of the burning of dung. In Nauje, Sherpas habitually burn dung, especially in the higher villages above the forest zone, and even distinguish between the seasonal variations in quality of dung as a fuel – autumn dung being best (Stevens 1993). However, other studies have demonstrated that the pattern of forest use is more complex. In Teri Gahrwal, villagers tend to conserve supplies of forest resources such as fuelwood near to the village for use during times when labour is scarce, or hard pressed, thus reducing collecting times significantly (Moench 1988). During periods of relative labour or time availability, resources more distant to the village are used. Issues such as the effects of development (tourism, use of stoves, etc.) are discussed later with respect to fuelwood conservation.

Fuelwood itself is also poorly defined. It may comprise a seasonal mixture of dead, seasoned and green woods, crop residues, or dung. Some is collected as deadwood from the forest floor, other wood lopped from specific species in a coppicing arrangement. Not all fuelwood consumption therefore represents damage to forests. Coppicing is an effective way of maintaining productivity, and the clearance of deadwood allows ground cover for grazing to develop and also reduces dry season fire risks and disease.

In the same way, estimates of consumption of fodder also vary – partly as a result of the variable definitions of fodder – from anything fed to animals to specifically tree-derived sources. Fodder is usually a combination of young, fresh leaves, dead leaves, crop residues, free grazing and cut and stall-fed grass. These vary seasonally, according to availability of the whole range of possible fodder sources. In the context of forests, therefore, perhaps only the leaf element is significant, and that may be further restricted to leaf fodder cut from forests rather than from other trees on private land. However, the balance actually fed to animals will not only vary according to seasonal availability but also according to access, community restrictions and changes in the balance of the whole range of material available. Some of the strategies adopted by farmers in relation to tree product use are discussed later.

Scale and extrapolation

The second issue relating to the reliability of data is that of scale. Where uncertainties exist in the basic facts and figures, this is increased significantly if these data are applied to large regions and taken to be representative of them. Problems arise when the casual observation of localised areas, such as those near to villages, roads and trekking trails, which tend to be cleared and intensively used, is employed as a general representation of Nepalese conditions. Too often, these have been used as the basis of eco-concern about deforestation. At the same time, limited observations of well-preserved, dense forest stands have led to the reverse conclusions. What is missing in these earlier accounts is an understanding of the underlying reasons for the differences observed. Consequently, the application of these observations to the whole of Nepal gives rise to huge misunderstandings and false interpretations of the reality of the state of the forests and the landscape.

Equally misleading are reports of landslide incidence based on observations made during the monsoon period. As indicated earlier, landslides and erosion are an inherent part of the Himalayan landscape, especially during the monsoon period. The farmers' strategies, which include leaving the land to stabilise and reducing the intensity of use of landslide scars until they can be re-terraced, are interpreted as careless abandonment of damaged slopes. The continual, small-scale maintenance operations of individual farmers in the course of each day's work are often overlooked, but are the most effective and crucial factor in the maintenance of slope stability of terraced land.

Most studies of physical parameters and processes are based on detailed analysis of small catchments or plots of land. Test plots of land cannot be treated as closed systems, and the context in which they lie in terms of the whole slope or valley may be underplayed. For example, sediment delivery rates may amount to 90 per cent per hectare, but this may be equivalent to 50 per cent for 80 hectares and 30 per cent for 500 hectares (Hamilton 1987). The inappropriate extrapolation of data therefore gives highly unrealistic information. But it is on this information that the external perception of crisis in the Himalayas is based. When confronted with observations of burning, woodcutting, hungry animals and people, and the erosion of agricultural land, it is difficult not to become alarmed. Nonetheless, it must be stressed that these observations need to be placed realistically into the wider framework of the Himalayas, or a region of it.

The effect of scale is not only relevant to the interpretation of observations, however, but also to the interaction of processes in operation. The natural susceptibility of the Himalayas to erosion is intrinsic to its nature. The presence of population and the variety of activities undertaken are also fundamental parts of the region. Ives and Messerli distinguish between macro-scale, meso-scale and micro-scale in considering the interaction between these two elements, natural and anthropological.

At the macro-scale, for the Himalayan range as a whole, natural control on rates of weathering, erosion and sedimentation is by far the most significant factor. The tectonic instability, earthquake frequency, relief, hydraulic head provided by relative relief, monsoon rainfall and influence of meltwaters are all processes which operate without reference to human activity. At the micro-scale, however, the processes themselves remain the same, but the human activity within the landscape can greatly influence the rates of erosion (for example by terraces and soil conservation measures). Changes in land use can affect the runoff, erosion, infiltration, water balance, etc. within the valley. It is at the meso-scale – the valley/watershed scale – that the greatest interaction between the two extremes operates. Here the reforestation or deforestation of whole hill slopes can affect the hydrological balance of adjacent lake, river or reservoir levels. The conversion of forests to terraced agriculture on a catchment scale can significantly change the runoff and erosion balances, particularly where conversion is from open, degraded surfaces to productive terraces. At the same scale, the influences of meltwater, monsoon floods, etc. are also significant, these occurring with a force that human activity can in some cases control and in others not. The interaction of these different processes and their influence at different scales act to both enhance and prevent the degradation process. Environmental processes also interact with politics, economics and development interventions. First, however, it is important to consider the historical context of changing land use strategies and the impact these have had on forest cover, subsistence, and erosion.

CONTROLS OVER EXPLOITATION:
HISTORICAL CHANGES

The development of indigenous management strategies for mountain resources is the product of the long-term evolution of needs and opportunities, and is expressed in a variety of individual and community behaviours. Such strategies apply not just to forests, but to water and irrigation supplies, rangelands and land tenure. These are discussed elsewhere, but bear similarities in that they require the co-operation or coercion of all community members in order to function effectively.

With respect to forests, we have already mentioned that rights of use and patterns of use may be controlled to a greater or lesser extent through different institutions – households, village elders, a major landlord, or local and national government. Religious institutions, such as the Buddhist monasteries, have played a central part in the preservation of forests, and in the management and control of their use. Local farmers perceive the forest to provide a diverse range of products requiring sustainable exploitation and therefore some degree of conservatory management. Alternatively the forest is viewed as a convertible resource that can be used as cultivated land. These attitudes are not mutually exclusive. They may be instigated or changed as a result of internal or external pressures, which will not only transform the attitude and therefore status of the forest but also the boundaries which are identified as belonging to individuals or communities.

In the Teri Gahrwal, a traditionally hazy definition pertains of the distant boundaries of forest land considered as belonging to the village (Moench 1988). Watersheds roughly determine these boundaries, but whereas an incursion into distant forest by neighbouring village people is at least tolerated, such incursions near to the village are strongly resisted. Rights of access are determined by use, and there are no hard and fast rules pertaining to quantities of products harvested. In such instances, the quantity gathered is generally determined by need. Conflicts tend to arise during times of increased demands and stress, and are settled between the communities concerned.

In other Himalayan communities, particularly those having experienced significant population growth and therefore pressure on resources, strict rules are enforced by village representatives pertaining to who can gather what, how much and when. Systems of fines are imposed on offenders, and peer pressure in traditional communities can be effective at keeping villagers within the laws agreed by the community. Conflict resolution generally occurs by negotiation, and agreed settlements, law adjustments, etc. within the immediate community.

Historically, leading figures, such as princes or major landlords, play a more central role. The ownership of forest, and in some cases all land, is claimed by the sovereign or landlord. This person can therefore dictate the pattern of exploitation and demand enforcement of rules imposed. In communities where all land effectively belongs to the ruler, a feudal system may exist whereby the tenants have very little redress to imposed rules, and little basis on which to establish security. Conflicts arising from the imposition of rules therefore tend to be limited to local protest.

Finally, state government and foreign colonial powers have had a significant influence on forest management and indeed on their extent and quality today. During the colonial era, the British relied on the Himalayan states for forest products, especially timber, particularly during the nineteenth-century railway development. They also observed problems of deforestation and took action to prevent it. This involved the enclosure of the forest lands, and exclusion of local people from this part of their livelihood. Inevitably, pressure increased on the areas remaining open and led to deforestation, this corresponding to the significant degradation of forests observed by Fürer-Haimendorf (1975). On the other hand, the reserved and protected areas appeared dense and healthy, corresponding to the observations of Houston (1982). The apparent disparity is therefore a function of forest history and ownership claims, rather than present abuse by local farmers due to wilful neglect. Work by Byers (1987a) is much less alarmist, and suggests that, over longer periods and wider areas, slower rates predominate.

Few authorities were inclined to appreciate the diversity of conditions in forest land, and the 'deforestation argument' prevailed both in the colonial period and through into independence. This reinforced the view that the farmers were incapable of managing commons resources effectively and therefore should not be permitted more control over them. However, as we have seen already through numerous examples, commercial logging was usually to meet national (usually lowland) demands for timber. This obviously hypocritical position, and the clear felling and great damage caused by commercial logging, caused substantial resentment in the local communities, leading to protests such as that of the Chipko movement (Guha 1989).

Overall, it is important to state that deforestation is a problem, but there is great geographical variability. Large areas of good, healthy forest remain – a far cry from the predicted total clearance by 2000. Erosion and landslides and floods are an established part of the natural landscape of the Himalayas, and it is primarily on a micro- to meso-scale that human activities can operate to enhance or reduce the intensity and magnitude of some aspects of these processes.

The importance of forest lands to subsistence livelihoods has since been increasingly appreciated. In addition, an acknowledgement of the existence of community forest management strategies has led to the rise of 'Community Forestry' as the most effective approach to local participation and involvement in the preservation and managed exploitation of forest resources. This may be based on the existing traditional rights of access and controls. Nonetheless, it has also been pointed out by Stevens (1993) that the greater and different pressures faced by communities as a result of capitalist penetration and economic integration with the wider world have meant that these traditional institutions alone are not able to cope. However, they form an effective starting point to the development of efficient forestry policy. This is discussed further on pp. 304–308, but first it is useful to consider the indigenous responses to shortage of forest resources. For example, having converted even degraded forest into cultivated terraces, how do they make up the shortfall in fodder, fuelwood and timber?

Afforestation by farmers

Trees are normally perceived as static features in the landscape. This arises from their relative longevity and apparent stability. However, various studies (Gilmour 1988, 1995; Gilmour and Nurse 1991) have maintained that, in order to appreciate the role of trees as a multiple crop source for subsistence farmers, trees need to be considered as much more dynamic elements of the landscape. Mountain livelihoods require a balanced exploitation of a wide range of resources, which are closely integrated. Any constraints in the supply of any one element within this system need to be compensated by alternatives. For example, lack of fodder reduces animal numbers, which reduces manure and therefore soil fertility. Communities have developed sophisticated systems to maintain productivity. Some responses are at an individual level, whilst others are community based. Responses may involve more distinctive boundary delimitation, or regulations regarding what may be gathered when and where.

One response of individual farmers to a shortage of fodder or fuelwood has been to plant their own supplies on their cultivated land. This either takes the form of deliberate planting where preferred species are selected, or the nurturing of natural regeneration. Gilmour (1988, 1995) describes this as the 'redistribution of trees in the landscape'. This is a relatively under-appreciated aspect of farmers' inherent ability to cope with change in different but often sustainable ways. Trees tend to be grown on the *bari* (rainfed) terraces, rather than the *khet* (irrigated rice terraces). The latter produce a much higher yield, and a more valuable crop. Rice is also much more susceptible to shading than wheat, maize or millet.

The effect of shade can be positive – for example in shading crops from the heat of midday – but other problems such as mildew may cause problems. The direction of shade is also considered important – the morning sun is thought to be warming and good for crops, but they benefit from shading from the burning midday heat (Carter 1992). In the Hunza valley, Pakistan, strict rules apply to tree planting in order not to cast shade on neighbouring terraces that belong to others (Whiteman 1988). The trees planted tend to reflect the preferences for species that produce good quality fodder and also good fuelwood and timber for construction. At the same time farmers also plant species that they believe to be under greatest stress in the remaining forests. Rusten and Gold (1995), for example, show that in Nepal alder is grown for timber and fuelwood because it is relatively fast growing, productive, and responds well to coppicing, despite the fact that it produces inferior fuel and fodder compared to other species such as oak. Likewise, fodder trees are favoured which give the best quality fodder for animal health, milk production and provide drier dung (which is easier to transport to fields). There is also a gender bias in the favoured fodder trees – women prefer a species of *Ficus* over *Quercus* because it is comparatively

302

easy to harvest. However, the men prefer *Quercus* as it produces a higher quality fodder and it is available during winter when other supplies are of poorer quality. Carter (1992) argues that while all year round availability is highly valued, it is the quality of the fodder that ultimately determines the choice of species planted.

A second incentive for tree planting is for cash cropping. Timber may be sold from private land to generate income. In areas where good timber is in short supply this may be very lucrative, but it only feasible, as in all cases where large stands of trees may be planted, for larger landowners who have sufficient land to spare beyond their subsistence needs. In Nepal, Rusten (1989) estimated that between 20–80 per cent of timber and other forest products were supplied by private planting. Planting was more dominated (45 per cent) by fodder in the lower altitudes (1,000–1,775 metres) of eastern and central Nepal, whilst non-fodder species dominated (68 per cent) above this altitude. This reflects the different shortages being met, as in the upper zones access to pastures would have lessened the fodder but increased the fuelwood needs.

Fruit trees (for example, apples, apricots, cherries and plums) represent a substantial capital investment for farmers, but can be highly lucrative in areas where a ready market is available for produce. The development of transport, particularly air cargo, enables fresh fruit to be transported more rapidly to distant markets. Previously, crops such as apricots in the Hunza would have been dried for sale elsewhere. The case of apple production in the Atlas Mountains, Morocco has been discussed in Chapter 7, but is typical of the growth of horticulture in many mountain regions. Fruit trees generally permit undersowing with barley or fodder crops, and an efficient agroforestry system prevails in these areas. Where substantial orchard planting has occurred, the landscape is effectively reforested as a consequence of cash crop production.

It has been estimated that the extent of indigenous planting in Nepal amounts to a 357 per cent increase in tree cover from 1964–8, equating to a productivity of between 65 and 298 t/ha (Carter and Gilmour 1989). This has occurred primarily on the *bari* terraces and uncultivated land and represents the net increase after usage. Although this only represents a small proportion of the total lost in thirty years, the estimated 16,000 hectares planted is still significant, if only because it is unprogrammed, unanticipated and unfunded by external agencies.

The role of the state in afforestation has generally taken the form of commercial plantations of single species. Plantations growing tea, coffee, etc. specifically for external markets are common, not only in the foothills of the Himalayas but throughout south-east Asia in particular. This evolved as part of the colonial exploitation of the resources of these regions. Plantation forestry also involves the planting of trees for timber production, paper and pulp and other industrial processes. The favoured species tend to be softwood conifers and eucalyptus, which grow quickly and are suitable for pulp production. Many of the forests are closed to subsistence farmers but in certain cases these users may still be permitted to gather litter and other products by agreement with the owner or by state licence. However, farmers recognise the poorer quality and unsuitability of pine litter compared with deciduous hardwood litter. Another effect of plantations is that the range of medicinal and food herbs, plants, and tubers tends not to survive the clearing and replanting process and, therefore, species are lost. This, of course, is the fundamental issue related to loss of biodiversity by monoculture.

Economic development can also help forest regeneration. Ohler (1999) reports that following forest management planning in Nepal, which has attempted to integrate livestock, fodder and timber production, there has been a substantial reduction in the number of unproductive cattle and an increase in milking buffaloes. The development of a lucrative milk market in Kathmandu has encouraged farmers to reduce the number of cows kept from around twenty per household to just two buffaloes, which are more amenable to being stall-fed with cut fodder. This significantly reduces the grazing pressure and allows fodder trees a better chance of regeneration. However, it also reduces the supply of dung (although this is now collected in the stall rather than off the grazing lands), but the cash gained can pay for chemical fertilisers to maintain arable productivity.

Policy implications

The intervention of the state, or NGOs and development agencies in implementing forest policies, has not been entirely successful. This may result from the inheritance of colonial measures, such as wholesale exclusion, protection or planting. In other cases, failure has arisen from a number of misunderstandings and inappropriate approaches, and from conflicting perceptions of the value of forests as a global resource (biodiversity and as a carbon sink countering global warming) and as a local subsistence or cash resource. This dichotomy is explored in Chapter 9.

Plantation afforestation schemes may put trees back on the hillsides but they also commonly cause resentment and protest from local farmers. Eucalyptus, a common species planted, is also highly resisted by local farmers because of the unpalatability of the litter and leaves for livestock. In both cases, the farmers do not object to the principle of tree planting but to the species, methods and appearance of plantations (Romm, cited in Thompson and Warburton 1985b). In the same study, Thompson and Warburton also discuss the problems of implementing different forestry approaches at different scales, noting that 'project blindness' is a particular scourge. The agencies often lack any awareness of the complexity of the context, and fail to relate the potential impacts of implementation to the diverse needs, institutional flexibility and dynamic attitudes to resource. Simplistic models of both context and project are, therefore, not appropriate.

Many other examples exist, such as in Hopar, Pakistan, where an incomplete understanding of the underlying religious and other belief systems has led to 'inappropriate' planting strategies. Although the state has planted forests for local use, they have been located at sites (high on the mountain sides) which are inaccessible to women, who do most of the gathering of fuelwood. The taboos in force prevent women from going to the high mountain areas (related to their safety and 'unclean' presence on high peaks held sacred). Thus the women have been unable to gain access to the resources so thoughtfully provided by the development initiatives (Hewitt 1989).

In Thailand, there is an effective stand-off in forest policy due to the national political agenda. Politics plays a significant role in Thailand's forests (England 1996). There is considerable international interest in forests for their significance in biodiversity protection and global climate change. In addition there is the local need for cultivable land for rice terraces to feed a growing population. In between, forest policies are caught in the battle between the need to exploit and recognition of the need to protect. There are several key actors in this scenario. The commercial interests of timber based enterprises have significant financial clout, small-scale farmers are sources of votes, and the military have their own agenda (against the pro-democracy commercial sector) and a propensity to overthrow the government. Forest policies therefore get stifled between the international demands for co-operation in global climate change and biodiversity policies, and the need to maintain national sovereignty over forest policies. The political games, playing factions off against each other without necessarily wielding absolute power, mean that policies frequently never come to fruition, particularly when the government is relatively unstable and subject to deposition.

Conflicts arising from forest policies that undermine the diversity and sustainability, from the farmers' point of view, are a common source of grass-roots community protest. The activity of the Karen in northern Thailand is but one example of the many protests, and demonstrates the real appreciation that farming communities have for the quality of their environment and the resources that forests can offer (Ganjanapan 1996).

Many of the conflicts and failures mentioned above arise from an incomplete understanding of the complex role that forests play in the subsistence livelihood of farmers, and in the wider natural environmental processes operating in mountains. A greater understanding of these complexities, the realities of these linkages and an acknowledgement of the uncertainties associated with the degradation of mountains could go a long way to the implementation of more sensitive and effective programmes for environmental protection. It is insufficient to invoke 'Community Forestry', whereby local communities may be facilitated in the regeneration and protection of their forest lands, as this does not always solve the problems. The traditional institutions have, in many cases, undergone considerable change as a result of the shift from subsistence to cash economies. Particularly where such institutions have been lost due to state intervention, they cannot always be resurrected. Such indigenous management strategies cannot always cope with the external demands (in quantity and type) thrust upon these communities.

However, it is possible to build on what is already happening in many areas – such as the farmers' own initiatives to supplement diminishing commons resources with their own supplies. Fisher (1989) indicates that not all traditional indigenous strategies for the management of commons resources are actually old, but are the products of a long evolution in dynamic response to changing needs. This evolution may be cut off by state intervention, as in the case of the nationalisation of Nepal's forests in the 1960s (Stevens 1993). They may not, therefore, currently exist to be built upon, or be in such poor, moribund state as to be ineffective.

Structured regeneration programmes such as that in the Kumaun foothills, a Swiss-funded project (the Tinau Watershed Management Project 1980), have been shown to be effective. Here land is handed back with conditions attached relating to ownership and use – for example a requirement to keep trees, and to cut and stall-feed fodder in order to permit natural regeneration. Flexibility is also needed, to allow for failures and also for farmers to choose from a range of options. These might include not only the choice of species to plant but also permitting land titles to be passed to the next generation. This provides an incentive to invest and protect the land, while maintaining some degree of control over land use to ensure that it satisfies the subsistence needs of the community.

Environment and development policies may be offered with the best intentions but still fail for apparently incomprehensible reasons. The introduction of cookstoves in Nepal to conserve fuelwood consumption has met with mixed responses. Household cooking fires need to burn very hot for shorter times, whereas stoves tend to be more efficient at producing steady heat for longer periods. Householders therefore found them unhelpful for rapid cooking, but teahouses found them highly efficient at producing a constant warm atmosphere.

Policies which restrict the gathering of forest products, and the cutting of fuelwood, may be effective in a number of ways, providing they do not cut off access to certain resources essential to the complex mountain life without providing alternative supplies. The growth of tourism is often associated with growing fuelwood demand, and the instigation of policies that require kerosene or other supplies to be provided by trekkers can reduce the pressure, or the need, to cut fuelwood. In the Khumbu region, for example, measures were implemented in 1979 by the Sagarmatha National

Park to ban felling for timber and the prohibition of the sale of fuelwood to trekkers and its use by such groups. A fuel depot supplying kerosene and other fuels was set up at the entrance to the park (Stevens 1993).

The complex interaction of the many variables in the Himalayan degradation 'problem' means that in fact there is not one single problem but a coagulation of many (see earlier comments). In order to solve these problems, a detailed understanding of how each variable has become, or contributes to, a particular problem is required. Consequently, a holistic response is required which needs to be flexible in order to allow adaptation throughout the system. Indigenous institutions provide a useful starting point to develop new structures but are not always appropriate, particularly in their traditional form. Much depends upon the specific manner in which they have evolved and the extent to which they can create new forms of socially acceptable management and, at the same time, form productive partnerships with other, wider scale institutions.

The greatest hindrance to the effective management of environment and degradation, population and development, lies in the failure to integrate the contributions of many different disciplines, approaches, government departments and countries. For the Himalayas, Thompson and Warburton (1985b) suggest six solutions – predictable ones such as tackling the erosion upstream, and dealing with the flooding, and more extreme measures such as allowing the system to collapse, and reclaiming the sediment islands deposited in the Bay of Bengal. Of course, the most effective strategies will be those which combine upstream and downstream initiatives, but any solution will be extremely complicated where this division is also marked by political boundaries. In the case of the Himalayas, upstream solutions apply to Nepal, Sikkim and the northern Indian states; downstream to the Indian plains and Bangladesh. The whole region is a finely balanced political chessboard, but unfortunately co-operative interaction is not always the primary motive of the respective governments.

The successful integration of different national agendas requires a legal framework. We have already mentioned the way that effective environmental action in Thailand is strangled by the contrary objectives of different factions. Robinson (1987) suggests that environmental law may be harnessed to provide such a co-operative legal framework. He provides two examples of existing treaties that could conceivably be used as starting points for bilateral action. These treaties are centred on a related issue, the sharing of Indus water (Pakistan–India) and of Ganges water (India–Nepal and India–Bangladesh). But within each country, the ministries and departments which are involved in the different facets of the degradation problem in the mountains are often independent and not given to free communication. Each may be operating a different agenda, and focus on different issues. A comprehensive reorientation of both the internal ministries and the international context would dissolve the fundamental characteristic of stern political gaming. Such changes do not occur spontaneously, quickly or at the behest of western environmental hysteria.

What is possible, however, is the modification of existing institutions and processes. Thompson and Warburton advocate this in preference to imposing 'grand designs'. They also argue that a simple 'bottom-up' approach does not really exist. Even sensitive, small-scale, flexible development options are usually just grist to the mills of national bureaucracy, and regularly emerge highly modified to meet government ends. The response to fuel and fodder shortages by planting trees is, effectively, a 'bottom-up' approach, but tends to be limited to fulfilling immediate (temporal and spatial) needs. In order to be effective, hill farmer strategies need to be able to take on the markets and political/development agendas on their own terms, but can only do this if facilitated by external agencies. These may be individuals or organisations, as is illustrated in the Andes by Bebbington (1995). It is the meeting of the 'top-down' with 'bottom-up' that appears the most promising approach, but actually achieving this is problematic (see pp. 370–371).

SUMMARY

This chapter has examined the discussions surrounding the Theory of Himalayan Environmental Degradation, noting its tenuous empirical foundations. Particular attention has focused on the role of forest development in areas that have become increasingly integrated into the capitalist economic arena. Overall this process is associated with an increase in the pursuance of private, individual gain at the expense of community-wide benefits. However, this has not always been the case, nor has it necessarily led to the general transformation of whole communities. Nonetheless, it is enough even in small doses to upset the traditional peer pressure policing systems which lie at the foundation of community resource management. In the examples above, it has been shown that such systems were not always egalitarian, particularly where ownership of land lay in the hands of a few, but that it often worked for long periods of time. The growth of private enterprise, from the time of the nationalisation of forests in Nepal by the state, has often caused irretrievable breakdown of community institutions. However, new institutions can always develop, and have always done so in response to changes in order to manage needs.

9

CONSERVATION IN MOUNTAIN ENVIRONMENTS

INTRODUCTION

This chapter examines conservation in a wider context, particularly its relationship to the debate concerning biodiversity. In particular it will focus on the differing perceptions of the nature and need for conservation strategies between the 'North' and the 'South'. These are in a state of considerable flux, currently changing from an exclusionist and protectionist stance to one in which people and environment are closely integrated. In the latter case there is a growing recognition of the importance of indigenous knowledge to inform these strategies. Although biodiversity, water resources and recreational demands have perhaps been the most powerful factors driving the demand for protection of the mountains and the conservation of their resources, a more comprehensive list is given below (IUCN 1990; Thorsell 1997):

- Mountains have high rates of biodiversity, endemism and ecosystem diversity; many species occur at their ecological limits and so are vulnerable to change.
- Mountains are fragile, sensitive to climatic and environmental changes and under increasing pressures from human and economic activity.
- Mountains have important downstream effects – erosion, water resources, timber supplies, etc. Thus, conservation in mountains may be beneficial to a much larger area in the lowlands.
- Mountains have long had sacred status to local cultures, high aesthetic value for visitors, and a growing recreational role.
- A highly diverse culture, traditional knowledge and livelihoods, language, arts, and religious beliefs are increasingly recognised as resources in their own right.
- Mountains often form borderlands, marking boundaries or buffer zones between nations, and are also areas of conflict, tribal strongholds with a strong identity associated with access to land.

ATTITUDES TO CONSERVATION

Attitudes to conservation differ globally and are shaped by concerns of a different nature and on different scales. Conservation itself is a highly controversial issue. Most agencies and actors, such as governments, farmers, scientists and a significant proportion of the populace, would agree that conservation is generally a good idea. However, the manifestation of conservation policies and strategies is highly variable, as is the motivation for them and physical expression in the landscape.

Two broad categories can be identified, North and South, using terminology which in the general literature roughly equates with 'developed' and 'developing' economies. These are born of, shaped by, and manifest in differing attitudes to resources, political, social and economic environments and differing abilities to cope with conservation demands, and requirements to implement such policies and strategies. Both North and South generate self-preservationist conservation strategies within their own boundaries, and also participate to differing extents in international concerns for issues such as biodiversity protection.

The relatively wealthy North comprises nations with economies and political environments which reflect the growing concerns over pressure on resources, not just in their exploitation but in the increasingly common process of 'ring fencing' of resources using formally defined limits surrounded by legal barbed wire. The allocation of responsibility, blame, protection and justice is the product of democratic societies to which governments, industries and the public are held accountable. The general public of western European countries, for example, illustrates a growing awareness of the degradation and pollution of mountain environments, particularly in the quality of recreational landscapes. This, together with their growing empowerment by the very accountability of governments, industry and related agencies, makes demands for controls, regulation and safety a much greater issue.

The greater wealth and recreation activity of many Europeans means that their real and expected access to environments such as mountains is much greater (Plate 9.1). With an increasing choice in tourist destinations and competition in the industry, travellers are becoming more selective in destinations, seeking the 'purer' experience and less spoilt resort. Money talks and resorts become increasingly discerning as to the importance of preserving the capital of tourism – the landscape itself.

The rise of corporate environmentalism (Williams and Todd 1997) reflects the greater environmental responsibilities for pollution prevention and the regulation concerns of industries. These concerns are derived internally from the need to tap into environmentally friendly market niches, and also to protect themselves against possible litigation by ensuring 'adequate' monitoring and controls over their activities. They are also generated externally

Plate 9.1 Cirque in the Pyrenees, now part of a protected area

by public demand, government and action group pressures and by the litigation that is set against environmentally damaging events.

Conservation in this respect evolves into a 'keep your own nose clean' issue of self-preservation, rather than a wider concern for the environment. In addition, NGOs bring to public attention more global issues such as tropical forest resource use, biodiversity and wildlife preservation; these then involve governments and, in time, local industries and activities. These strategies and concerns can then become imposed upon other countries through international protocols and actions in which a resurgence of colonial arrogance may figure. This is also a response to the increasing internationalisation of industry and service sectors and the resulting regulatory pressure to standardise health and safety provision, along with environmental controls. Enforcement is generally considered more effective in the North – the resources and access to legislation exist to assist in this. In the EU, and the USA for instance, not only is there a regulatory framework, with bans, subsidies and other instruments, but it is reinforced to some extent by political pressure groups, which at times have proved to be very effective.

In contrast, the South is approaching conservation from a different perspective, with a different agenda. Many developing nations may recognise the issues of erosion, deforestation and biodiversity concerns, but often have more immediate problems. These include civil conflict, natural disasters, crippling debts, internal corruption, and issues associated with population growth, which they usually consider more worthy of immediate concern than international pressure to embark on conservation programmes. This has, of course, changed with the evolution of conservation attitudes away from exclusion and towards inclusion of people in the plans and activities of conservation areas. However, conflict with remote, autonomous tribal groups, disputed national boundaries and the importance of political gaming points (such as Kashmir) are often hindrances to conservation policies in these areas.

Conservation itself can become bound up with political games – Qomolangma National Park in Tibet is an important political point for China (Williams and Todd 1997), although local Tibetan objectives were more concerned with resource conservation *sensu stricto*. Some of the initiatives recently presented by Preston (1997), such as Debt for Nature swaps, may in fact potentially become important political bargaining points between countries having the resources which international bodies seek to conserve.

A sea change in conservation strategies has evolved as a result of the increasing integration of people into conservation activities (i.e. conservation for/with the people). This results from the realisation that in densely populated mountains, people cannot be excluded from conservation areas without causing problems elsewhere or even altering the nature of the landscape being conserved. In the planning stages of many National Parks or Protected Areas people were often considered a problem, and plans developed which effectively excluded settlement, deliberately ignoring the legitimate needs of the resident population (Price 1996). Indeed, even in some recent literature (Huang and Yao 1992) the terminology used to describe traditional activities in the Hengduan Mountains, China, is 'primitive', without apparently recognising the potential value of these activities, as is increasingly the case elsewhere. Recent UNEP/UNESCO/IUCN documentation, such as that relating to Pakistan (IUCN 1996, 1997) openly encourages and advocates integration of local populations and livelihoods into conservation strategies.

There are therefore three different approaches to conservation arising from different political and socio-economic climates:

1   *Litigation, corporate environmentalism, and standardisation.* The motives are for self-preservation and accountability at corporate/national level. The imposition of agenda by national or international bodies, and by public pressure to increase momentum. Usually public funding is available.
2   *National concern for environmental conditions and sustainable uses of resources for subsistence agricultural activity, in countries where a large proportion of the population rely on such resources for their livelihoods.* This agenda is directed towards optimal use, intensification, longer-term

sustainable use and may involve conversion rather than preservation. In many countries poor organisation, co-operation and co-ordination, funding and conflicts between different users' agendas hamper such objectives.

3 *International concern for global issues – biodiversity, climate change – which are major transboundary problems.* NGOs and the scientific community develop the agendas. Designations and impositions usually placed by richer countries upon the poorer developing nations, with external controls, monitoring and funding. They may actively oppose internal development/resource exploitation of the country concerned and relate to issues that are considered marginal by the host government but of global significance by international organisations.

The approach of IUCN to conservation

Thorsell and Harrison (1992) and Thorsell (1997) examined data from the World Conservation and Monitoring Centre to describe the conditions that have been adopted to form the basis for including or excluding particular relief characteristics, including mountains. To be designated mountainous a landscape must have relative relief higher than 1,500 metres and cover an area larger than 10,000 km². This immediately excludes many of the European mountain areas such as the Pyrenees and the Scottish Highlands. However, many of these areas are already protected by National Park status of some kind, and therefore considered to be at less risk than other regions without such protection. In both cases, the national interest in landscape, recreation resources and the scientific value of such areas has long been recognised.

The categories identified by IUCN comprise:

1 Strict Nature Reserve (ecological importance; science, education uses).
2 National Park (scenery; science, education, recreation – primarily natural landscapes).
3 National monument/landmark (small size; significant landscape feature).
4 Managed Nature Reserve/Wildlife Sanctuary (nationally significant species/communities/landscape; some controlled harvesting/use).
5 Protected Landscape (nationally significant landscape; human–environment interaction; recreation, education, etc.; traditional land uses).

Three other categories were also identified initially, but were later dropped:

• Resource Reserve (resource development and control).
• Anthropological Reserve (traditional cultures and livelihoods).
• Multi-use Management Area (providing sustainable use of water, timber, wildlife, etc.).

These last three were largely incorporated into wider, more integrated conservation with development approaches of the late 1980s–1990s.

Internationally recognised Biosphere Reserves (protecting present and future communities, ecosystems, species) and World Heritage Sites (natural features of outstanding universal significance) are the last two which reflect more global significance of particular aspects of mountain environments and are designated by UNESCO with advisory inputs from IUCN and other bodies. Price (1996) points out that the main focus of international activity in mountain conservation, particularly the coverage of Biosphere Reserves, has been to protect gene pools and genetic resources and diversity from overexploitation and projected future climatic and environmental changes. However, whilst the designation of such reserves is internationally recognised it is not legally binding, in many cases instilling only hopes and opportunities for protection rather than concrete obligation. This is characteristic of many internationally 'designated' phenomena, and indeed environmental law as the enforcement of 'soft law' is the greatest weakness in effective international environmental conservation and sustainable development.

Coverage

Thorsell and Harrison (1992) analyse the status of Mountain Protected Areas, which is the broad designation of the state of protection of mountains throughout the globe. The World Parks Congress in Bali (1982) recommended that 10 per cent of each of the world's biomes should be protected. However, the species diversity is highly variable and greater proportions of more diverse biomes should be represented – Myers (1983) therefore recommends 20 per cent of tropical mountains, 10 per cent of temperate mountains and 5 per cent in the high latitudes. It is immediately apparent that there is a demand for greater protection and coverage in areas where the stress is greatest, the population dependent upon the resources for life is highest, and also where the most pressing needs for protection of environmental quality and biodiversity are most immediate.

This is reflected in the actual coverage. According to Thorsell's (1997) survey there are 473 sites, covering 264 million hectares, which fall within the categories he used. In addition to this there are many other sites – for example, they may be smaller, with lower relative relief, or not be primarily set apart for conservation. There is a bias towards the larger tracts of land being set aside in the Antarctic, the Arctic and in North America where large areas are relatively unused and with low indigenous pressures. The huge Greenland National Park boosts this figure. Conversely, coverage of the Palaearctic, which includes the European, Himalayan and south-east Asian mountain chains, is much more limited (14.8 per cent of the total area).

In theory, future extensions involve both increasing coverage, and modifying the rationale in different areas to account for the growing conflicts between development resource use and nature conservation, particularly in densely populated mountain areas. This is considered below in the discussion of the 'myth of multiple use' of mountain areas. Protection also needs to be reconsidered in some areas where the 'island' concept means that a limited area is protected, which may be too small to preserve a gene pool sufficient to maintain species diversity and viability. There is a particular bias towards protection of high-altitude zones, whilst the lower zones continue to be intensively used. There is also the problem of accounting for future climate change – a general warming of climate is predicted to cause movement upslope, resulting in loss of species and ecosystems, with only anthropogenically maintained species lower down the mountain to replace them.

The myth of multiple use: a 'northern' perspective

The problem of reconciling a multitude of uses and demands is common to all mountain areas. Although the nature of the demands varies, both developed and developing countries have to face the issue of how to integrate them. In both cases, conflicts emerge as a result of the differing evaluations of economic benefits and costs, and the restrictions which conservation-oriented regulations place on resource use and development opportunities.

In the Swiss National Park (*Unterengaden*) the actual economic benefits of the park for the communities are relatively small (Elsasser *et al.* 1995). The park is actually sited on land previously occupied by villagers, who were resettled elsewhere much in the style of a US National Park (M. F. Price, personal communication). Activity and development is severely limited by Swiss federal law, which prevents transport, permits access on paths only, and sticks close to its primary objectives of strict nature conservation and scientific research. Direct employment and earnings from the park are minimal and in fact financial losses are incurred as a result of the constraints of development. As a result, there are few opportunities for locals to be directly employed or involved in the park either now or in the future, unless federal law is changed.

The state may alter access rules in response to particular claims, for example the change of ownership of land in the Yukon, where land was returned to the natives, resulted in the exclusion of non-native populations. This meant that the latter were no longer able to maintain mining, trapping and hunting activities (Slocombe 1992). This raises important issues concerning entitlements and usage rights, and whether long established indigenous populations have greater entitlement to land and resources than later arrivals with conflicting rights, or governments who desire to exploit

or, equally, protect/conserve the resources in the same area. This fundamental issue underlies many such conflicts which now exist in mountain
areas as well as in other zones. The new draft Declaration on the Rights of
Indigenous People (1994) confirms the increasing international recognition
of indigenous peoples' rights to self-determination, land, and other resources
and could be an important step forward in the protection of people–environmental relations in many areas.

In the last decade one move towards greater flexibility in conservation
strategies concerns the UNESCO Biosphere Reserves, and a good example
of the debates about this framework concerns the proposed designation of
the Kluane–Wragnell–St Elias National Parks and Preserve region as a
Biosphere Reserve in Alaska and the Yukon. On the US side the Wragnell–
St Elias area was designated a National Park in 1980 and consists of
4.4 million hectares of wilderness in which lie some of the highest peaks in
North America. The Kluane National Park in Canada was established in 1972
and consists of 2.2 million hectares of land with the world's largest non-
polar icefields, glaciers and extensive wildlife set amongst Canada's highest
mountains. The designation issue is discussed by Slocombe (1992) and at
the time of writing (2000) still remains to be resolved. The principal functions of biosphere reserves are research and education, but they include the
relatively innovative approach of creating 'buffer' zones. These consist of
areas where human activity is controlled but not prohibited, whereas the protected zones exclude all human activity. Slocombe argues that the proposals
have implications for both native and non-native populations, but the need
to co-ordinate activities within each zone may stimulate much greater awareness of the possibilities for regional collaboration across national boundaries.
Moreover, the biosphere initiative could be used to be proactive in involving local populations in the process of the design and development of the
reserve. Although there is no guarantee that this would happen, Canada has
recently been attempting to improve this arena of its domestic policy and so
this mountain area could serve as a real test of political initiatives.

The decentralisation of federal law and the increasing regional responsibility for environment and development also require increased integration
of action. However, in the US, federal law can still override state sovereignty
where habitat destruction contravenes laws relating to conservation. This
can occur in federal lawsuits, for example both the case of *Palila* habitat
destruction in Hawaii (Juvík and Juvík 1984) and that of the spotted owl
were brought under the US Endangered Species Act, 1973.

In the *Palila* case, a honey creeper whose habitat is limited to forest
in Hawaii, had become endangered as a result of the expansion of feral
sheep and goats, whose grazing habits prevented regeneration of forest
(Juvík and Juvík 1984). The feral animals were introduced by European settlers as domestic animals, but escaped and thrived. They became
important economically with the rise of game hunting at the end of the

Second World War. Controls on hunting imposed to protect these species resulted in increases in their numbers and hence encroachment on the forest habitat of *Palila*. Additional pressures to the landscape have arisen from the development of the astronomical observatories since 1969. Eventually the conservation lobby brought a case to court, and as the bird was by now endangered the Federal Court ordered protection measures to be instated, and a removal of all feral sheep and goats, as required by the US Endangered Species Act, 1973. This proved highly controversial due to the dependence of the population on hunting for their income. There are twenty-nine further endangered species on the island of Mauna Kea, and this judgement sets a precedent for further environment over economy cases. Although the honey creeper habitat is limited to 2 per cent of the area, how much more might be needed to protect these other twenty-nine species?

A more recent but similarly controversial issue arose from the conflict between protection of the endangered northern spotted owl versus employment in the logging industry in Oregon. This featured prominently in the presidential elections between Bush and Clinton (Yaffee 1994). The owl is dependent on old-growth conifer forests. Opposition to logging predated the discovery of the owl as endangered but it lent crucial weight to the environmental case for suspending logging. The fact that an Environmental Assessment of the effect of logging on the owl, required by the Endangered Species Act, had not been completed resulted in suspension of logging activities. A hot debate ensued about the relative justification of protection as opposed to use of resources, or particular parts of those resources, which reflects the constant underlying conflict in conservation issues. The Endangered Species Act provisions now protect much of this area, but the loss of income to local residents continues to be a vocal loser in the battle over resources.

Integrating development and conservation: a 'southern' perspective

In the developing world, and specifically the Himalayan range, similar conflicts of use arise as a result of the imposition or introduction of conservation measures. A closer dependence on the environment for subsistence agriculture and the higher densities of population are critical here. The original imposition of US-style nature conservation approaches, which effectively excluded activity or severely limited human activity, has been modified significantly in recent years. However, the problems of exclusion from benefits and the effects of increasing costs (burdens) on the indigenous population are similar to those above – exclusion from local needs and future development opportunities (Kharel 1997).

We have already noted the rising profile of indigenous institutions in development planning and this has also become apparent in the conservation world. The process is not, however, straightforward. Whereas official attitudes to conservation tend to alter along with new ideas and approaches, this may not be a rapid process where traditional institutions are involved. For instance, access to and use of particular resources may be governed by traditional practices but these may not be consistent with the currently proclaimed conservation standards. Moreover, as we have noted earlier, some of these institutions would fail to meet the social norms dictated by external agencies – for example in the treatment of women. Consequently, conservation strategies relying upon existing institutions, whilst very appealing, may be less successful than imagined.

Ramble (1990) discusses the importance of the headman of the village of Te in Nepal as being the force for cultural conservation. This contrasts with adjacent villages where the authority of the headman has been reduced, and cultural change has occurred. The issue of whether such cultural stasis is in reality fossilisation as opposed to adaptation to changing conditions, and whether this undermines the capacity of such communities to cope with increasing external pressures, is also called into question (Funnell and Parish 1999). Similar cases of differences between adjacent villages in their attitudes to change and the upholding of individual and cultural identity occur elsewhere. In Hunza, Pakistan, for example, the oldest village, Ganesh, differs in its Islamic sect to adjacent villages, which has affected the villagers' attitudes to development opportunities. This may have slowed their 'development' in terms of externally funded projects, but has also engendered a spirit of entrepreneurship in individuals who help fund similar projects through their own efforts (Parish 1999). Strong indigenous authority, therefore, can be both a driving force for conservation initiatives and also a source of strong protest and non-co-operation with perceived alien objectives.

In the south of Quinghai Province, China, the importance of the Buddhist monasteries in the protection of the local environment has led to significant differences in environmental quality and species protection between monastery controlled lands and adjacent lands (Harris 1991). This provides a complex mosaic of effective protection over the wider area, with areas of relative sanctuary interspersed by areas of species decimation. Not all communities have such conservation 'legislation' within their culture. For example, Buddhist culture does not support killing, but in western Yunnan Lisu hunters have no taboos on hunting. Harris and Shilai (1997) report efforts to instil a hunting 'ethic' in such communities to avoid killing certain species or destroying their habitats. Whilst this operates for some species such as the gibbon, which poses no threat to their livelihood, there is a tendency towards retaliating by killing wild boar which destroy their crops and farm animals. There is also the feeling that if they didn't kill, others would and they would lose out, and an indication of some incomprehension as to why they should not kill. Animals are considered to be a gift of deities, and whilst they could in principle grasp the idea of longer-term preservation in order that killing could continue, albeit on a smaller scale, the reality seemed rather different. Similar situations occur elsewhere, especially if crops or livestock have been damaged. In the Langtang National Park, Nepal, the extension of yak grazing displaces the red panda. The reduction in wild prey for carnivores leads to increased predation on domestic livestock, and consequently increased retaliation by farmers (Kharel 1997). In their eyes, it makes little sense to protect the carnivores at the expense of their own beasts.

A number of important recommendations concerning multiple use management were made at an intergovernmental meeting of European ministers at Trento, Italy (Backmeroff *et al.* 1997). These recommendations (common to many regions) are presented under a number of headings: Protection and development; Cultural landscapes; Biodiversity and human activities; Agriculture, forestry and nature conservation; Tourism and recreation; Transport; Energy. These advocate the development and promotion of sustainable practices, compensation, impact assessments, and stimulating changes to infrastructure development and efficiencies in energy use. However, little evidence is provided of realistic assessments of the potential for implementing such measures, let alone giving more than broad statements as to what should be done. This is characteristic of many policy documents which are developed in response to complex, interrelated issues such as environmental development strategies.

Lawrence (1999) describes an interesting development in conservation strategies in the Philippines. She refers to studies of the increasing tendency of farmers to look to their own farms for tree products. This builds upon the previously discussed Nepalese examples of rearranging the trees on the landscape, but also recognises that greater biodiversity survives on farms than in the surrounding forest, as this is selectively exploited (Halladay and Gilmour 1995). She offers an alternative view of colonisers of forest edges. Traditionally viewed in the Philippines and elsewhere as evil encroachers on diminishing resources, Lawrence indicates that they are also tree planters and that such communities can develop important, sustainable, tree-based and biodiverse systems on their farms. They may seek as afar afield as 160 kilometres to find species to plant, and they maintain species which are valued higher than, but less used than, exotics. In six communities she identified 135 species of tree, 70 per cent of which were native. The ban on logging in their areas prevents them selling and thereby profiting from native hardwood species, even if they planted them. She also reiterates the prevalent issue of considering what is being conserved, for whom, and why.

Experience from Nepal

In Bishop's (1998) study of Melemchi, Nepal, she discusses some aspects of the impact of the formation of the Langtang National Park upon the local community. The Park was established in 1976 with the original emphasis on conservation and protection of natural resources and the development of tourism. Those living outside the Park's borders were no longer permitted access to its resources. This promptly provided an incentive for those people to relocate and live within it, despite the costs that external regulations might have on their lifestyles. There was some confusion and lack of understanding over the rationale for nature conservation, in the same way that there was misunderstanding between the policy-makers and the local population concerning the nature of trees as a convertible or renewable resource. There was also considerable fear that resources would be taken from them and not returned. In addition were the rumours and fear of the enforcement agencies – the Park wardens and the Nepalese army.

Melemchi was incorporated into the Park in 1989, with these real and anticipated problems. The local community was reported to have a 'vague sense of National Government' as a result of relative isolation and limited intervention. With the inclusion of the village into the National Park, the population suddenly found themselves to be controlled on all sides by central government administration. Regulations prohibited or severely constrained the collection of fuelwood, fodder and other forest products, and of grazing and timber use. Interestingly, as the local officials were not based in the village itself, the original practices continued to some degree because enforcement was relatively weak and it was a nuisance to trek a long way to request permission just to cut down a tree!

Johnston (1990) refers to the issues of rights of ownership, access, control and cultural expression under Park rules and questions the real benefits to locals, and changes in quality of life. Certainly in Melemchi, the incorporation into the Park instituted changes in grazing strategies and livestock husbandry, but the Park was in fact only one of several contemporary factors operating to change use patterns. For example, shifts towards increased middle-altitude grazing were in response to changing attitudes to the hard life of transhumance, labour availability, changing markets and demands for different goods, as well as the regulations imposed on grazing and fodder by the Park authorities.

The Park is committed to the maintenance of traditional ways of life, and may in fact permit certain cultural traditions to continue even in the face of changing economics and attitudes that are affecting culture outside the Park. Whether this is an advantage or disadvantage for either is a moot point. Certainly the promotion of literacy and education in the Park in order eventually to enable local people to take much more effective and prominent roles in park management is already under way. This should also open other opportunities for locals to take up more demanding and commanding roles in tourism, as well as giving them enhanced prospects in seeking employment outside the mountains.

Again, however, there is a parallel with examples from Switzerland with constraints on tourism and other economic development within National Parks. The Park authorities may control the tourism activity, pricing structures, infrastructure, etc., and the locals may not be, but perceive themselves to be, exploited and constrained to more demeaning tasks such as porterage. However, the increase in literacy and education is beginning to change this.

Bishop relates three areas where she recommended change and flexibility in the Park regulations in order to allow more flexibility and practicality for the Melemchi villager. First is the deliberate preservation of the hybrid herding tradition of transhumance. This has considerable tourist appeal, and any relaxation on grazing restrictions would encourage more people to maintain this tradition. In order to allow this, the herders' options and flexibility need to be incorporated. Herders could then continue to change and adapt their practices in responses to changing environmental and economic conditions. This is also a significant arena of indigenous expertise, and an effective and sustainable use of the high-altitude pastures.

The second point was the extension of the education programme to develop local involvement at higher levels in the future, and give the children a greater sense of responsibility for, understanding of and involvement in the Park as a whole. Finally, she made a plea for the Park authorities to be sensitive to the political history of the area. Melemchi, in Yolmo, only represented one history, and the whole location of the Park incorporates areas with different histories. This might affect the understanding of conservation objectives and emerge as a vitally important factor explaining different attitudes and responses to the imposition of regulations.

Stevens (1993) also referred to the issue of changing the attitudes of local people to their environment. This takes time and sensitivity. His study has been noted earlier, but in this context the issue centres around management strategies. Early management tactics involved responding to Sherpa needs and incorporated local consultants, acknowledgement of their rights and the importance of subsistence strategies in supporting populations and in maintaining environmental productivity (Lucas *et al.* 1974, cited in Brower 1991). In the 1980s confrontations occurred as a result of growing hostility to the removal of the right of access to lands, and after some fourteen or more years, no sign of their reinstatement for subsistence use. Both the local population and the Park authorities recognised the need for conservation, but had different attitudes and perspectives as to how this might be achieved. Traditional Khumbu strategies would combine conservation with use, contrasting with the exclusionary policy of the Park.

The imposition of regulations is not only a source of conflict in terms of rights of access but also in enforcement. As in Melemchi, distance from the central offices can dilute the efficacy of regulations. Inconsistency also occurs through rule bending by certain groups or permissiveness in one aspect, such as grazing, but enforcement of others such as forest protection within the same region. In Yunnan, Harris (1991) reports the forest guards enforcing felling or hunting bans, but cutting trees or practising hunting themselves. The Thai case of illegal felling 'permitted' by commercial operations or by mixing cut timber with imported sources has already been mentioned.

Inconsistency breeds confusion, mistrust and then conflict. The reinstatement of responsibility for resources, possibly with conditions attached, can enhance co-operation and clarity within both Park authorities and indigenous populations. Stevens (1993) refers to the potential of the reinstatement of traditional forest guardian systems of *shingyi nawa* as a way of developing this. This strategy is not a change in the decision or policy, but in the way it is enforced. Agreement is always preferable and more peaceable and effective than enforcement, especially when enforcement can introduce further confusions.

The IUCN in Pakistan instigated a series of initiatives for conservation incorporating rural development in 1995. These were located in the Northern Areas and North West Frontier Province (IUCN 1996, 1997). Various projects are involved, including wildlife preservation, concentrating on species such as ibex, and medicinal plant identification and monitoring. The concept of stakeholder participation at grassroots is incorporated, by which the local communities become an integral part of the conservation process. As part of this hunting fees are paid into the local communities and local people are used for identification, monitoring and protecting wildlife. They act as guides for recreational hunters, who pay fees for the sport. Thus the initiatives offer local employment, use the existing knowledge of species, haunts and terrain, and the communities can benefit from the hunting fees paid by visitors. It builds on existing village organisation institutions set up under the Aga Khan Rural Support Programme whereby representative groups of villagers come together to initiate development projects funded by AKRSP.

SINGLE SPECIES AND ECOSYSTEMS

Much of the early impetus for conservation was centred on single species, particularly large mammals that caught the imagination of western conservation lobbies. The significantly high rate of endemism in mountains has meant that these species have received substantial attention with respect to endangered species. Mammals such as the snow leopard, spectacled bear and mountain gorilla continue to attract considerable interest, funds and even fanatical support. Conservation centred on single species has proved effective in restoring habitats as an indirect consequence of interest in the selected species itself. The *Palila* example of Hawaii is a case in point. However, the current trend in conservation is much more holistic, in the preservation of entire ecosystems. This acknowledges the complex interactions of individual species with their environment and all aspects of their ecological surroundings. Indeed, conservation needs to take the complexity of ecosystems into account in order to be effective. The extension of this to include human activity is one aspect.

It is often the case that individual endangered species of insects, birds, mammals, and reptiles are endemic to faunal or floral associations. Cloud forests, for example, contain a number of endemic species. Conservation, therefore, needs to be centred on the habitat rather than the species *per se*. It is also often the case that within one habitat more than one endangered or threatened species will be involved. Ladakh, for example, is perhaps less aesthetically appealing than neighbouring parts of Tibet and Nepal. However, its relatively 'unspoilt' status means that it has become the last stronghold in the region of the snow leopard, two species of wild sheep (the Ladakhi *urial* and the Tibetan *argati*), brown bear and *kiang* (wild ass) (Fox *et al.* 1994). Thus the approach to conservation should consider the interactions between carnivores and herbivores, domestic and wild species. This also extends much further down the food chain to incorporate invertebrates such as butterflies and the flora on which they depend.

Heinen and Yonzon (1994) discussed the necessity for rethinking the existing approach of Nepal to conservation of species as part of a Biodiversity Action Plan. Within the existing legislation, the primary focus is on large mammals, even in Nepal, which is considered the leader in conservation due to its early implementation of wildlife protection legislation in 1972. However, a number of species no longer exist despite appearing on the list, and others which are currently endangered are not included on the lists of threatened species. Several species traditionally used for medical purposes are not monitored, along with those species often demanded as tourist souvenirs and hunting trophies. Consequently there is a pressing need for an evaluation exercise of these different species and a redrawing of the schedules.

This emphasises the point that conservation strategies should not be set in stone, but need to be flexible. Conservation strategies need to be constantly revised to reflect the changing status of a species resulting from natural or human induced factors. Heinen and Yonzon also emphasise the point that many studies in single species conservation are locally conducted and only short-term – usually less than three years. Such research makes for a disjointed approach comprising many small studies and initiatives that would perhaps be more effective if combined in a more holistic way. They also point out that in Nepal the schedules listing species and families that are threatened may not distinguish all species at the species level. Thus it may be possible to harvest whole families of birds and small mammals within the existing legislation without differentiating between those more or less at risk.

The importance of cloud forests has been mentioned previously with respect to their contribution to the hydrological balance of otherwise barren areas, and their importance in having a highly endemic flora and fauna. Hamilton (1995) reviewed the status of cloud forests in mountain regions. The biodiversity of cloud forests is comparable to that of tropical rainforests, but within a much smaller area. The reliance on cloud capture of water has meant that a specific flora has developed, rich in epiphytes, herbs, ferns and shrubs, which tend to increase with altitude in the humid tropics whilst tree species and lianas decrease. In Peru an estimated 32 per cent of fauna were endemic to cloud forest habitats, and in addition large mammals such as the spectacled bear (in the Andes) and mountain gorilla (Central and East Africa). An estimated 90 per cent of cloud forest in the northern Andes has been lost as a result of conversion into agricultural land. This arises from the expansion of lowland and low-altitude plantations pushing the traditional swidden agriculturists upslope into the forest zone. Clearing for opium in Thailand and forest depletion in the Tatra Mountains of eastern Europe by air pollution are other causes.

The role of ecotourism

Ecotourism is another way in which conservation is becoming integrated into other activities on which communities depend. Tourism is the fastest growing business in the world, and mountains are increasingly important destinations (see Chapter 7). Ecotourism is only a small and relatively new part of the industry, but is rapidly expanding. In some respects it is closely connected to participatory development initiatives. In a similar way to Community Forestry, the emphasis is upon the control of tourism by local populations, and of entitlements to the benefits arising from these enterprises. These come with responsibilities to conserve resources, the environment and their own heritage, but it is more appropriate and 'politically correct' that these communities should determine these issues for themselves, particularly with respect to their cultural heritage, rather than experience the imposition of a paternalistic view by external authorities. Naturally, local populations can benefit from education and assistance in developing management plans, technical and economic skills and exchanging information with similar enterprises elsewhere – the recurring theme of facilitation and capacity-building.

The burgeoning of micro-enterprises along trekking routes, together with the wider-scale community of family monopolies (Parish 1999) and competition between villages, may evolve into an *ad hoc*, competitive and uncoordinated development of the industry. More emphasis is emerging on co-ordinated and integrated approaches, whereby the whole community can benefit and the balance of power can remain within the community rather than with selected members. Substantial discussion on community-based tourism was one of the subjects of a Mountain Forum e-conference in 1998 (Godde 1999). An example from Ladakh (van Beek 1999) demonstrates the need for co-ordinated approaches and a convergence of government, local elite, external authorities, and local populations, instigating co-operation rather than assuming shared values and aspirations. Ladakh suffers from unregulated expansion of tourism-related activities, and at present some 13,000 visitors arrive each year, primarily between July and August. Some have a 'volunteer' status and act as catalysts to local initiatives.

The objectives of ecotourism are to contain tourist activities within sustainable limits with respect to the environment and economic capacities of the region. It also serves to control unfettered development and to ensure more equitable involvement and benefit. However, despite the widespread general appeal of ecotourism within some development agencies a strong note of caution is advisable. Many mountain areas are now the subject of publicity offering opportunities for extensive recreation based upon ecotourism, but the plain fact is that many of the arrangements fail to meet even the most weak definitions of the subject (Honey 1999).

The international scale

International incentives for forest protection arise from a need to protect biodiversity, and also to counteract global warming. Recent developments under the 1992 Convention on Climate Change signed at Rio UNCED have included the Kyoto Protocol (1997). This incorporates legally binding targets for the reductions in emissions of greenhouse gasses, especially carbon dioxide. In order to meet these targets, a number of flexible mechanisms (or loopholes) have been developed. These include 'joint-implementation' whereby an industrialised nation can implement afforestation programmes in another country which can offset part of its own emissions. Such a scheme operates between Norway and Costa Rica (Chaverri-Polini, 1998) where the value of the project amounts to US$45m. The general idea is to protect forests as carbon sinks to counteract the effects of carbon emissions. This 'sink-accounting' (Krorick 1999), however, is subject to considerable controversy. In the first instance there are arguments about whether it actually reduces the need to cut emissions. Second, doubts are raised on the basis that the actual contribution of trees to counter-balance emissions is subject to uncertainty with respect to the lag time (for trees to grow) and the effectiveness of planting forests in the face of eternally continuing emissions. It is seen by industrialised nations as an easy alternative to cutting emissions, and is another escape hatch along with emissions trading and similar flexible mechanisms.

A second initiative is the concept of Debt-for-Nature swaps, whereby donor countries write off a debt owing to them, or a third party (government or NGO) buys up a proportion of debt owed by a country in return for a commitment to protect specified areas of land. This concept, although simple, has been fraught with difficulties, such as the degree of coercion in giving up sovereign control of resources in return for much-needed debt relief (Sher 1993). This raises issues of who has control over the resource and enforcement of the agreement. Protection of forest lands under early agreements caused native populations to be excluded, and more recent developments have begun to take into account the role of these indigenous populations in maintaining the value of the resource by staying put, and a more co-operative and equal balance of power in management. Such an initiative is that operating between Norway and Costa Rica. International NGOs have proved important in the role of mediators and in the effective negotiation of agreements, but it requires the political will of governments to take the conservation part of the bargain seriously, and the co-operation of local populations, NGOs and environmental organisations to realise it (O'Neill and Sustein 1992).

Incorporating climate change

The predicted response of global warming is a rise in global temperatures, causing an upslope movement of ecological zones. As such belts move up slope, the available land area decreases. This causes a loss in species diversity due to competition, but may also include speciation by adaptation of existing species. The impacts of climatic change have been discussed in earlier sections, and are at present only relatively vague scenarios. Attempts have been made to model the effects of climate change on mountain ecosystems, such as that by Halpin (see Figure 5.1). However, these can only remain conjecture. With respect to conservation, however, the fact that climate changes and that vegetation responds to it poses serious problems in the planning of protected areas. It is difficult enough to estimate what might be the optimum areas to be protected in order to maintain sufficient area for diversity and the long-term viability of the gene pool under current conditions, without incorporating the added uncertainty of future climatic changes.

Halpin discusses the issue of core and buffer zones, whereby the central core is strictly protected whilst the surrounding buffer zone may incorporate varying degrees of human activity. This leads to the so-called 'island' phenomenon, whereby a series of protected, isolated islands develop, widely separated. If these are large enough, internal genetic viability may be sustained, but it is inevitable that some losses will also occur as a result of the cessation of genetic exchange with the wider area. Under the conditions of changing climate, where the buffer zone is heavily used by human activity, additional demands such as human population growth will make cultivation at higher altitudes not only potentially possible (given the capacity to construct terraces, etc.) but also desirable. This, of course, eats into the central protected core. Conflicts arising from this 'nibble' effect pose serious ethical questions about the maintenance of genetic diversity for possible future gain (for example through new agricultural and medicinal products) as opposed to the immediate necessity of providing food for populations where the lowlands may be experiencing less favourable agricultural conditions.

There is a growing emphasis on connections between protected areas by networks or corridors. These may become increasingly important physical routes of species movement as well as serving to provide regions with coherent, linked networks of protection. Such networks not only link mountain ranges (for example the Corredor Biologico Mesoamericano) but also, perhaps more importantly, encompass both highland and lowland. This potentially allows for response to changing climate with its altitudinal shifts. Such networks as Natura 2000 in Europe are established with the vision of providing a coherent ecological network of special areas of protection. In Europe, however, such initiatives are still sidelined by concerns of 'overriding public interest' such as transport networks, agricultural or other economic needs (Scott 1998), and thus protection or conservation can never be considered as absolute.

SUMMARY

This chapter has examined the particular issues of conservation and protection as they are revealed in mountain regions. The focus has been towards forest resources and particularly biodiversity, and there is now a wide range of instruments – National Parks, Wildlife Reserves, Biosphere Reserves, etc. – at all geographic scales used to seek some acceptable equilibrium between exploitation and conservation. One important dimension that the chapter has highlighted is that of the different perspectives between the 'North' and the 'South' about this issue. For mountain areas, whilst by no means all-embracing, this tends to highlight the dichotomy of Birnbaum, made in Chapter 1, between those mountain areas with sparse communities and those where there are relatively dense, usually agriculturally based populations. In this instance the slow but increasing momentum of recognition of the need to embrace 'people' in the conservation programmes marks a major shift away from the decades of exclusion which both damaged and alienated mountain communities. Nonetheless, conservation programmes in mountain areas, as elsewhere, should be seen as essentially dynamic as the balance between utilisation and conservation is constantly reformulated.

10

POLICY AND DEVELOPMENT INITIATIVES IN MOUNTAIN AREAS

INTRODUCTION

This chapter begins with a brief account of the main development models that have had an impact in mountain regions. This is followed by a more detailed exploration of the framework in which mountain policy initiatives have been developed, particularly at the international scale, which involves an appreciation of the wider political agendas. This builds upon the short synopsis provided in Part 1 and is followed by an examination of some specific examples of policies developed for mountain regions.

DEVELOPMENT AND MOUNTAIN REGIONS

The historical evolution of a particular society is a long-term process of economic and social transformation. The discourse of development, whilst at one level part of this transformation process itself, has had a somewhat separate evolution. In western literature it has been closely allied with the concept of progress which is, in turn, a culturally and therefore politically sensitive issue. A striving for material improvement can be contrasted with the search for nirvana, though increasingly the globalisation of a Eurocentric model of progress has tended to assume that the two are almost identical! The more pragmatic 'development discourse' has been dominated by the belief in the validity of agency intervention into social structure and behaviour to address deficiencies and problems that are defined by prevailing social norms. Thus today, internationally set goals emphasise sustainability and the alleviation of poverty, particularly in less developed countries.

The discourses of development are controversial and form the topics of many excellent books (Preston 1994; Corbridge 1995; Grillo and Stirrat 1997). For simplicity we will look briefly at two major themes, recognising that there are many variants. First, and historically the most powerful, is the approach which associates development with modernisation. This involves the application of technology and the construction of institutions that will deliver progress, usually defined in terms of economic growth and social 'improvements'. As understood by conventional economic analysis modernisation involves not only technical change and the application of formal (western) scientific knowledge but also more fundamental shifts in social structure. Work patterns, family organisation and occupational and geographic mobility are altered. For instance work arrangements are increasingly separated from domestic activities and become more rigid and hierarchically controlled. Proponents of this perspective emphasise the fact that an underdevelopment or marginality can be attributed to the failure of a specific society or area to transform itself. Consequently, development strategies involve the conscious removal of these 'blockages' to modern development. These themes are associated with both capitalism and various forms of state socialism and have generated their own vocabulary that has become either the language of the state or of capitalist enterprise as the principal agent of change. At least from the middle of the nineteenth century, development has become associated with nation building and is therefore embedded in the prevailing political rhetoric.

Second, there are strong critiques of the 'modernisation' model, arguing that its Eurocentric and now global rhetoric is associated with exactly the same processes that create and perpetuate divisions between rich and poor. The development of the underdevelopment school which flourished in the 1970s placed the explanation for global patterns of poverty squarely on the international division of labour and the framework of unequal exchange that ensued (Kitching 1980). As we noted in Chapter 7, marginality is created by a specific set of economic and social relationships following the introduction of new socio-economic activities, particularly those under capitalism or state socialism. A good example of a more recent critique is that of Escobar (1995). He argues that the process of development, which is the key element of contemporary modernisation, has served to create underdevelopment in a form that can be easily controlled by the state. In particular, the technical and political strategies, which are the cornerstone of the modernisation discourse, serve to perpetuate social division as a means to uphold the essentially western model of civic society. He also maintains that there has been a distinctive growth of social movements that are struggling to produce alternatives to the development model built around modernisation. These are usually grassroots movements that focus on specific localities and attempt to redefine the process of transformation in terms of predominantly local cultural norms and thereby often generate active opposition to the state.

Mountains have not featured explicitly in the vast bulk of the practical or theoretical literature on development. This is not surprising because conventional concepts of development have been constructed by a lowland, mainly urban-based elite whose principal interest is in maintaining and enlarging its own power base. In terms of practical policies this means that urban based initiatives and 'high potential areas' have dominated attention. Development interventions in mountain areas have been constructed as part of more general policies and have therefore followed the prevailing rhetoric. This is evident from the discussions in Chapter 7, most of which may be labelled as 'modernisation'. The attempts to use 'Green Revolution' technology for improving agricultural output, for example in the Himalayan foothills of northern India or in parts of the Andean range, have generally been of limited success. Most of the agricultural technologies promoted in development programmes require the shift to monocropping and the increased use of purchased inputs. This has had its successes when measured by strictly financial criteria. In the highlands of Thailand, some individuals are successfully exploiting the market for cabbages and other fresh vegetables, building on state support for such enterprises. Similarly, logging has become a critical part of the economy of some mountain areas, despite the environmental lobby discussed earlier.

In these examples, the development projects have built upon the comparative advantage of mountain areas; that is to say, the development logic is founded on the presumption that the local population will benefit from exploiting a resource which has a market principally outside the mountain areas. This of course is not a new feature. We saw that, for many centuries, mining enterprises in mountain areas have been temporary features of the economic landscape. When either the resource was mined out, or the market changed, so the community died. But often the scars remain, both social and physical, and it is this element that has led to a raft of policies, at least in rich countries, which attempt to maintain jobs and a community in these areas. On the other hand, as Part 3 explored, many other technical developments have proved very significant for mountain areas, for example the development of transport and communications that have been instrumental in opening up mountain regions.

The modernisation ethic has had its most powerful support during the period of state socialism in the former USSR and similar countries. The development of many mountain areas was explicitly geared to the provision of commodities for industrialisation. In addition to the well-known mining, forestry and water investments, many of which were large-scale and a hallmark of the period from 1917–91, agricultural enterprises were reorganised according to the tenets of state socialism. In some cases full-scale collectives were created, although peasant farming persisted in some remote areas. An example of this is Tajikistan (Badenkov 1992), where much of the landscape lies above 3,000 metres. From the 1920s development policies were orchestrated from Moscow and were designed to provide raw materials for the industrialisation of the USSR. In addition, the mountain resources have been utilised to provide power and irrigation water for extensive state-run farms producing cotton. On the other hand, what was once extensive, high quality grazing in the mountains has been severely damaged by persistent over-grazing. Another feature of centralised planning in the Soviet period was the forced migrations. In Tajikistan they occurred mainly in the 1930s, the 1950s and the 1970s. To operate the new agricultural enterprises producing cotton, large populations were forcibly removed from some mountain zones that were considered 'unpromising', and settled in the irrigated valleys and lowlands. Today, population growth in Tajikistan is one of the highest in the world at about 4 per cent per annum. In a recent environmental evaluation of the region (ERIN 1998) it is clear that much of the present development policy is geared to coping with the enormous pressures on local resources and the recovery of the natural resource base. Unfortunately, one of the consequences of Soviet policies, apart from increasing the marginalisation of the mountain districts, was to generate considerable ethnic mixing. Since 1991, development initiatives have been severely disrupted by the outbreak of civil strife between the various groups within the region, each seeking to recapture the resources left behind by the retreat of the Russian state. The capital Dushanbe became the centre of violent clashes, which makes the establishment of a long-term recovery programme very difficult.

In many former Soviet mountain areas the problem has been one of serious depopulation, very similar to that found in other parts of the world. The Caucasus is a case in point, for instance in Ossetia where there has been large-scale migration from the highlands to the lowlands since the nineteenth century. Most of this movement occurred as a response to the earlier industrialisation of the lowlands and was reinforced by the Soviet emphasis on industrialisation and urbanisation. Today less than 1 per cent of the population of this area live in the mountains, which comprise about 20 per cent of the land area. These communities remain very marginalised, as much for their ethnic composition and cultural status as for direct economic conditions, but the mountain areas have suffered from both heavy exploitation for industrial resources and a decline in careful landscape management practices by those remaining.

A considerable proportion of government investment has focused on the 'modernisation' approach and consequently often bypassed mountain areas, but international aid programmes have increasingly contained an explicit demand for a poverty focus. The programmes and projects that have arisen under this guise have often used the same logic as that more commonly associated with 'modernisation', but mountain regions have often been included as recognition of the need to tackle poverty issues specifically. In the Andes, Himalayas, Karakoram and many low-latitude mountain areas a wide range of projects of this type can be found. As might be expected, the success of the projects and the institutional frameworks in which they have been implemented have been closely related to the political status of the areas. Those zones that are considered as 'enemies of the state' for example, the Kurdish mountain regions of Turkey, receive very different treatment to those which are part of the prevailing political culture.

It is also important to recognise that the precise manner in which development policies have followed the modernisation approach has varied considerably. In the 1960s and 1970s World Bank programmes, especially for rural development, passed through a whole range of different 'fashions', including Basic Needs, Getting the Prices Right and others which have had an influence on the nature of individual projects. For example, under the influence of the Basic Needs approach to rural development, some projects in the Andes emphasised a much greater level of public sector provision of infrastructure. Many of these initiatives were (and remain) designed as part of the Integrated Development Plans which aim not only to change the economic structure of a community but to recognise that this involves a broad range of interlinked initiatives (Plate 10.1).

By contrast, from the 1980s, under the regime of Structural Adjustment, policies have emphasised the retreat of the state and withdrawal of subsidies, some of which had been of considerable benefit to mountain areas. This has been the case in Morocco where Structural Adjustment policies, particularly through their impact on agriculture, consequently affect mountain producers (Kydd and Thayer 1993). The price support programmes that encouraged relatively remote producers into markets have disappeared, primarily because of the excessive benefits lowland producers were alleged to have accumulated. On the other hand, encouraging local initiatives, perhaps minimising bureaucratic interventions, theoretically provides mountain residents with a chance to experiment with new enterprises. However, even this has proved problematic for many communities.

Sustainable development

Much of the recent interest in mountain environments has arisen as a result of the widely supported swing towards the notion of sustainable development. Although many of the ideas surrounding sustainability were present long before, the public acknowledgement of this concept came with the release of the Brundtland Report in 1987 (World Commission on Environment and Development 1987). Escobar (1996) suggests that sustainable development represents a rethinking of the relationship between nature and society and attempts reconciliation between growth and environment. It also firmly establishes the fact that environmental management at the international scale has become a crucial element in the construction of a global political economy.

The key attraction of sustainable development for mountain areas lies in the fact that, at least at the level of public rhetoric, sustainability explicitly recognises the significance of environmental variables in the development process. Consequently, the sustainability agenda appeals to those who are

Plate 10.1 The Kalam Integrated Development Project, Pakistan. Note the wide-ranging framework of the project

concerned about environmental change, an issue in which mountains play a major role. Sustainable development has been subject to extensive critical analysis and the basic elements can be examined elsewhere (Redclift 1987; Kirkby *et al.* 1995; Moffat 1996). However, the work in political ecology, and its reconstruction in a postmodern idiom (Peet and Watts 1996), provides a useful approach to examine the contribution of sustainability arguments to mountain development.

At the outset there are two general issues. For some, the essence of sustainability is the fact that it has been the principal mechanism by which the environment has been 're-invented' for political discourse. Others note that it is also a politically acceptable framework for continuing 'business as usual' and not a search to define new kinds of civil society. For the main agencies, such as most national governments, NGOs and the World Bank, it has become a tag assigned to almost all development initiatives, certainly if these are to gain acceptability in the global context. In the case of mountain regions, therefore, we must guard against the fact that today almost all policy initiatives are labelled 'sustainable', with little regard for the deeper implications of the concept. Many of the problems emerging from the transformation of mountain areas highlighted in Part 3 are now embedded in policies labelled 'sustainable development'. It is the linchpin of the work of Messerli and Ives (1997), who, whilst admitting to the problems concerning practical definitions of the concept, do not themselves embark upon a critique or evaluate the implications of 'sustainability' beyond seeing it as a major step forward from previous frameworks. In many ways the content of many of these policy initiatives has remained very similar to the approaches already discussed, namely highly technocratic but with the recognition that the economic demands may lead to long-term environmental damage. Soil degradation, deforestation and pollution are part of the prevailing production imperatives and this raises the very fundamental question of whether technical solutions can be found which do not question the prevailing institutions and social norms. This becomes particularly apparent when we note that the 'crisis' mentality surrounding much of the discussions of sustainability reinforces the desire to maintain particular lifestyles (i.e. development as 'modernisation') but does not seek to reconsider the sustainability of local cultures struggling to retain or develop alternative social models. This can be seen in the global attack on peasant agriculture by many environmental scientists seeking the 'causes' of degradation, and the problems of communities attempting to renegotiate their relationship with nation-states.

Sustainable development in mountain regions

If we return to the various components of the transformation process examined in Part 3, such as agricultural development, tourism and the management of forest areas, the last two decades have seen the introduction of sustainability criteria to individual project design and evaluation. In the discussion on Himalayan Environmental Degradation in Chapter 8, the current thinking is to design sustainability into projects aimed at the mountain population, and to ensure that projects initiated primarily for lowland benefit also recognise the environmental dynamics of highland–lowland linkages. Thus the key point about sustainable projects for mountain areas is that they are adapted to the supposedly highly diverse and fragile nature of mountain resources. Accepting the existence of diversity and actively attempting to respect it are two very different issues. Much of the involvement of agencies (governments, NGOs, international organisations) is predicated upon the existence of a framework of governance and an acceptance of bureaucratic attitudes that may not always correspond to local cultural diversity. For the most part the general attitude to diversity by these agencies is instrumental; it is recognised as a step towards managing this community, usually to ensure that it meets the demands of some goals external to the community.

In his extensive review of the potential for sustainable agriculture Jodha (1990, 1997) argues that projects have been successful in mountain areas when they involve the careful selection of innovations which match the requirements of diversity and resource limitations. He cites cases of off-season vegetable farming, stall-feeding of dairy cattle and angora rabbit breeding in Himachal Pradesh as examples of profitable diversification. These innovations are designed to exploit comparative advantages offered by the mountain environment. The work by Bebbington (1997), already noted, is another example where the enterprise and capabilities of local organisations are stressed as key factors stimulating successful intensification in the Ecuadorian and Bolivian Andes. As noted earlier, intensification is a hallmark of the conventional approach to agricultural innovation but often falls foul of the charge of 'mining' the environment. In the examples examined by Bebbington it is the targeting of high value markets that has been of considerable importance – for example, horticultural products, coca, dairy and timber products. The key factor has been the ability to add value to these products within the locality itself, rather than the normal response to export the raw products. Thus cheese, dried vegetables and cocoa and cocoa powder are the traded products which now command a good market price. It will be recalled that in Chapter 7 we noted the problem of the production of drugs in mountain areas and the difficulty of finding suitable replacement products. These examples indicate that, at least in some mountain areas, commercial success can be achieved with farm products. In the long run, however, sustaining the market for such products will depend upon continued innovation and probably further investment into processing facilities.

Three other elements are also relevant in these examples. First, at the local level the effective returns to production have been obtained only because the specific trading relations have been renegotiated, often avoiding traditional 'middle men'. This has occurred through the mediating assistance of various NGOs as well as the local initiative. Second, the family producers and their organisations have been able to develop the crops within the framework of traditional resource management procedures, which in the main are relatively sensitive to long-term environmental requirements. Finally, the success of these initiatives is still predicated on the existence of satisfactory demand that can easily melt away in the face of national or international economic policies as well as simple consumer taste.

As noted above, sustainable development has become a major influence on mountain policy initiatives. However, since the late 1980s the practical application of sustainable development has led to subtle shifts in the way the concept has been articulated. The current sustainability debate treats the environment much more in terms of resource vulnerability and is therefore one step beyond the viewpoint that treated the environment only in terms of resource availability. Nonetheless, the World Bank has gradually moved its perspective further, linking resource vulnerability and degradation to poverty (Martinez-Alier 1990). This presents some problems in so far as the development discourse presents both greed and poverty as key issues in the environmental problematic, but this at least recognises the variety of circumstances in which environmental problems may emerge. The inclusion of a poverty focus has also emerged as a reaction to the overemphasis on the specifically physical properties of the environment that characterised much of the early work on sustainable development. Currently, the concept of 'sustainable rural livelihoods' is now being applied to policies aimed at eradicating rural poverty. This tries to amalgamate concepts such as agro-ecosystems and farming systems into a practical framework for sustainable poverty alleviation. The concept draws heavily on the work of Scoones (1998) and Ticehurst and Cameron (1998) and attempts to provide a more holistic approach to development initiatives in rural areas. Much of the recent thinking of the UK Department for International Development is moving in this direction (Carney 1998).

At the same time that agencies have been reformulating their activities to embrace sustainability a number of other, arguably equally important, issues have moved centre stage. These are the interrelated themes of empowerment, governance, participation and decentralisation. These themes directly confront the problems of scale which we have highlighted throughout this book and which are of particular relevance for tackling the diversity of mountain regions. These themes are also intrinsically linked to the sustainability question. As more and more attempts are being made to generate environmentally sensitive programmes for mountain (and other) areas, it is increasingly evident that success depends on the carefully co-ordinated involvement of all stakeholders at all levels. This may seem obvious, but it has certainly not been the practice in mainstream development programmes.

On the one hand, local knowledge and enterprise require active participation at the 'grassroots level'. This itself is fraught with problems despite the fact that almost all projects now include 'participation' as a key element (Mosse 1995). Moreover, despite the enormous literature that has emerged on this topic, this is usually insufficient, especially given the strong inter-relationships between mountain and lowland and beyond. This was noted in the example from Bebbington (1997) discussed earlier, and it is the form of this relationship and nature of governance that can prove critical. For instance, the move to develop 'ethical trade' (Welford 1995), where the emphasis is on goods obtained from environmental and socially responsible production systems, is as much a function of the nature of the market-place as the local scale production environment. In their review of the possi-bilities of sustainable development in selected mountain areas, Rieder and Wyder (1997) conclude that mountain areas face a range of difficult economic pressures both internally and externally. How these are mitigated depends crucially on the structure of trade relations, between household and village, between village and state, and between the state and the rest of the world. In turn this depends on the structure of governance and regula-tion. We saw this in Chapter 7 where the decentralised authority of the Swiss system provided powerful checks on the centralised state, providing a politically constituted 'voice' for mountain regions. Many development programmes have now included elements of 'capacity building', but this often falls well short of an effective revaluation of the bureaucratic and political structure.

Paying for mountains

Recently, some researchers have begun to advocate another approach to sustainability in mountain areas, namely 'Paying for Mountains'. The concept is derived from neo-liberal thinking and the argument that sustain-able development of any kind requires that all initiatives must have a sound economic base, a point made forcibly by Rieder and Wyder (1997). This requires a number of conditions to be met. First, that the natural and social resources of mountain areas must be assigned their full value in society and, second, that these mountain communities should be able to ensure that a realistic portion of that full value is returned to the highlands. Without these, mountain communities will not escape poverty and continue to invest in the future of mountains. But these conditions require a suitable structure of economic regulation, in turn demanding appropriate modes of gover-nance. It is interesting that none of the foregoing material has contained a quantitative presentation of the flows of resources (goods, services, people and finance) into and out of a specific mountain range. The Sierra Nevada Ecosystem Project (1996) in the western USA has been one of the few attempts to measure this. It found that the whole ecosystem produces about

$2.2 billion in commodities and services each year. Despite the fact that water resources constitute 61 per cent of the total resource value, its share in the reinvestment of these resources within the area is zero. This is because the institutional framework for the extraction of water provides no means of taxing the returns from this resource. It is not considered to be 'property' as in some other regions. Interestingly, recreation uses, which only account for 21 per cent of the total resource valued at $470 million, have a reinvestment value of $10 million. It might be remarked that even here the local reinvestment is only 2 per cent. Nonetheless, it is evident that such an analysis can highlight relationships that are a crucial element in the marginalisation process and which underlie the many examples of political unrest in mountain regions (Rangan 1996).

This has a number of implications. At the level of development agendas, new initiatives are needed for financing mountain-based enterprises. These will be examined below. At a conceptual level there is a need to examine the framework in which the mountain environment itself is 'valued'. Since the 1980s an important component of the discussions relating environment and development has been the notion that 'correct' allocative decisions would be selected if resources were assigned financial values. One strand of environmental economics promoted by David Pearce (1993), Turner *et al.* (1994) and many others has developed an artillery of methods which seek to bring environmental variables 'into the market'. One advantage of this approach, which has been strongly reinforced by the pervading market ideology of the last two decades, is that techniques of financial and economic decision-making such as benefit–cost analysis have been brought into programmes of environmental management.

This approach has continued with the inclusion into decision-making models of a much larger socio-political dimension. One example of this is an increasing emphasis on stakeholder analysis where the first stage consists of the identification of each of the persons/agencies likely to be affected by a given initiative. Moreover, the analysis of partners and participatory relations introduces a direct awareness of the institutional framework in which resource evaluation is made. Needless to say, this has provoked a considerable debate that is relevant for mountain regions. In particular, the problems associated with assigning realistic monetary values to questions of landscape beauty or the valuation of peace and quiet cannot realistically be reduced to an exercise in hedonic pricing or modelled through the travel cost method. Nonetheless, these approaches do focus the debate onto the question of alternative resource use, which for practical purposes confronts all potential users.

Unfortunately, although these methods begin to indicate how mountain resources might be valued they do not provide answers to questions about how to generate or capture income. In this respect the electronic conference held under the auspices of the Mountain Forum entitled 'Investing

Table10.1 Mechanisms for financing sustainable development in mountain areas

Category	Mechanisms	Examples
Tenure rights	Property rights	Forest user groups, Mexico
	Transferable development rights	Mountain protection, USA
	Tradable water rights	Tradable water rights in Chile
User fees	Royalties	Mountaineering royalty Sagarmatha, Nepal
	Entrance fees	Gorilla viewing, Rwanda
	Tour operator fees	Pippen system, India
	Environmental taxes	Lodge tax, Nepal
Market Strategies	Green marketing	Ecotourism, Sikkim
	Co-operatives	Co-operative movement, Trentino, Italy
	Micro-finance	AKRSP, rural banking
External funding sources	Foreign aid	Global environmental facility
	National Trust resources	Snow Leopard Trust, Tibet and Mongolia
	Debt-for-nature swaps	National Trust Fund for Protected Areas, Peru
Private sector funds	Investments in conservation	Shore Trust Bank, Washington, DC

Source: Simplified from Preston (1997)

in Mountains' provides very interesting information and ideas (Preston 1997). Apart from the relatively novel form of the conference, where participants conducted their discussions by submitting short papers and commentaries through electronic mail, a number of interesting examples emerged which indicate the rich variety of initiatives currently in hand in mountain areas. This issue is likely to receive much more attention in future years.

Table 10.1 illustrates the range of mechanisms for generating finance in mountain areas. Broadly the categories include specific applications of tenure rights, the use of fees and taxes, the encouragement of market support, and external sources of funds. Some examples can give a flavour of what is involved. For instance, we have already noted in Part 2 that in many mountain areas resources have been managed by institutions that use a wide variety of tenure conditions. Although communities once managed these directly or allocated the rights to individuals, a combination of state and market pressures have meant that local control has disappeared. Often it remains unclear just who has formal tenure rights; in other cases the state has abrogated these rights. In the National Parks of Nepal the state has 'returned' these rights and communities are expected to assist in conserving the resources of the park. The allocation of ownership has also meant that income from fees from tourists and other sources can be channelled back to these groups for specifically local investment. In some cases the solution might lie in the privatisation of resource rights, an approach which currently finds strong support in many international agencies. This may be inadequate, however, because private rights are rarely constituted in a manner that takes account of other users. This is a classic case for regulation that has traditionally been supplied by local institutions (for example, the case of water management in the High Atlas) but today increasingly demands new powers and perhaps new institutional arrangements. As Pratt and Preston (1998) note, the sustainable use of mountain resources needs more than just entrepreneurial skill. All stakeholders must be brought together to negotiate the appropriate patterns of behaviour in an atmosphere conducive to mutual respect and a willingness to renegotiate if resource use is leading towards unfavourable consequences for some stakeholders. This is a tall order.

In the case of market support, this has involved the promotion of co-operatives that can provide enterprises with assistance in the face of global marketing pressure. The example of Beaufort cheese has already been noted and the success of the Trentino co-operatives in Italy also illustrates the way in which mountain agriculture can survive (Backmeroff et al. 1997). The co-operative movement handles about 80 per cent of the agricultural produce of the Province and linked institutions provide credit, social welfare and technical advice. Such programmes are a key part of the Alp Action strategy examined below. Finally, the work of the AKRSP (Aga Khan Rural Support Trust) has been instrumental in encouraging local initiatives in new crops and non-farm enterprises, particularly in the Karakoram (Kreutzmann 1993; Parish 1999).

This electronic conference highlighted a number of crucial characteristics that are generally essential for the success of these initiatives. For poorer mountain areas seeking to stimulate local enterprise, the preconditions include appropriate local institutional capacity and equitable access to education and credit availability, precisely the conditions that characterise a richer region. For the more affluent areas, although also requiring some of these infrastructural or institutional arrangements, it was the mechanism of redistribution that seemed to be most important. Many of these factors were elaborated in a later Mountain Forum Electronic Conference on Mountain Law and Peoples, which emphasised the role of community-based property rights associated with sustainable development (Kynch and Maggio 2000). In particular this conference examined the nature of devolved authority and the allocation of rights to manage resources, and it highlighted the overtly political nature of this process. Better analytical methods might indicate the nature and direction of resource flows, and interesting management or innovative ideas for enterprise have considerable potential. Nonetheless, it will be the character of both the local civil society and the structure of the politics at all levels that will determine just how these innovative ideas will be used.

MOUNTAIN DEVELOPMENT AND INTERNATIONAL REGULATION

While Chapter 7 pointed out that the populations of mountain areas have become increasingly enmeshed in the long-term process of capitalist transformation, we have illustrated above how part of this process concerns the development of formal interventionist policies under the umbrella of development initiatives. The rationale and form of these policies has shifted over time, largely in tune with the influence of international organisations. Where mountain areas form a large part of the landscape of a country and are politically important (for example in Switzerland and Ethiopia), policies have been designed with the specific mountain conditions in mind. Elsewhere, whatever has been currently the 'fashion' for general development policies has been applied to mountainous regions as part of the ongoing agenda. This can be combined with the fact that since the earliest formal recognition by international agencies of the problems facing particular environments, such as coastal zones or tropical forests, these have received special consideration. From a policy viewpoint, at least at the international level, mountains were non-existent! Consequently, both research funding and development assistance for mountains had no special claims on national and international agencies. The process by which mountain areas have become one of the key areas of international policy is therefore an important issue in our understanding of mountain environments and communities.

International support for mountain research

In order to understand how mountains have been placed on the agenda of international organisations it is necessary to consider the position of mountain research as it appeared in the 1960s. Although there has been a long tradition of geomorphological and geo-ecological work, some of which has been discussed earlier, little of this linked with the entirely different traditions of studies of mountain communities undertaken by anthropologists, geographers and others. Moreover, much of this work remained in the hands of a few individuals and, in the case of mountain research, was heavily linked to exploration and colonial occupation. Textbooks in English available at that time were heavily biased towards the physical environment, and only relatively few articles published in research journals discussed specific policy issues. A stronger tradition of mountain research existed in other European countries, with writing on the Alps and Pyrenees. In Italy, mountain policy dates back at least to the period of National Socialist rule in the 1920s and 1930s.

In Part 1 we noted the development of theoretical perspectives on mountain regions, but the work of Ives and Messerli (1990) and Price (1998) describes the establishment of the institutional capacity for much of this research and policy formation. In 1968 the International Union, under the influence of Carl Troll, established the Commission on Mountain Geo-ecology and Resource Management, and in 1971 the International Potato Centre was established in Peru and began to stimulate work on the agro-ecosystems of the Andes. In 1973 a major step forward occurred with the inauguration of the UNESCO Man and Biosphere Programme, which included (MAB-6) entitled 'Impact of human activities on mountain and tundra ecosystems'. This built upon the work of the International Biological Program, particularly the Man at High Altitude and Tundra Biome work. Consequently a key element of the MAB project was the recognition of the need to examine biosphere zones, taking account of the impact of human activity. Research projects examined these relationships in the Andes, Himalayas, the Pyrenees and the Alps. Price (1995) provides a detailed review of the European and former USSR component. The European programme focused heavily on the impact of tourism as this represented a major issue for applied research in these areas. Most of the projects tended to have a natural science bias, although the French and Spanish projects offered some exceptions and included strong input from social scientists. However, despite the plethora of research activity under MAB-6, the difficulties associated with multidisciplinary research remained and it is interesting how many of the objectives listed under the project remain as important goals twenty-five years later. Nonetheless, the MAB-6 research programme has influenced subsequent research and policies, and a series of mountain-based initiatives emerged elsewhere.

Ives and Messerli (1990) note that whilst the MAB-6 programme was in operation other scientists working in mountains began to link up with the initiative. It was from a meeting of the 'Munich Group' in 1974 that Eckholm published his famous book *Losing Ground* (Eckholm 1976). As we have seen in the discussion of the Himalayas, this had a considerable influence on the environmental movement in general and on the general perception of a crisis in mountain areas. In 1977, the United Nations University (UNU) established a research project examining Highland–Lowland Interactive Systems which itself began a series of initiatives involving the integrated study of both physical and human systems. Another important institutional step occurred in 1980 with the creation of the International Mountain Society that has acted as a focus for the dissemination of mountain issues. Probably one of the most significant developments of the 1980s has been the creation in 1984 of the International Centre for Integrated Mountain Development (ICIMOD) based in Kathmandu, which was initially sponsored by UNESCO and funded mainly by GTZ (German Development Agency) and the Swiss Development Corporation. This itself indicates that development agencies were beginning to respond to the call for greater recognition of the particularities of mountain problems.

The establishment of many of these organisations often grew from proposals developed at research workshops that gathered together researchers whose work focused on mountain regions. Ives and Messerli (1990) and Price (1998) list these, but the main point for this discussion is the fact that a loose network of interested individuals and agencies had been established. The research interests increasingly overlapped with a changing policy agenda, which has already been described. There was a strong representation of those primarily interested in the dynamics of the physical environment, which by the 1980s had become increasingly engaged with problems associated with climate change. Then, as noted in Part 1, there were those researchers who were developing new approaches to the analysis of human activities in the mountains, particularly as a response to the debate about Himalayan Environmental Degradation. Finally, there was a very strong recognition that the international agencies were adopting the notion of sustainable development as a political banner and increasingly providing funds to support policies and research in this field. Together, these developments led a group of researchers to recognise that the long-term success of mountain research and development depended upon achieving the same level of international recognition for mountains that had been granted to other ecosystems such as deserts or tropical forests. From this position developed the Mountain Agenda.

The Mountain Agenda

This originated as a result of the joint efforts of Dr R. Hoegger, formerly of the Swiss Development Corporation, and Peter Stone, who had considerable experience of international organisations. Drawing upon the technical expertise of Professor Bruno Messerli from the University of Berne, a team of co-authors was assembled to produce a public statement that charted the problems facing the world's mountains. Based at the Institute of Geography in Berne, the Mountain Agenda team set their sights on providing a case for mountains to be included as a distinctive environment at the UNCED (United Nations Conference on Environment and Development) at Rio in 1992. Their statements were published as *An Appeal for Mountains* (Mountain Agenda 1992) and *The State of the World's Mountains: A Global Report* (Stone 1992). This latter book contains many regional accounts as well as attempts to synthesise the key problems involved. The Agenda highlighted four principal themes:

1   Mountains as watertowers for human consumption: this idea is based upon the key role of mountain systems, particularly the Himalayas and the Andes, but is also evident elsewhere such as in the Atlas Mountains and the Rockies, in which large lowland populations depend for their livelihoods on water flowing from mountains.
2   Mountains as weather makers for large parts of the world: mountain massifs affect the circulation of global airmasses and, equally, are important influences on local climates.
3   Mountains are important repositories of global biodiversity, especially in tropical areas.
4   Mountains as privileged locations for spiritual and cultural recreation.

These characteristics of mountain areas were chosen because they were seen to be under serious threat arising from the following interrelated factors:

- the growing contradiction between marginalisation and integration within the national and global economy;
- the problems posed by the development and increasing use of mountain resources for agriculture and forestry;
- the impact of new activities in mountains, especially tourism;
- the impact of climate change.

The proposals contain recommendations for action that emphasise improving the level of knowledge about mountain systems, encouraging existing mountain communities in their efforts to develop sustainable livelihoods, and to improve international co-operation in research and policy design.

At the time of the preparations for UNCED-92, it was certainly clear that the research and policy framework for mountains was extremely diverse. The extensive work in the Andes, particularly its socio-political implications, was rarely accessible to workers in other areas, partly due to language difficulties. As we have already noted, the 1980s saw the development of various critiques of the approach associated with Eckholm and the Himalayan crisis. Members of the Agenda team had published widely with their concerns about how to tackle mountain problems (Ives and Messerli 1989). However, the Agenda discussion side-stepped one of the most radical critiques of development in mountains, containing no direct reference to the seminal work of Thompson et al. (1986) and its implications, discussed in Part 1. This is perhaps not surprising given that the essential purpose of the Agenda was to convey a clear message about mountain issues rather than question the whole approach to policy construction.

As a result of the enormous work put in by the Mountain Agenda, and appropriate lobbying in the byways of international agencies, mountain areas were given their own special section: Chapter 13, under the heading of

'Managing Fragile Ecosystems: sustainable mountain development', a term which itself raises interesting questions!

Agenda 21: Chapter 13

The concept of a fragile ecosystem was current within the global environmental community and had been the original designation for other areas such as deserts. As a result of the discussions at Rio, two programmes were agreed:

1 Generating and strengthening knowledge about ecology and sustainable development of mountain ecosystems.
2 Promoting integrated watershed development and alternative livelihood opportunities.

The first programme was directed towards improving the knowledge base of mountain ecosystems and socio-economic activities through networking of information and research. Not only were national and international official agencies to be involved, but also ICIMOD, UNU and other regional NGOs. The second programme emphasises the concern for environmental degradation, a theme the conference organisers had already identified prior to the inclusion of a specific mountain focus. The background to this programme introduced many of the themes concerning sustainable development and population growth which were incorporated into the ideas for mountain watershed development and which place the emphasis for action on the mountain community itself. Land use management systems, alternative sources of income and appropriate hazard management were considered to be of major importance.

These two programmes are interesting in so far as they also represent different aspects of the prevailing thinking about mountain development. The first subscribes to the view that the highly vulnerable mountain environment cannot be adequately protected without a substantial increase in knowledge. In fact, this is not quite what is being argued here. Certainly, our knowledge about the vulnerability of the environment is limited but the real issue concerns the provision of information that can justify specific political action. In other words, how much do we need to know about the problem and the efficacy of particular strategies before we are willing to take action! This itself is a political question and is contingent upon the prevailing pattern of social norms and political power. Put in this way, it opens up a considerable debate about the efficacy of improved knowledge *per se* in an essentially political question. In a similar vein, the agenda for the second programme, as read from the official text, indicates close affinity with the FAO promoted strategies of land use management. It continues to stress the role of smallholder activities in the soil erosion process despite the evidence from the Himalayan debate, and many other sources, that responsibility often lies elsewhere.

It is also worth noting that one area that could profit from improved information concerns the population of mountain areas. Almost all statements on mountain development, including Chapter 13 and subsequent documents, begin by asserting that 10 per cent of the world's population live in mountain areas and at least half of the world's population are heavily dependent on mountain resources. Despite some attempts to check this (Denniston 1995), no one appears to acknowledge that this same statement was published in Messerli (1983) which cited a source dated 1974! Can we assume, given changes in global population, that the proportion residing in mountains and elsewhere has remained stable for twenty-five years? It has been shown in Part 3, that mountains have been subject to both depopulation and high levels of population increase and so, at least at the regional level, the proportions living in mountain regions may have shifted markedly. The recent publication of the WCMC global map of mountain areas is a step in the right direction as it may prove possible to develop a spatially referenced demographic database that could be linked with the topographic database used by WCMC. At the time of writing, however, this had not been achieved and therefore this book has repeated the now rather suspect statements, though recognising them to be only 'best guesses'.

Following the Rio conference the FAO was designated as Task Manager for Chapter 13, and there have been a number of significant developments including the publication of a revised policy brief by the Mountain Agenda (1997), along with the detailed studies (Messerli and Ives 1997). Probably the most significant change in the Mountain Agenda material is the increased focus upon the problems and possibilities for mountain communities and the recognition of considerable cultural diversity. *Mountains of the World: Challenges for the Twenty-first Century* (Mountain Agenda 1997) highlights socio-economic stresses as being a crucial factor in the pressures on mountain environments. In addition, much greater emphasis is placed upon the input from mountain communities themselves, reflecting the donor driven models of participation. A number of evaluations of the progress achieved on Chapter 13 are available. In 1995 and 1997 reports to the Secretary-General of the Commission for Sustainable Development provide early indications of achievements in the first few years after the Rio Conference (FAO 1995, 1997) and a more detailed review has recently been produced for the FAO itself by Price (1999a).

Price describes the series of conferences at regional, national and international level that have followed up the initiatives from Rio and which have been initiated by FAO. In 1994, in Rome, a group involving official agencies, research institutions and some NGOs effectively became a consultative body on the implementation of Chapter 13 and subsequently met in Lima (Peru), Aviemore (UK) and Trento (Italy). One of the most noticeable characteristics about these meetings was the fact that very quickly the number of non-governmental representatives increased rapidly as organisations who had interests in mountain zones wanted to participate. In fact, a twin track approach resulted with a series of regional intergovernmental consultations. For instance, conferences were held in Asia at Kathmandu in 1994 and in Africa at Addis Ababa in 1996, whilst at the same time non-governmental organisations met in India at Dehra Dun in 1994, and in Europe at Toulouse in 1996.

As a result of the consultations at these meetings there is no doubt that a new international awareness of the role and problems of mountain regions was evident. This has been particularly noticeable with the inauguration of The Mountain Forum, a new form of participatory network with regional and global nodes largely funded by the Swiss Agency for Development and Co-operation. At the centre of this Forum is an Information Server Node, based at The Mountain Institute in the USA. A number of agencies were designated as regional co-ordinators, for instance ICIMOD (Nepal) is responsible for the Asian region and CIP (Peru) for the Andean region. This has facilitated Internet exchanges, e-mail conferences and discussions, and provides an online library. This Forum allows participants from all parts of the world and from widely diverse backgrounds to supply information and experiences, to initiate discussions and to promote policies and activities for development in mountain areas. Altogether five electronic conferences have been held, ranging from discussions on enterprise and investment through conservation and tourism to the most recent on Mountain People, Forest and Trees. Other e-conferences have been held within the Latin-American network. In the case of the Mountain People and Forests conference there were 850 subscribers, and just over one hundred made contributions. Nonetheless, the Forum still has much work to accomplish to encourage private entrepreneurs and many more government organisations to partici-pate. This is not just a question of access to Internet resources – it often requires a wholesale change in the approach by the aforementioned sectors to public discussion. It is also notable that there remains a huge gap in terms of the lack of involvement of those individuals living in mountains who have no organisational affiliation. The potential, even in relatively poor countries, for participation at this level remains to be explored.

As we noted earlier, the 1990s has been the decade in which sustainable development programmes have come to dominate agency thinking, and so attempts to evaluate the success of the Chapter 13 initiative are difficult because many other programmes include projects which involve mountain areas (Price 1999a). Of particular importance has been the Global Environment Facility (GEF), which concentrates upon promoting the sustainable use of natural resources and promotes biodiversity. The GEF was initially established in 1991 and then reorganised following UNCED-92 as an agency which mobilised financial resources to deal with four key threats to the global environment – namely, biodiversity loss, climate change, the degradation of international waters and ozone depletion. The GEF has its own operational programme (OP4) for mountain ecosystems, which states that mountains are a 'storehouse of diverse endemic and endangered biological diversity of global significance' (http://www.gefweb.org/html/operational_programs.html). This programme falls under the biodiversity focal area and is targeted on mountains in Mesoamerica, the Andes, East Africa, the Himalaya–Hindu Kush–Pamirs, as well as the highlands of Indo China and mountains on tropical islands. In the initial phase only three of fifty-seven projects dealt explicitly with mountain regions. The projects are normally implemented by national agencies and include assistance to Uganda for work in the Bwindi Impenetrable National Park, to Bhutan to help the management of the Jigme Dorji National Park, and in some of the former states of the Soviet Union in the promotion of a Transboundary Biodiversity Project. The current level of GEF commitments (April 2000) under the biodiversity programme has reached $1 billion, to which must be added a further $1.7 billion from co-financing agencies. Unfortunately it is not possible to specify clearly just how many projects would fall under a mountains label, in part because appropriate activities stretch much wider than OP4. However, a brief examination of the GEF Council documents for April 2000 (http://www.gefwb.org) suggests that although there are increasing sums directed to mountain-based projects, these are still dwarfed by allocations to other environments.

Another feature of the consultations that took place under the aegis of Chapter 13 concerns the recommendations for action. Beyond reaffirming the importance of raising the profile of mountain communities and the need for greater research and intervention, the recommendations emphasised different aspects of mountain problems and, most importantly, introduced new concerns to the mountain debate. For example, it became clear from some of the discussions that representatives wanted to see more effort directed towards the delicate issue of encouraging cultural diversity and enhancing the heritage of mountain peoples. Mainstream development thinking consistently undervalued sacred and spiritual values in mountain areas, but it was recognised that not all mountain communities wanted external involvement in these issues. Another important area was that of developing catchment-based sustainable development programmes. The particular emphasis on watershed catchments ties in neatly with the discussions in Part 2 of this book and once again highlights the role of scale. Many scientists believe that developing institutional frameworks that mimic the structure of the natural environment will enhance the sustainable use of the resources. It is also interesting to note that participants often felt that Chapter 13 failed to emphasise the need for poverty eradication, and, especially in the case of African countries, the need to focus on food security and the reduction in vulnerability to drought. As has been mentioned earlier, participants at the European meetings at Trento were keen to emphasise the need for government support, but at the same time wanted to see greater administrative autonomy for mountain communities. Finally, in a number of meetings, Chapter 13 was criticised because it failed to give sufficient priority to the improvement in status and conditions of women and children and tended to hide behind the umbrella term 'community' in its agenda.

An example of the outcome of one of these meetings is provided by Mujica and Rueda (1995) who report the conclusions and recommendations of the intergovernmental consultation concerning sustainable development of mountains in Latin America. This meeting took place at the International Potato Centre in Lima, Peru. This document is interesting because its conclusions place far greater recognition on the historical background to mountain problems and the role of the political economy of development than others of its kind. Moreover, unlike many of the other statements, which are full of 'agency talk', this report states quite directly the confrontational nature of the problems. It is worth quoting at length because it summarises, perhaps more succinctly than most other reports, some of the key issues that Chapter 13 fails to confront openly.

> That conflicts exist in the achievements of the objectives of sustainable development in the mountains of Latin America, as outlined under the guidelines of Chapter 13 of Agenda 21. We cannot fail to recognise that after 500 years there continue to be cultural clashes in systems of production between local forms of mountain region development, and outside forms of development, a result of market globalisation and the employment of foreign technology.
>
> The goals of these two development styles [are] such that the people as well as the governments themselves must look at the mountains from two different perspectives; on the one hand, it is seen as a zone of great cultural richness and biodiversity, but with a poor and underdeveloped population; while the other perspective is based on a vision of large-scale productivity which some may see solely as a source of wealth.
>
> The consequences of this situation are manifest in some of the following ways:
>
> a. The conditions of free access to a vast amount of the natural resources, especially those which have been extracted at much higher levels than would be considered socially desirable and ecologically sustainable;
>
> b. The absence of mechanisms that allow ways to account for negative externalities which many productive activities in mountain regions generate;

c. The poverty and low levels of education, characteristic of a majority of the populations living in mountain ecosystems, results in their intense consumption of natural resources, and does not give them easy access to technology which could improve their economic situation, without damaging the environment; in addition, the poorest people are those most vulnerable to contamination, environmental degradation and disasters caused by the inadequate management of natural resources;

d. Those who consume the most, particularly the richest of the population, are known for their inefficient use of renewable natural resources.

(Mujica and Rueda 1995: 2)

The document goes on to argue that the crucial task for sustainable mountain development is to design strategies which harmonise the two development styles to reinforce any complementarity between the two. In doing so it stresses the need for the state to be more involved (contrary to the current paradigm of reduced state intervention) in a regulatory role, acting as a mediator between the conflicting elements. Like other conferences the Lima meeting also proposes specific strategies pointing to the need for catchment-based initiatives, for strategies aimed at conservation and protection, and the formulation of regulatory markers by which agencies can evaluate the sustainability of projects. But, finally, and somewhat different again in emphasis, this meeting points out quite clearly that the challenge is also directed at global-level stakeholders:

To demand from first world countries to assume in proportion to their responsibilities, the goal of mitigation of so called global environmental liabilities, which particularly affect now and will do so more drastically in the future, sensitive areas, or extremely sensitive areas, such as mountains.

(Mujica and Rueda 1995: 3)

Following the review of all initiatives under Agenda 21 at the UN General Assembly Special Session in June 1997, agencies are now placing more effort on integrating existing programmes so that it is now possible to obtain a clearer picture of all the activities taking place in mountain areas. Price (1999a) reviews some of these areas and illustrates how often mountain problems emerge as part of sectoral activities. More importantly, he notes the fact that there is an urgent need for all participants in the mountain debate to establish clear sets of priorities. Whilst the original twin objectives of Chapter 13 were appropriate in the light of the political circumstances of UNCED 92, the subsequent plethora of discussions has meant that the mountain initiative is in danger of losing its focus. He argues that the FAO

Task Manager's main objective must be to ensure that mountain issues remain on the global agenda. This involves greater efforts to ensure that all stakeholders contribute to the task. In particular, he highlights the fact that few national governments have yet established specific programmes of sustainable mountain development at local, regional or national levels and that there is enormous scope for persuading the private sector that they have a part to play in many of the initiatives. In his view the future of the mountain agenda lies in moving the initiative away from a small group of scientists and development professionals to a much broader public. Whilst initiatives such as the Mountain Forum point the way in some respects, and both governments and the scientific community are aware of the need for greater public participation, there are entrenched attitudes to be overcome. It is at this point that educational programmes designed for mountain inhabitants play an important role in producing the level of informed, confident participants that are needed for such an enterprise.

Overall, therefore, the work of the Mountain Agenda team has been a major success in the process of politicising mountain issues at the global scale by linking localised or national concerns about the management of mountains to global policies about sustainable development. With sponsorship from the SDC (Swiss Agency for Co-operation and Development), the Mountain Agenda has prepared annual reports since 1997 which have been submitted to the Commission on Sustainable Development (CSD). Each year has focused on a specific topic: in 1998 water was the focus, in 1999 tourism, and in 2000 the topic will be forests.

Although *Mountains of the World* (Mountain Agenda 1997) sets out the objectives for the future, there remain considerable questions about the precise way forward. This has as much to do with more general questions of science policy and environmental management as with the problems of mountains themselves; and also a focus has not yet been articulated that does not leave the documents littered with managerialism. It is at this point that some questioning of the Agenda approach is called for. Aside from the tasks of generating a better knowledge base, there seem to be few queries asking 'Whose knowledge, whose development?' What after all is the prime purpose of the endeavour? It is to raise the consciousness of the regional, national and global communities to mountain issues so that more resources will be provided for improving the knowledge base and creating initiatives for sustainable development. The increased influence of the 'mountain lobby' means that its policy recommendations may become significant in long-term funding programmes. This approach is understandable but does leave certain assumptions unstated. For example, bids can be made for financial resources through the official channels of the UN and intergovernmental agencies. This assumes explicitly that the present structure of the state continues to retain legitimate control of mountain resources, and hence the focus on state managerialism is unproblematic (Escobar 1995, 1996). Even when NGOs are involved, there remains considerable doubt about their role as 'agents of the people'.

These problems are clearly recognised. As Ives *et al.* state in their review of future plans for mountain development, optimism must be tempered by cold reality: 'When the World Mountain Balance Sheet is reviewed as a whole, military assault on mountain peoples, legitimised by the state, surely greatly offsets all progress that has been made' (Ives *et al.* 1997: 456). It also begs the question as to whether the Agenda should follow the suggestion of Peet and Watts (1996) and put at least as much attention into the developments in civil society in mountain regions, involving civil rights and other matters not discussed in many of the policy documents.

MOUNTAIN POLICY INITIATIVES: SOME EXAMPLES

Throughout this book we have used many examples drawn from projects at all scales and located in many different mountain areas, and in the previous section we examined the process through which mountain issues have become absorbed into mainstream policy developments, particularly by international agencies. For example, in some countries national level institutions now exist which focus on sustainable development (for example, Vietnam, Slovenia) and where mountain issues play a major part in policy formulation (Price 1998). In addition, mountain initiatives are now in the process of being increasingly co-ordinated between countries under transboundary provisions. This draws attention to the fact noted earlier that many mountain ranges are divided by national boundaries but their utilisation and management involves cross-national links. A good example of this process can be found in Europe where the development of the EU, on the one hand, offers transnational political and economic structures through which to implement mountain policies; but equally, some mountain regions within the EU are themselves claiming increasing autonomy. The next section explores the European example in some depth because it contains most of the salient issues facing the development of mountain policy, at least in affluent countries.

Mountains within Europe

In the previous chapters of this book there has been frequent mention of the issues associated with the long-term development of the mountain regions within Europe. A considerable literature exists outlining the specific conditions surrounding problems and policies in individual countries, many of which are now members of the European Union. A complete issue of *Revue de Géographie Alpine* (80, 4) in 1992 provides some interesting examples from within the European Union. In addition, Euromontana, an organisation originally founded in 1974, brings together some EU countries and others, particularly from eastern Europe, which are currently outside the EU but have a common interest in mountain problems. We shall examine formal Community policies, those of selected countries, and initiatives that focus exclusively on the mountain regions *per se* regardless of national boundaries. Like many of the main mountain regions of the world, some of those in Europe such as the Alps and the Pyrenees also pose important transboundary issues, as each of the countries involved has its own legacy and attitudes towards the mountain regions. Furthermore, although Switzerland and Slovenia are not currently members of the EU they have a critical role to play in questions of future development of the Alps. It is possible in the European context to identify a 'politique montagne' (Broggio 1992) which has played a major role both in individual countries and, since 1958, in the increasingly significant context of the EU.

As the EU has enlarged over the past forty years so has the area of mountains contained within it. In 1992 mountain areas represented about 17 per cent of the land surface but have greatly differing roles in the politics of each country (Lowe 1992). In France 20 per cent of the territory is classified as mountainous, but Denmark or the Netherlands have virtually no high land. With the recent expansion of the EU and, for example, the addition of Austria, this figure has increased. According to a study carried out by the International Centre for Alpine Environments (ICALPE (1997), changes in the composition of the Community have meant that while mountains now represent at least 20 per cent of the community area, there has also been a considerable increase in the area classified as eligible for community assistance.

The development of policy for mountain areas is closely connected with the key role of the Common Agricultural Policy (CAP). Although other aspects of intervention associated with tourism – non-agricultural, commercial exploitation and latterly environmental issues – have become increasingly involved, the problems of agriculture originally dominated community thinking. It is in this political arena that the somewhat esoteric discussions of the definition of mountain areas, noted in Part 1, become particularly important. Many countries had already developed their own operational definitions; for example, the French had already adopted the principle of compensation for ecological hardships in mountain areas. However, in 1975 when the UK, Ireland and Denmark joined the then EEC, there had to be considerable debate about the harmonisation of special policies under the CAP. By far the largest amount of assistance for mountain areas was directed to the agricultural sector under CAP. Under Directive 75/268/EEC the Community provided for financial assistance to 'less favoured areas' (LFA). This established the fact that mountain areas, along with other areas which had special handicaps for agriculture, should receive funds to enable the continuation, and if possible the modernisation, of agriculture.

This policy builds upon one of the principal issues in the then EEC: namely that of agricultural modernisation. The Community had to cope with a highly diverse set of farming structures ranging from the large, commercial, mechanised farms (for instance in northern France) to small, very poor, technically backward units in some mountain areas. Depopulation was a serious problem in some LFAs and so the policy combined economic, environmental and social issues. One of the interesting features of the LFA programme was the fact that the Commission issued a Directive rather than a Regulation setting out the terms and conditions under which the LFAs would operate. This subtle distinction is important as a guide to how mountain policy issues might emerge in the EU. A Regulation would effectively leave little freedom for individual states to work out how best to apply this policy, whereas a Directive is more flexible in this regard. Given the wide diversity of existing practices, and the nature of mountain environments, this approach recognised political reality. The LFA definitions fell under three categories: mountains (600 metres and slopes over 15 per cent), other areas where production deficiencies reflected the difficult terrain and where output fell below the national average with problems of depopulation, and finally other areas where specific ecological problems could be identified (such as marshland). Financial assistance for farms in these areas came under four headings: annual compensatory payments reflecting higher production costs, aid for the completion of development plans, aid to investment schemes, and specific funds already allocated by national governments.

Within the broad guidelines, each country implemented its own criteria for an LFA, so that mountains and uplands were variously designated. In France, for example, several categories existed which included high mountain areas; in Britain most of the upland areas of Scotland and Wales were designated. However, it is not possible to state, from the data available, just how much of the compensatory payments for difficult conditions were directed to mountain regions alone. There is no doubt, however, that these payments, and additional benefits for capital expenditure available to farmers in such areas, did keep many in business. Over the Community as a whole, payments constituted 36 per cent of farm incomes, and, in the case of the UK, sometimes as much as 90 per cent of farm income was generated through these subsidies, which effectively maintained livestock farming in the highlands. This is well illustrated in a study by Bone and Seton (1987) of farm holdings in Speyside, where, even allowing for the subsidies on livestock, disposable incomes before interest payments barely allowed survival. However, it is also evident that there were large discrepancies between the beneficiaries and that in general the payment system did little to even out the differences in farm incomes across the Community. This is an issue of more general regional policy rather than just compensatory payments for difficult environments, but it was part of the debate that ensued which led to the reforms of the latter part of the 1980s and early 1990s.

In the 1980s, the crisis in Community financing, especially under CAP, and issues associated with regional development in the Community, led mountains to be considered as part of the general regional problem rather than just an agricultural issue. Mountain areas as such were no longer recognised as target areas, although payments could be made under objectives 1, 2 and 5b – particularly to mountainous areas in countries bordering the Mediterranean. More attention has been paid to the development of enterprises that stimulate the local economy, including tourism and local artisanal activity and assistance to co-operative activities. One of the most significant has been the use of 'Appellation d'Origine Controllé' to provide protection for high quality mountain products. The idea was initiated most successfully in the 1950s in the Beaufort valley, France, where subsequent expansion of traditional cheese production has resurrected the local dairy industry. In a community beset by depopulation and the loss of pasture through flooding for dam construction, subsequent community action involved negotiating for better rent payments for the dam. The farmers capitalised on their traditional knowledge of dairying and increased output of high quality milk from 600 to 3,000 tonnes, which commands a 25 per cent premium over other types of milk. At the same time, new technology such as portable milking sets have been introduced, enabling milking to take place on high pastures, so facilitating supplies. The producers have developed 'micro-markets' which provide some measure of success in economies becoming increasingly dominated by large food manufacturers (Vivier 1992; Warsinsky 1996).

From 1991 Community policy also began to incorporate specific measures aimed at maintaining the natural environment. In particular, it was formally recognised that payments could be made to farmers in mountain areas to continue methods of cultivation or livestock use which were thought to prevent excessive erosion. However, from the production side, the renegotiated terms of the CAP posed problems for some upland areas, although lowland grain and dairy farmers felt much of the initial pressure. The overall objective of these renegotiations was to reduce the cost of the CAP and bring farm gate prices nearer to world market levels, but by doing so it has exposed mountain farmers to increasing pressure in the long term. Certainly since 1992 two major shifts in policy within the EU have affected mountain areas: first, the much greater awareness of environmental issues, so that policy initiatives must be 'sustainable'; second, the increasing stress on reducing state involvement, leaving the Commission in a regulatory role rather than emphasising just the payment of public funds. Whilst the first has generally raised the profile of mountain areas because of their ecological role, the second has raised increasing doubts about the willingness of the EU itself to maintain and improve the socio-economic conditions of mountain populations.

Various agencies within the Community have been developing a more cohesive mountain policy. The study by the International Centre for Alpine Environments for the European Parliament (ICALPE 1997) represents the most important attempt at the present time. Within the framework of the debate on Agenda 2000 this report from the Parliament calls upon the Commission to strengthen existing measures to mountain regions, including the formulation of a specific Community Plan for the mountains. In turn, this should address cross-border issues, provide Community-wide provision for improved access and associated transport problems, as well as give increasing emphasis to the development of renewable energy resources through financial or tax incentives. Specific attention should be addressed to maintaining employment in local production through various incentive packages to attract young enterprises and especially those based on information technology.

This proposal remains 'on the table', but is also problematic. Attempts to harmonise approaches across the EU become highly contentious because of the fact that, with enlargement, more countries already having mountain-oriented policies might be pressurised to reduce the level of support that they offer. In turn, this has raised again the long-standing debate about the social and other benefits arising from maintaining population in the mountain regions and continues to be a key issue in the current EU discussions concerning Agenda 2000. This is perhaps most clearly indicated in the fact that at a recent conference arranged in 1998 by Euromontana (see pp. 363–364) Dr Franz Fischler (the EU Commission Member for agriculture and rural development) spoke about the approach of the EU to mountain issues. He claimed that the EU Agenda 2000 package offered excellent prospects for mountain regions. These included better premiums for livestock and protection for designations of origin (locality labels), citing cheeses such as Gailtalet Almkase and meat products such as Tiröler Speck. In addition, under rural development proposals specific attention would be given to encouraging community initiatives and participation in project design and implementation. However, and this is the interesting point, he declined to support the ideal of a separate development initiative for mountain areas. He claimed that Member States had unanimously agreed to concentrate resources on areas of very high unemployment and that existing and planned measures were quite adequate to deal with specific mountain problems. Thus at the level of the EU there are more important priorities than mountains.

National policies and the 'politique montagne'

Whilst these policies have emerged at the European level in the last forty years the 'politique montagne' has a much longer history at local and national levels, reflecting the differing significance of mountain regions in each country. In France, for instance, various legislative measures for the maintenance of mountain slopes and forests date from 1882 and 1922 respectively, whilst controls on pastoralism and provision to assist mountain zones were formalised in 1973 with the creation of L'Indemnité Spéciale Montagne. In 1985 la Loi Montagne was promulgated, which establishes mountains as a political entity and set up the Consueil National de la Montagne et des Comités Massifs. In the case of Switzerland, which remains outside the EU, we have seen in Part 3 that the structure, and history, of the state is built upon the need to incorporate diverse mountain areas. The status and power of the communes and cantons *vis-à-vis* the federal government is a recognition of the diversity of culture and environment. This contrasts markedly with the French example, where the foundations of the modern state are associated with the principle of uniformity across the territory and hence it is only in more recent times that spatially focused measures have been developed for mountain areas.

The Italian experience is interesting because of the long public recognition (if not successful action) of the need to provide support for mountain areas. In the Fascist period (1920s and 1930s), legislation was enacted which attempted to set up an institutional framework to provide special help to the mountain valleys. Legal provisions of 1923 and 1948 (Law 1102) defined mountain territories (Broggio 1992). After 1952, further legal provision was made introducing 'consorzii e comprensori di bonifica montana', and later the Comunita Montana were established. These are administrative agencies for a mountain region, created by the aggregation of mountain communes into an organisation with responsibility for the development of the locality. According to Casabianca (1967) by the mid-1960s there were approximately 352 such units involving 8,000 communes and 10 million people, and thirty years later Romano (1995) reported that total government financing for this initiative had reached US$1.25 billion, apparently just for administration. These *comunita* have absorbed funds both through EU programmes and other government initiatives for improvements to roads, land reclamation, irrigation and afforestation.

As an example of the impact of local, national and community policies it is interesting to consider the view of Barberis (1992) who examines some of the more recent research on Italian mountains. He concludes that whilst it is now reasonably certain that most mountain *comunita* have stabilised their populations, this has been at the expense of agriculture which has generally shown a marked decline. Between 1981 and 1990, of 3,524 *comunita* surveyed, about 80 per cent showed either some growth in population or a stable condition. The road building programme, dating from the Fascist period, has meant that by the 1980s many mountain communes are accessible, and approximately two-thirds of the *comunita* show signs of positive developments. Those that do not are usually the smallest and contain the most difficult landscapes for development, along with seemingly intractable local institutions often associated with land ownership. Moreover, reporting on work carried out by Merlo and Zaccherini (1992), Barberis (1992) notes the following features:

- disposable income of the mountain areas is only about 10 per cent below that of the Italian average;
- there is almost the same degree of entrepreneurship (measured by registered business initiatives) as the national figure.

Overall, other indices suggest that the mountain areas are today much less backward than popular imagination maintains. Whilst some of this advance may be illusory, resulting from distortions of the statistics (particularly for mountain locations), there can be no doubt that the policies discussed above have played their part. Recently, much of the emphasis has encouraged 'pluri-activity', of which tourism is a critical element, and, despite the portfolio of initiatives, the decline in agriculture has continued.

On the other hand, the analysis by Romano (1995) suggests a less favourable view, as he argues that success stories are not to be found in all communities. At the macro-scale, the decline in the proportion of population in some mountain areas has continued, whilst it is evident that the mountain areas in southern Italy have shown far less success in the implementation of *comunita montagna* than areas in the north. This difference is attributed to the persistence of clientalism that serves to undermine any new initiatives that conflict with old alliances and institutions. Romano feels that relatively recent legislation (1991) which established new national parks may offer some hope. For example, the Abruzzo mountain area contains both attractive landscapes and also small settlements of historic interest. Whilst many residences are now abandoned or only used seasonally, as in other areas of the French and Italian Alps, Romano suggests that the framework of the national parks provides a new format for rethinking the way in which a local economy can be stimulated. He emphasises that 'typical' tourist investment, as seen elsewhere in the Italian Apennines, is not desirable, in line with the comments made in Chapter 7. Most of the benefits leave the area, and massed tourist influx in mountain regions often produces either a damaged landscape or one so highly protected as to be more like a 'living museum'! In his view, the funds available should ensure that the relevant agency offices are located in existing mountain settlements and encourage much greater use of existing residential capacity for tourists. He further argues that much available investment (from European funds) is misguided because it aims to create 'elite' tourist centres entirely out of keeping with the social structure of the existing community. Thus any new initiatives must spring from local involvement, in line with recent IUCN suggestions. Developing family businesses and ensuring that 'soft tourism' predominates are just two such strategies which have had success elsewhere. Consequently, the onus lies with the enabling institutions to provide the right stimulus for locals to react. In this sense, his argument foreshadows that now being proclaimed by EU officials.

Transboundary agreements

As was noted above, follow-up work under Chapter 13 was produced by the European Intergovernmental Consultation on Sustainable Mountain Development at meetings in Aviemore and Trento. These produced a policy statement, 'Toward Sustainable Mountain Development in Europe' (Backmeroff *et al.* 1997). The recommendations from this meeting emphasised the need for co-ordinated action to promote cultural diversity and the empowerment of mountain communities in decision-making. However, whilst also listing a range of specific policy objectives for tourism, transport, and multiple land use management there is little detailed strategy discussed. For instance, in one section it is proposed that financial adjustments should be made so that mountain areas are compensated for services provided. This goes to the heart of the political economy of mountains but is not elaborated on. Presumably this is based upon the Swiss model in which cantons receive money from hydro stations for the use of local resources, a practice which if applied in other areas would cause considerable strife (see p. 362). However, although there is often much emphasis placed upon the highland–lowland linkages, it must be said that the increasing economic peripheralisation of many mountain regions, except for recreation, suggests that the balance of transfers would be to the lowland advantage. This remains an interesting issue to be explored.

An important dimension of political action within Europe's mountains has been the result of the work of NGOs. This is not something new. In France, the Fédération Française d'Economie dates from 1913, and the Association National des Elus de la Montagne from 1984. Similarly, the Bayer Arbeitsgemeineschaft Für Bergbauernfragen dates from 1985. Recent developments, particularly stimulated by action under Agenda 21 initiatives, including Chapter 13, have brought these organisations into the limelight and stimulated further action. At a meeting in Toulouse in 1996, co-ordinated by ARPE (Agence Régional pour l'Environnement Midi Pyrénées) and CIAPP (Conseil International pour la Protection des Pyrénées), a large number of organisations were brought together to discuss the results of a survey of NGOs, and to comment on the findings of a parallel intergovernmental meeting which was held at Aviemore, UK. The findings of the survey were interesting and these can be found at the Mountain Forum's online library (http://www2.mtnforum.org/resources/library/arpex96a.htm).The results are not easy to interpret statistically but they suggest that many organisations see the approach to mountain development as being multi-focal, with no clear priorities. Most also maintained, as noted earlier in the book, the need for greater participation by the public in the planning process and effectively the application of the subsidiarity principle. Local areas would be given much greater capacity to implement and manage projects, with higher levels of state involvement only if the lower levels are unable to proceed.

The European Charter of Mountain Regions

Another interesting initiative is that of the draft European Charter of Mountain Regions which originated in a series of conferences under the auspices of the Congress of Local and Regional Authorities of Europe (CLRAE). (The details can be found at http://www.coe.fr/CPLRE/eng/etxt/einstrjur/emontagne.htm) A draft Charter was approved at the CLRAE conference in Chamonix in 1994, but so far the Council of Europe has not ratified it. The Charter incorporates most of the issues discussed in this book, but several points are particularly interesting. It emphasises the principle of subsidiarity; that is, maintaining the key role of local authorities in the management of mountain areas. They need suitable financial and legal powers to act in this capacity. This is not surprising given that the Charter is a CLRAE initiative, but it does highlight the recognition of diversity inherent in mountains – as much from their historical and social experience as from environmental characteristics. In this sense it also reflects 'grassroots' opinion from mountain areas. Second, the Charter requires not just the formal recognition of mountains as a specific territorial category in the administrative structure but also maintains that populations should have the 'right' to remain living in these areas (Preamble Para 7). The corollary to this is, of course, that there is a specific 'duty' on the part of the appropriate authorities to maintain an attractive and viable working environment. In many other political manifestos or proposals this may have been implicit, but usually has been addressed through the argument that such populations are necessary to maintain the landscape as a valuable asset. The emphasis in this Charter raises questions of human rights that are only just beginning to be explored in the mountain context.

A number of organisations have focused on the co-ordination of development initiatives for particular mountain regions, most notably the Alps and the Pyrenees. As a contribution to the discussion on future developments in the Alps, the *Green Paper on the Alps* (Pils *et al.* 1996) is an important document which highlights the problems facing the region and sets out some important policy markers. The Alpine region today has a population of around 13 million and stretches across nine countries (including Liechtenstein and Monaco) and is highly diverse. Current development policy is both highly variable, reflecting the policies of the different countries, and also subject to the principal control of EU regional and financial policies in which the Alpine areas are peripheral. This is exacerbated by the fact that currently Switzerland is not a member of the EU. However, the

360

Green Paper argues that the current problems, principally stemming from the decline in employment on the one hand and environmental deterioration through overuse on the other, ought to be tackled by building a co-ordinated Alpine strategy that highlights the specifically mountain character of the region.

This study seeks to politicise the issue by arguing that the Alps could serve as a strongly autonomous region within the EU model. In doing so, the discussion highlights the problems facing many mountain regions in that in addition to the specific ecological, economic and social problems mountain chains often cut across existing political boundaries, a point noted at the beginning of this book. It remains to be seen if the Alpine framework is sufficiently strong to carve out a new political constituency within Europe. The *Green Paper* contains many suggestions about the achievement of sustainable development and accepts that many initiatives are needed in areas outside of agriculture, especially with regard to tourism, transport, energy and conservation. It advocates the careful integration of highland and lowland activities and recognises the fact that there are marked regional disparities within the Alps which reflect historical and cultural differences. Nonetheless, the critical suggestion that the foundation of an autonomous region should be based upon a physiographic entity places this proposal well beyond the current thinking of the EU and, indeed, of individual states. Of the many proposals made in the *Green Paper*, once again there is a call for a special EU Aid category for mountain regions, which, as we have seen, has been rejected so far by the Commission.

The Alpine Convention

Perhaps the most interesting long-term initiative for the Alps concerns the Alpine Convention which is discussed in detail by Price (1999b). This originated in discussions carried out several decades ago in 1952 under the auspices of CIPRA (Commission internationale pour la Protection des Alpes), but little ensued until the 1970s and 1980s when there were several calls for the establishment of a transboundary agency for Alpine protection. For example, in 1974 the International Symposium of the Future of the Alps was held at Trento. Progress was slow and in 1987 CIPRA again reinvigorated the campaign for an Alpine Convention. This led in 1989 to a meeting in Bavaria and the development of a Convention that was signed in 1991 at Salzburg by six of the seven Alpine states and the EU. In subsequent years the signatories changed as the Republic of Slovenia replaced Yugoslavia and Monaco was included. It has been ratified by all states except Italy.

Its principal objective is to recognise the Alpine region as a whole as being in need of special development and conservation action, necessitated by its central role in the European landscape and its fragility in the face of changes in the economic development of the various regions. The Convention maintains that its members will protect the Alpine environment by using legislative means and by adopting a 'polluter pays' framework. This will apply to the problems of land use management, mountain agriculture, transport and tourism handled though a series of protocols.

The Convention is maintained by a Conference of Contracting Parties which has established a Standing Committee at which most operational matters are discussed. Also many NGOs, such as Club Arc Alpine and Euromontana, have observer status. Despite the fact that the Convention establishes only broad-based principles and uses protocols as the main instrument of implementation, it has only been signed recently (1999) by Switzerland, which of course contains a substantial proportion of the Alpine area. The problem here lay with the fact that, unlike other territories, the cantons already had significant powers and they were not prepared to agree ratification until more details of protocols were established. In particular, many canton members felt that the Convention was likely to be biased in favour of environmental protection rather than the stimulation of economic activity. An agreement was finally reached in return for the Swiss federal authorities' willingness to increase the rent paid to cantons for hydroelectricity (Price 1999b). At the time of writing there are protocols on nature protection and landscape management, mountain agriculture, regional planning and sustainable development, and mountain forests. Not all parties have signed each protocol.

The Alpine Convention provides a good example of the opportunities and problems posed by mountain regions. It is multi-state, involving transboundary agreements, but there are difficulties at agreeing on the working languages. It is a Framework Convention, thus allowing individual states freedom to co-operate or not on specific issues raised in the protocol procedures. Each new protocol tends to demand new resources and states are wary of entering into such arrangements, especially if, like Italy, the mountains involved form only a small part of the nation-state and could easily lead to demands from other mountain regions within the country. However, unlike the European Charter, which is far more prescriptive and geographically general, progress has been made.

One practical result of this Convention has been the establishment, in 1996, of a working group entitled 'Monitoring the Alps' which has designed a methodology for monitoring entitled ABIS (Alpine Monitoring and Information System). Following this, a programme labelled ALPMON has been developed to determine the condition of the Alpine environment, to assess the pressures on this environment and to evaluate the state of the Alpine economy and population. With these goals in mind the first programme is under way. This is utilising a variety of remote-sensing and GIS based sources to provide information on land-cover; specifically, vegetation and residential areas will be matched with terrestrial data to produce

a GIS for monitoring and planning purposes. In particular, it will emphasise hydrological modelling and runoff, the production of avalanche and
landslide risk maps, environmental quality assessment information and the
provision of information for national parks. A series of sites within the Alps
are currently under detailed investigation. For example, a test site at Tarvisio
is being used to evaluate remote sensing and GIS methods for tourist
research, whilst in Engadin, Switzerland, a project has begun which will
establish a monitoring programme in a National Park (ALPMON 1997).
The report highlights immediately the paucity of data available for appropriate monitoring procedures, and suggests ways in which a GIS base can
be used to begin the assembly of information.

The Pyrenees

Turning to the Pyrenees, CIAPP produced a Charter for the Protection of
the Pyrenees which, whilst having certain features in common with the Alps,
made clear the specifics of the Pyrenean context in the light of possible EU
action. It noted that the Pyrenees had been particularly badly hit by depopulation and, at the same time, attempts to bring employment to the range
had resulted in inappropriate developments. It is interesting that this document considers that the previously mentioned French Mountain Law of
1985 is a retrograde step compared with the 1975 EEC directive, in part
because of bad wording and the likelihood that the benefits will rapidly accumulate to the richer communities. Moreover it claims that development
control provisions are likely to see a further encroachment of urban settlement into the mountain zones to the detriment of the landscape. Above all
CIAPP claim that there is public support for the protection of the Pyrenees
environment and a strong case for encouraging only the low intensity form
of tourism described in Chapter 7. The point to note here is not so much
the detailed proposals but the fact that it is an NGO agency that is mobil-
ising public opinion and working as a pressure group to ensure that its image
of a mountain environment is maintained. On the Spanish side, a number
of academic studies have also documented the changes described above
(Puigdefabregas and Fillat 1986; García-Ruiz and Lasanta-Martinez 1990).
Since the 1950s the traditional, highly diverse land use pattern has undergone radical change. On the one hand depopulation has changed the social
structure, and labour use in agriculture has been severely affected. At the
same time improvements in transport and price effects have meant that
cereals for livestock feed can now be brought into the high valleys. The new
labour force is just as likely to be employed in tourist related activities, or
new light industries encouraged to move into some of the valleys – for
example on the route through Vic and Ripoll.

So far we have looked at institutions that focus specifically on the Alps
and the Pyrenees as mountain ranges, and on mountains within the purview

of the EU. However, in 1974 the European Confederation on Agriculture established a group, which was called Euromontana, specifically to look at socio-economic problems in the mountains. The scope of this organisation was widened in the early 1990s to also include countries in central and eastern Europe and this led to a 1995 meeting in Cracow from which the current activities stem. In 1998, another major conference was held at Ljubljana. We have already noted elements from the speech by Franz Fischler but the meeting produced yet another set of recommendations. This meeting treated with some scepticism the willingness of the EU to promote mountain areas directly. With the enlargement of the EU to include eastern and central Europe, most of the mountain ranges of the area would, in principle, be amenable to concerted action through one authority. It also requested that Euromontana become a 'favoured negotiator' with regard to mountain policies at regional, national and European levels, including resources to put this into effect. This objective indicates a vigorous drive for command in the 'politique montagne'.

MOUNTAIN POLICIES IN OTHER REGIONS

The extensive revue of the European experience should not overshadow the fact that similar activities are taking place in the Andes and in the Himalayas. Though with perhaps less intensity, a series of conferences have taken place under Chapter 13 to stimulate regional co-operation. Some initiatives have been mentioned earlier, particularly CONDESAN (Consortium for the Development of the Andean Ecoregion). This organisation was established in 1992 to co-ordinate research and policy discussions throughout the Andes. Research projects have been initiated which focus on long-term change in the region, for example on production systems and biodiversity. A number of benchmark sites, usually where existing institutions have research capacity, have been chosen to provide data across the diverse conditions of the Andes. In the Himalayas, ICIMOD provides this co-ordinating role and has been responsible for initiating research projects and conferences. Many of the general issues discussed in more detail in the European case also appear in these regions. The big difference lies in the greater focus on agricultural research and production-based strategies for mountain areas as, of course, a much higher proportion of the population is engaged in agricultural activities, and poverty is both deeper and widespread. Although there are various NGOs operating in these regions, especially in the Andes, in most cases they have yet to develop the degree of political influence that some have attained in Europe.

The preceding sections have provided an overview of the institutional arrangements associated with development interventions in mountain areas, especially the initiatives following Chapter 13. The purpose of the review

has been to illustrate the way in which mountain policy is not just something that emerges from academic research but has, as it were, a life of its
own. This is important if we are trying to understand the factors that
influence the direction of policies. In the European case it is clear that the
'politique montagne' existed long before the existence of the EU but has
remoulded itself to the requirements of the new political infrastructure.
The real problems of the region, in terms of improving economic and
social well-being and environmental maintenance, have been tackled
principally through sector programmes, especially agriculture. Despite
considerable campaigning, the EU itself seems unwilling to designate mountains as a special structural programme, mainly because most member states
recognise that, whatever the ecological importance of these areas, the bulk
of the EU population and their needs are located in the lowlands. Therefore
much of the intervention in mountain regions occurs as a result of general
programmes (for example in agriculture, social welfare or energy provision)
that apply to many different geographic areas, of which mountains may be
just one. Moreover, with the increase in the size of the EU and the added
area of mountains involved, special provision would be immensely costly.

The Commission response has been to initiate a more widely based rural
development programme, which includes non-agricultural activities for rural
areas, and in particular places much greater emphasis on environmental
management. At the present time there is no lack of policy proposals for
mountain areas, some well researched, others less so, but there is a major
problem of constructing an appropriate administrative umbrella for their
implementation. On the one hand, both national governments and the
EU operate their own interventionist approaches which either regulate activities (for example, land use rules) or provide financial assistance (subsidies,
grants, tax relief, etc.). However, at all levels there is the desire to reduce
the amount of formal state involvement so that many new initiatives are
directed to building capacity for local enterprise. Some of the NGOs recognise this and are quite willing to foster these initiatives, but these demand
much greater empowerment at all levels of the policy-making and implementing process. This is further complicated by the claim that mountains
as a geographical entity require their own autonomous treatment which
would cut across existing political boundaries, as is proposed, for instance,
by the *Green Paper* on the Alps and, to some extent, Euromontana. Apart
from obvious political objections this approach also fails to recognise the
significance of the linkages between highland and lowland which are the
basis of the livelihoods in both areas. It will be interesting to see to what
extent an increasing application of the subsidiarity principle will work in
these situations, particularly when many mountain regions still remain highly
dependent upon financial support from the state.

MOUNTAIN LAWS

The discussions about institutions and mountain development have, so far, neatly sidestepped the nature of legal provisions for mountain areas. Of course, mountain communities are subject to the prevailing legal code of the country concerned, but in many regions this is complicated by the fact that several different legal frameworks may exist. For instance many mountain communities still operate under traditional law despite the existence of national legal provision. Some governments accept this, others actively attempt to impose the national legislation. In Morocco, an extensive discussion of water law and irrigation practices shows how this has evolved from a combination of Islamic practices and French colonial notions of water rights. Although all water resources and rights to use are held by the state, in the mountains traditional rights and procedures are still recognised (Roche 1965). The problem arises when new developments, for example the construction of a dam, require significant changes in riparian rights. Legal provisions operate at all scales, and there are international conventions that are signed by national governments which serve as major legislative frameworks for action. Much of the recent debate on environmental protection and trading relationships has been about preparing appropriate legal instruments to further the underlying policies involved. We have already seen how national policies on land use, conservation and development control can be very influential in mountain areas, but many other issues are relevant, in particular inheritance and property rights and tax law. Basic questions on human rights and empowerment for political action may also play a major role in any positive action by mountain citizens. At the local level, the laws and by-laws play a key role in determining land use practices on individual plots, or in the licensing of specific activities. What varies is the capacity of these local administrations to make and modify these laws.

We can see therefore that legal regulation reflects the prevailing political viewpoint and can be used to protect particular individuals or groups, geographical areas or specific objects such as trees and landscapes. Few materials exist on the subject but the Mountain Forum organised an electronic conference in April 1997 that provides some interesting points of view (Lynch and Maggio 2000). For example, traditional laws relating to landholding or access to water operate within a formal national legal framework and operate at all scales. In Switzerland new federal legislation is being drafted for the protection of mountain landscapes and control of development, but the cantons play an important part in the regulation process. In Nepal, forest protection laws play an important role and these have slowly changed to empower some of the local user groups, defining rights and responsibilities. In many countries, as we noted earlier, National Parks established by legal directives influence the way tourist activity and local people interact. These illustrate one of the principal functions of legal provision – namely, to act as a framework for managing conflict. It will be recalled that in a statement from the Andean Mountain group it was made clear that conflict was an inevitable element in the process of achieving sustainable development. It could be argued that conflict is always present and the legal provision may or may not assist in its resolution. For example, the main thrust of many debates is towards seeking the regulation of mountain resource use, emphasising especially the protection of biodiversity and the environment in general. At the same time, legislation that makes it simpler to establish new commercial enterprises or change land use may well lead to the exploitation of the environment.

Another important component concerns intellectual property rights. This is a particularly controversial issue with respect to biodiversity management, especially in tropical mountain areas. The increasingly global reach of commerce, coupled with the astonishing degree of bio-technical advance in the last few decades, has meant that mountain communities are effectively stakeholders in a global resource. However, their claims both to the raw materials and indeed to some of the basic ideas involved in their use (because they have used products, for example, as medical treatment) may be lost without the legal recognition of their rights (Brandolini 1997).

Therefore, the practical effect of much of the discussion about new developments since UNCED 92 depends on the extent to which existing or new legislation can regulate and/or enable activities in mountain areas. The enactment of legislation is a political process. Whatever the merits of particular projects for greater environmental protection or the encouragement of sustainable development, their successful implementation needs the support of both those who will be affected and those who have a critical role in the political structure.

WHERE TO NEXT: COMPLEXITY AND ALL THAT?

It is obvious from the discussions above that policy-makers have to confront the difficult problems posed by sustainable development in mountain regions. Whilst the history of intervention in mountain regions indicated a wide range of initiatives, a successful approach to this problem is unlikely through traditional frameworks of policy formulation and implementation. The cog diagram (Figure 1.4) illustrated the fact that the traditional development of policy agenda and the implementation of specific projects have usually operated within a defined geographical area, and also within a discrete time-frame. It will be recalled from Part 1 that these conditions rarely correspond to the framework of socio-economic transformation or to the far more fluid evolution of natural ecosystems. Most of these systems experience what Clark et al. (1995) term 'co-evolutionary change' and evolve under their own dynamics, for example the population life cycle of particular species or the cycles of economic activity, and also interact at different scales and frequencies. This produces new elements and combinations that would have been difficult to predict. Moreover, sustainability produces problems that are different from traditional policy concerns involving the recognition of irreversibility in natural ecosystems, the question of ecological limits and a range of interrelated socio-economic issues. Existing approaches have kept quite strict boundaries between these arenas, typically through sectoral policy formulation. New approaches to environmental policy are required which foster links between ecological systems, the socio-economic framework, institutions and communities.

The problems of managing sustainable development in mountain areas are inherently difficult because they have all the features of complexity discussed earlier. For many relationships there are a number of possible conceptual frameworks, with uncertainty being a central characteristic. Individuals and institutions struggle to minimise uncertainty, but policy-makers in their political arena have been reluctant to admit publicly that this is a permanent element of human existence. For mountain areas these issues are of particular importance because, perhaps more than in any other environment, there is a high probability of natural hazards and many of the changes that have taken place expose communities to socio-economic uncertainty in a manner never previously experienced. Innovations are needed at all levels: with research methodologies which increasingly require a high degree of multidisciplinarity; with new management tools that allow for 'self-reflexivity'; and with institutional arrangements that empower all elements in society, not just through formal political channels but more directly in policy formation.

At the level of management tools one increasingly popular approach is that of 'adaptive management strategies'. This concept was originally developed in Canada and draws upon the work in ecology by Holling (1978) and others. Its principal starting point is that of imperfect knowledge and uncertainty. Coping with the uncertainty and the unknown becomes a central part of both research and policy formulation. It is based on the principle of management as experiment, an attempt to establish a framework in which precise objectives are framed, tested and revised in a continual process. According to Lee (1993), the important feature of adaptive management is that policy interventions are 'experimental probes' rather than confident predictions. This is described in more detail in Mitchell (1997) and is critically examined in Dovers and Mobbs (1997). Most early uses of this management model utilised quantitative models, especially with regard to planning forest areas and river catchments. Several large-scale experiments have been conducted in a North American context.

Much of this work has deliberately gone beyond the purely quantitative system modelling to explore the capability of such approaches to absorb ideas about adaptive learning and reflexivity. Lee (1993) noted how most models have been prescriptive in the sense that the public policy resulting from such experimentation is part of the 'expertise model' rather than a collective decision-making process. It is interesting to refer back to the Obergurgl model discussed in Part 1, where efforts were made to involve the community, suggesting in some respects how mountain research and policy formulation were in the forefront of these approaches. However, as all commentators note, whilst there is an increasing number of projects with these participatory objectives in mind, we are a long way from achieving 'project sustainability'. The more integrated participatory process requires fundamental changes in the way interventions are established, budgeted and monitored. To be self-reflexive involves not only a change in the attitudes of individuals to uncertainty and failure, but also suitable support mechanisms. In essence self-reflexivity is a historical process and most programmes are woefully too short to allow the impact of re-evaluations to be monitored effectively within the usual project-based framework. It is also costly as it involves considerably increased 'negotiation time'.

There is also the danger that we confuse the construction of analytical methods with the management tools designed to provide the solutions to social and environmental problems. The increasingly large literature on science policy and sustainability indicates that there is now a wide range of analytical devices that attempt to handle the scenarios of complexity. Modelling systems such as AVS and VENSIM, the latter employed by Perez-Trejo et al. (1993), hold out interesting possibilities for work in mountain areas. However, as Shackley et al. (1996) remind us, there is a danger that theoretical models designed to improve our understanding of the way in which complex systems work soon become acceptable as heuristic tools for planning. Criticisms of Cultural Theory as an attempt to explain behaviour were noted in Part 1, but the ideas are valuable contributions to the debate. Nonetheless, Shackley et al. argue that problems arise when agencies slavishly follow these constructs and turn these helpful theoretical ideas into practical planning devices. Incorporating uncertainty and complexity into analytical models does not in itself make the models superior as planning tools. Such a procedure is tantamount to nullifying the basic principles of diversity and effectively cutting off real reflexivity by denying the validity of alternative perspectives. Thus the apparently objective practical tool is in fact being used in a highly political way.

As Thompson and Warburton point out (1998a, 1998b), the dichotomy between lay and expert participants, between knowledge-based and ethical approaches to environmental issues, and the recognition that all policy issues are contested in some way or another, means that we have to ask whether there are some institutional frameworks which are better at stimulating reflexivity than others. In this book the greater part of the information used is derived from formal scientific enquiry, but, as has been intimated at various points, this does not mean to say that this is the only perspective on the evolution of mountain areas. The real problem arises when particular issues become the basis for policy discussion, for example in the Himalayan Crisis debate, and then it is quite noticeable that governments, once interested, sought advice and policy assistance from a very limited constituency. The results are amply illustrated.

Clearly, the construction of appropriate mountain policies remains constrained, both by the ways in which science and social policy are formulated and by the mechanisms through which appropriate political agenda are discussed. This is an enormous topic that reaches well beyond the confines of mountain policy *per se*, but the development of sustainable mountain communities and environments hinges on this. Thompson and Warburton (1998a, 1998b) argue for an approach which looks at the discourses of policy; that is, not just the policy proposals themselves but an understanding of where they are coming from. To understand that no policy is itself likely to be objective (whatever scientists state), and to accept and incorporate the contested nature of the political terrain, is an essential prerequisite for self-reflexivity and resilience. In most instances, the resulting structures that emerge are non-hierarchic and non-exclusive in their political domain, more akin to traditional village committees. This is a point worth emphasising. One of the key characteristics of many traditional mountain societies (contrary to popular perception) was their flexibility, tempered by pragmatism. Each year and season was seen as different and their institutions and behavioural rules recognised this. In the main, real problems have occurred when rigid structures have been imposed by outside agencies. Although autarchy is not realistic today, a firm recognition of this point is a prerequisite for successful mountain policy.

The problems of self-reflexivity, coupled with the diversity of mountain environments facing uncertainty, suggest that rigid structures of governance may be inappropriate. It is not simply a case of enforcing rules, for example in relation to forest or slope management. It is a question of how these rules emerge and whether they are responsive to the dynamics of the mountain environment. The recent interest in pluralism has been noted in Chapter 7, but work in rural development has emphasised the key role of social capital (Bebbington and Kopp 1998). This refers to the interpersonal, inter-community sets of networks and relationships which serve to mediate the particular way in which other forms of capital (financial, natural, etc.) actually operate. It is argued that those societies that are rich in institutions and networks between the various factions are actually the most enterprising and able to respond to new circumstances. This is because, for any given problem or initiative, there may be several channels through which all members of the community can find a voice, which in turn facilitates effective conflict resolution. Bebbington and Thiele (1993) have examined numerous NGO and government relationships. They report that there is no straightforward answer in so far as social capital can be built both from the 'bottom up' (as is usually advocated) but also from the top down, the latter being particularly effective when wider-scale linkages are significant.

The debates on governance certainly suggest that innovative forms of political structure and behaviour need to be encouraged, but it is hard to say, and probably quite wrong to specify, which particular systems are suitable for mountain communities. While there is considerable recognition in the development literature about widening the constituency of decision-making, at least at the local or project level, the same does not always apply at higher levels. Moreover, the mountain debate illustrates again the geographical problem of setting boundaries. If the highlands are so critical to the well-being of large areas of lowland then the mountains cannot really be seen as an autonomous region, either physically or socially. Does this in turn suggest that they do not really constitute a sufficiently distinctive part of the globe for the purposes of policy formulation? More particularly, do regions based upon physical geography – mountains – really form a satisfactory basis in the twenty-first century for the construction of territorial units of governance and policy implementation?

The purpose of this book has been to suggest that it is wrong to consider mountains as just a physical entity, as they have social meaning as well. Furthermore, in the light of new discussions of governance and the increasing acceptance of local diversity, mountains provide an excellent example of how local values and conditions form a prerequisite for sustainable development. The real issue lies in how the larger-scale structures of political control can be made to contribute positively to these local requirements.

BIBLIOGRAPHY

Agnew, J. (ed.) (1997) *Political Geography: A reader,* London: Arnold.

Ajiki, K. (1993) Household composition and its reproduction process in remote mountain villages in the Kitakami Mountains, northeast Japan, *Geographical Review of Japan,* Series A 66, 3: 131–150.

Aldrich, M., Billington, C., Edwards, M. and Laidlaw, R. (1997) Tropical montane cloud forests: An urgent priority for conservation, *WCMC Biodiversity Bulletin,* 2: 1–14.

Alexandrian, D., Esnault, F. and Calabri, G. (1999) Forest fires in the Mediterranean area, *Unasylva,* 197, 50: 35–41.

Allan, N. J. R. (1986) Accessibility and altitudinal zonation models of mountains, *Mountain Research and Development,* 6: 185–194.

Allan, N. J. R. (1987) Impact of Afghan refugees on the vegetation resources of Pakistan's Hindukush–Himalaya, *Mountain Research and Development,* 7: 200–204.

Allan, N. J. R. (1988) Highways to the sky: the impact of tourism on South Asian mountain culture, *Tourism and Recreation Research,* 13, 1: 11–16.

Allen, B. J. (1988) Adaptation to frost and recent political change in Highland Papua New Guinea, in N. R. J. Allan, G. Knapp and C. Stadel, *Human Impact on Mountains,* Towota, N.J.: Rowman & Littlefield, 255–264.

Allen, P. M. (1994) Coherence, chaos and evolution in the social context, *Futures,* 26: 583–597.

Allen, P. M. and Lessor, M. (1991) *Evolutionary Human Systems: Learning ignorance and subjectivity,* Chur, Switzerland: Harwood Academics.

Allison, R. J. and Thomas, D. S. G. (1993) The sensitivity of landscapes, in D.S.G. Thomas and R. J. Allison (eds), *Landscape Sensitivity,* Chichester: Wiley, 1–5.

ALPMON (1997) *Report on WP1: Requirements of the customer,* Brussels: European Commission.

Amin, A. and Hausner, J. (eds), (1997) *Beyond Market and Hierarchy: Interactive governance and social complexity,* Cheltenham: Elgar.

Anton, D. J. (1993) *Thirsty Cities: Urban environments and water supplies in Latin America,* Ottowa: International Development Research Centre.

Arbos, P. (1922) *La vie pastorale dans les Alpes Français,* Grenoble: Armand Colin.

ARPE/CIAPP (1996) *Recommendations of NGOs and Mountain Populations to Governments and to the European Union,* Toulouse: ARPE/CIAPP.

Aulitsky, H. (1967) Lage und Ausmass der 'warmen Hangzone' in einin Quertal der Innenalp, *Ann. Met.,* 3: 159–165.

Avery, D. (ed.) (1989) *The Complete History of North American Railways,* London: Brian Trodd Publishing.

Backmeroff, C., Chemini, C. and La Spada, P. (eds) (1997) *European Inter-governmental Consultation on Sustainable Mountain Development: Proceedings of the final Trento Session,* Trento, Italy: Provincia Autonoma di Trento.

Badenkov, Y. (1992) Mountains of the former Soviet Union, in P. Stone (ed.), *The State of the World's Mountains,* London: Zed Books, 257–298.

Bajracharya, D. (1983) Fuel, food or forest? Dilemmas in a Nepali village, *World Development,* 11: 1057–1074.

Baker, P. T. (1969) Human adaptation to high altitude, *Science,* 163: 1149–1156.

Baker, P. T. and Little, M. A. (eds) (1976) *Man in the Andes: A multidisciplinary study of the high-altitude Quecha,* Stroudsberg, Pa.: Dowden, Hutchinson & Ross, Inc.

Bandyopadhyay, J. (1992) On the perceptions of mountain characteristics, *World Mountain Network Newsletter,* 7: 5–7.

Bandyopadhyay, J., Rhodda, J. C., Kattelman, R., Kundzewicz, Z. W. and Kraemar, D. (1997) Highland waters – a resource of global significance, in B. Messerli and J. D. Ives (eds), *Mountains of the World: A global priority,* Carnforth: Parthenon, 131–155.

Banskota, K. and Sharma, B. (1997) Case studies from Ghandruk, *MEI Discussion Paper,* 97/5, Kathmandu.

Barberis, C. (1992) La montagne ou les montagnes italiennes identité et civilisation, *Revue de Géographie Alpine,* 80: 65–77.

Barbier, E. B. (1987) The concept of sustainable development, *Environmental Conservation,* 14: 101–110.

Barker, M. (1982) Traditional landscape and mass tourism in the Alps, *Geographical Review,* 72: 395–415.

Barry, R. G. (1992) *Mountain Weather and Climate,* London: Routledge.

Barsch, D. and Caine, N. (1984) The nature of mountain geomorphology, *Mountain Research and Development,* 4: 287–298.

Bartelmus, P. (1994) *Environment, Growth and Development: The concepts and strategy of sustainability,* London: Routledge.

Bätzing, W., Perlick, M. and Dekleva, M. (1996) Urbanisation and depopulation in the Alps, *Mountain Research and Development,* 16: 335–350.

Baumgartner, A. (1960) Die Lufttemperaur als Standortsfaktor am Grossen Falkenstein, *Forstwiss Centralblatt,* 79: 362–373.

Bazin, G. (1992) Quel bilan de la PAC dans les zones de montagne et defavorisées?, *Revue de Géographie Alpine,* 80: 43–64.

Bebbington, A. (1995) Rural development: policies, programmes and actors, in D. Preston (ed.), *Latin American Development: Geographical perspectives,* Harlow: Longman, 116–145.

Bebbington, A. (1997) Social capital and rural intensification: local organisations and islands of sustainability in the rural Andes, *Geographical Journal,* 163: 189–197.

Bebbington, A. and Kopp, A. (1998) Networking and rural development through sustainability for management frameworks for pluralistic approaches, *Unasylva,* 194, 49: 11–18.

Bebbington, A. and Thiele, G. (1993) *NGOs and the State in Latin America: Rethinking roles in agricultural development*, London: Routledge.

Beck, G. (1995) Amenity migration in the Okanagan Valley, B.C. and the implications for strategic planning. Unpublished M.Sc. thesis, University of Calgary, Canada.

Behnke, R., Scoones, I. and Kervin, C. (eds) (1993) *Range Ecology at Disequilibrium: New models of natural variability and pastoral adaptation in African savannas*, London: ODI.

Belfanti, C. M. (1993) Rural manufactures and rural proto-industries in the 'Italy of the Cities' from the sixteenth through the eighteenth centuries, *Continuity and Change*, 8: 253–268.

Belfanti, C. M. (1996) The proto-industrial heritage: forms of rural proto-industry in northern Italy in the eighteenth and nineteenth centuries, in S. C. Olgivie and M. Cerman (eds), *European Proto-industrialisation*, Cambridge: Cambridge University Press, 155–170.

Bellaoui, A. (1996) Tourisme et développement local dans le Haut-Atlas Marocain, *Revue de Géographie Alpine*, 84: 15–24.

Bencherifa, A. and Johnson, D. L. (1990) Adaptation and intensification in the pastoral systems of Morocco, in J. G. Galaty and D. L. Johnson (eds), *The World of Pastoralism: Herding systems in comparative perspective*, London: Belhaven, 394–416.

Bencherifa, A. and Johnson, D. L. (1991) Changing resource management strategies and their environmental impacts in the Middle Atlas mountains of Morocco, *Mountain Research and Development*, 11: 183–194.

Berkes, F. (1989) *Common Property Resources: Ecology of community based sustainable development*, London: Belhaven.

Bernard, P. P. (1978) *Rush to the Alps: The evolution of vacationing in Switzerland*, East European Monographs 37, New York: Colombia University Press.

Bernbaum, E. (1997) The spiritual and cultural significance of mountains, in B. Messerli and J. Ives (eds), *Mountains of the World: A global priority*, Carnforth: Parthenon, 39–60.

Berque, J. (1953) Notes sur l'histoire des échanges dans le Haut-Atlas occidental, *Annales Economies–Sociétés–Civilisations*, 8 année: 289–314.

Berriane, M. (1993) Le tourisme de montagne au Maroc, in A. Bencherifa (ed.), *African Mountains and Highlands: Resource use and conservation*, Vol. Colloques et Seminaires Series No. 29, Rabat: Faculté des Lettres et des Sciences Humaines, Université Mohammed V, 291–404.

Bertelsen, K. B. (1997) Protestant Buddhism and social identification in Ladakh, *Archives Sciences Sociales des Religions*, 99: 129–151.

Bhatta, B. R. (1992) Table of major forest types, dominant species and their occurrence in the Himalayan region (Unpublished, ICIMOD), in P. Stone (ed.), *The State of the World's Mountains*, Norwich: Zed Books, 107.

Bhattari, A. (1999) Community forestry in Nepal. Contribution to Mountain Forum Electronic Conference on Mountain People, Forests and Trees, April (http://www2.mtnforum.org.mtnforum/archives/document/discuss99/mpft/mpft.htm)

Biddulph, J. (1880) *Tribes of the Hindoo Kush*, Lahore: Ali Kamran Publishers.

Bishop, N. H. (1989) From zomo to yak: change in a Sherpa village, *Human Ecology*, 17: 177–204.

Bishop, N. H. (1998) *Himalayan Herders*, Orlando, Fl.: Harcourt Brace & Co.

Böhm, R. (1986) *Der Sonnblick. Die 100 jährige Geschichte des Observatoriums und seiner Forschungstätigkeit*, Vienna: Osterreichischer Bundesverlag.

Bolgiano, C. (1999) Communal/private ownership: Appalachian mountains. Contribution to Mountain Forum Electronic Conference on Mountain People, Forests and Trees, April (http://www2.mtnforum.org/mtnforum/archives/document/discuss99/mpft/mpft.htm)

Bone, J. S. and Seton, J. R. (1987) The financial viability of farms in the Spey valley, in D. Jenkins (ed.), *Land use in Speyside*, Aberdeen: ACLU, University of Aberdeen.

Boserup, E. (1965) *The Conditions of Agricultural Growth: The economics of agrarian change under population pressure*, London: Allen & Unwin.

Boujrouf, S. (1996) La montagne dans la politique d'amenagement du terroire au Maroc, *Revue de Géographie Alpine*, 84: 37–50.

Bradley, R. F. and Jones, P. D. (1992) *Climate since AD 1500*, London: Routledge.

Bragg, K. (1992) Akha ethnobotany: the forest resources of a mountain people, in A. R. Walker (ed.), *The Highland Heritage: Collected essays on upland North Thailand*, Singapore: Suvarnabhumi Books, 145–162.

Brandolini, G. V. (1997) The protection of indigenous cultural patrimonies and global markets. Contribution to the Mountain Forum Electronic Conference on Mountain Policy and Law, April (http://www2.mtnforum.org/mtnforum/archives/discuss/discuss.htm)

Braudel, F. (1972) *The Mediterranean and the Mediterranean World in the Age of Philip II*, London: Collins.

Brett, M. and Fentress, E. (1996) *The Berbers*, Oxford: Blackwell.

Broccoli, A. J. and Manabe, S. (1992) The effects of orography on midlatitude northern hemispheric dry climates, *Journal of Climate*, 5: 1181–1201.

Broggio, C. (1992) Les enjeux d'une politique montagne pour l'Europe, *Revue de Géographie Alpine*, 80: 26–42.

Bromley, R. J. (1971) Markets in developing countries: a review, *Geography*, 56: 124–132.

Bromley, R. J. (1974) *Periodic Markets, Daily Markets and Fairs: A bibliography*, Monash University.

Brookfield, H. and Allen, B. (1989) High-altitude occupation and environment, *Mountain Research and Development*, 9: 201–209.

Brower, B. (1991) *The Sherpa of Khumbu*, Delhi: Oxford University Press.

Browman, D. L. (1990) High altitude camelid pastoralism of the Andes, in J. G. Galaty and D. L. Johnson (eds), *The World of Pastoralism: Herding systems in comparative perspective*, London: Belhaven, 323–352.

Brugger, E. A., Furrer, G., Messerli, B. and Messerli, P. (eds) (1984) *The Transformation of the Swiss Mountain Ranges*, Berne: Verlag Paul-Haupt.

Brunsden, D. (1993) Barriers to geomorphological change, in D. S. G. Thomas and R. J. Allison (eds), *Landscape Sensitivity*, Chichester: John Wiley & Sons, 7–12.

Brunsden, D. and Allison, R. J. (1986) Mountains and highlands, in P. G. Fookes and P. R. Vaughan (eds), *A Handbook of Engineering Geomorphology*, Guildford: Surrey University Press, 150–165.

Brunsden, D. and Thornes, J. (1979) Landscape sensitivity and change, *Transactions of the Institute of British Geographers,* NS 4: 463–484.

Brush, S. B. (1976a) Introduction to cultural adaptations, *Human Ecology,* 4: 125–133.

Brush, S. B. (1976b) Man's use of the Andean ecosystem, *Human Ecology,* 4: 147–166.

Brush, S. B. (1986) Genetic diversity and conservation in traditional farming systems, *Journal of Ethnobotany,* 6: 157–167.

Brush, S. B. (1988) Traditional agricultural strategies in the hill lands of tropical America, in N. R. J. Allan, G. Knapp and C. Stadel (eds), *Human Impact on Mountains,* Towota, N.J.: Rowman & Littlefield, 116–126.

Bryant, R. L. and Bailey, S. (1997) *Third World Political Geography,* London: Routledge.

Brzeziechi, B., Kienast, F. and Wildi, O. (1994) Potential impacts of a changing climate on the vegetation cover of Switzerland: a simulation experiment using GIS technology, in M. F. Price and D. I. Heywood (eds), *Mountain Environments and GIS,* London: Taylor & Francis, 263–279.

Burns, S. F. and Tonkin, P. J. (1982) Soil-geomorphic models and the spatial distribution and development of alpine soils, in C. E. Thorn (ed.), *Space and Time in Geomorphology,* Binghampton Symposium in Geomorphology 12, London: George Allen & Unwin, 25–43.

Butz, D. (1994) A note on crop distribution and micro-environmental conditions in Holshal and Ghoshushal villages, Pakistan, *Mountain Research and Development,* 14: 89–97.

Byers A. (1987a) Landscape change and man-accelerated soil loss: the case of Sagarmatha (Mt Everest) National Park, Khumbu, Nepal, *Mountain Research and Development,* 7: 209–216.

Byers, A. (1987b) An assessment of landscape change in the Khumbu region of Nepal using repeat photography, *Mountain Research and Development,* 7: 77–81.

Cain, M. (1982) Perspectives on family and fertility in developing countries, *Population Studies,* 36: 159–175.

Caine, M. (1975) An elevational control of peak snowpack variability, *Water Res. Bulletin,* 11: 613–621.

Caldwell, J. (1976) Toward a reinstatement of demographic transition theory, *Population and Development Review,* 2: 321–366.

Campbell, J. G. (1978) *Community Involvement in Conservation: Social and organisation aspects of the proposed resource conservation and utilisation project in Nepal,* Nepal: USAID.

Campbell, J. G., Shrestha, R. J. and Euphat, F. (1987) Socio-economic features in traditional forest use and management. Preliminary results from a study of community forest management in Nepal, *Banko Janakar,* 1: 45–54.

Campbell, J. K. (1964) *Honour, Family and Patronage: A study of institutions and moral values in a Greek mountain community,* Oxford: Oxford University Press.

Carney, D. (ed.) (1998) *Sustainable Rural Livelihoods: What contribution can we make?,* London: Department for International Development.

Carson, B. (1985) *Erosion and Sedimentation Processes in the Nepalese Himalaya,* Kathmandu: International Centre for Integrated Mountain Development (ICIMOD).

Carter, A. S. and Gilmour, D. A. (1989) Tree cover increases on private farm land in central Nepal, *Mountain Research and Development*, 9: 381–391.

Carter, E. J. (1992) Tree cultivation on private land in the Middle Hills of Nepal: lessons from some villages of Dolakha District, *Mountain Research and Development*, 12: 241–255.

Casabianca, F. (1967) Y-a-il place pour un développement des zones montagneuses en Méditerranée?, *Peuples méditerranéens*, 38–39: 209–218.

Chapin, F. S. and Körner, C. (eds) (1995) *Arctic and Alpine Biodiversity*, Ecological Studies 113, Berlin: Springer Verlag.

Chauverri, A. and Cleef, A. M. (1997) Las communidades vegetacionales en los paramos de los macizos del Chirripo y Buenavista, cordillera de Talammanca, Costa Rica, *Revista Forestal Centroamericana*, 17: 44–49.

Chauvin, C., Berger, F. and Courband, B. (1997) *A Research Programme for Mountain Forest Management*, Proceedings of the XI World Forestry Congress, Antalya, Turkey, 13 to 22 October 1997, Volume 2, topic 9.

Chaverri-Polini, A. (1998) Mountains, biodiversity and conservation, *Unasylva*, 195, 46: 47–54.

Chayanov, A. V. (1966) *The Theory of the Peasant Economy* (trans. D. Thorner), Homewood, Ala.: RD Irwin.

Clark, L. G. (1995) Diversity and distribution of the Andean woody bamboos, in S. P. Churchill, H. Balslev, E. Forero and J. L. Luteyn (eds), *Biodiversity and Conservation of Neotropical Montane Forests*, New York: New York Botanical Garden, 501–512.

Clark, N., Perez-Trejo, F. and Allen, P. M. (1995) *Evolutionary Dynamics and Sustainable Development: A systems approach*, Aldershot: Elgar.

Clark, S. P. and Jager, E. (1969) Denudation rates in the Alps from geochronologic and heat flow data, *American Journal of Science*, 267: 1143–1160.

Clark, W. M. (1986) Irrigation practices: peasant-farming settlement schemes and traditional cultures, in *Scientific Aspects of Irrigation Schemes*, London: The Royal Society, 229–243.

Cohen, E. (1989) Primitive and remote: hill tribe trekking in Thailand, *Annals of Tourism Research*, 16, 1: 30–61.

Cohen, R. (1996) *Theories of Migration*, Cheltenham: Elgar.

Cole, J. W. and Wolf, E. R. (1974) *The Hidden Frontier: Ecology and ethnicity in an Alpine valley*, New York: Academic Press.

Coleman, D. and Schofield, R. (1986) The state of population theory, *Population and Development Review*, 36: 159–175.

Collins, D. N. (1989) Hydrometeorological conditions, mass balance and run-off from alpine glaciers, in J. Oerlemans (ed.), *Glacier Fluctuations and Climate Change*, Dordrecht: Kluwer, 305–323.

Conway, G. (1987) The properties of agro-ecosystems, *Agricultural Systems*, 24, 2: 95–118.

Cooper, P. J. M. (1979) The association between altitude, environmental variables, maize growth and yields in Kenya, *Journal of Agricultural Science*, 93: 635–649.

Cooper, R. G. (1984) *Resource Scarcity and the Hmong Response: A study of settlement and economy in Northern Thailand*, Singapore: Singapore University Press.

Corbridge, S. (ed.) (1995) *Development Studies: A reader*, London: Arnold.

Cox, C. B. and Moore, P. D. (1985) *Biogeography: An ecological and evolutionary approach,* Blackwell: Oxford.

Crook, J. (1994a) The history of Zangskar, in J. Crook and H. Osmaston (eds), *Himalayan Buddhist Villages,* Bristol: University of Bristol Press, 435–474.

Crook, J. (1994b) Social organisation and personal identity in Zangskar, in J. Crook and H. Osmaston (eds), *Himalayan Buddhist Villages,* Bristol: University of Bristol Press, 475–518.

Crook, J. and Osmaston, H. (eds) (1994) *Himalayan Buddhist Villages,* Bristol: University of Bristol Press.

Crook, J. and Shakya, T. (1994) Monastic communities in Zanskar: location, function and organisation, in J. Crook and H. Osmaston (eds), *Himalayan Buddhist Villages,* Bristol: University of Bristol Press, 559–630.

Damianos, D. and Hessapoyannos, K. (1997) Greece and the enlargement of the European Union, *Sociologica Ruralis,* 37, 2: 302–310.

Danz, W. and Henz, H. R. (1979) *Integrated Development of Mountain Areas: the alpine region,* Regional Policy Series, No. 20, Brussels: Commission of the European Communities.

Daubenmire, R. (1954) Alpine timberlines in the Americas and their interpretation, *Butler University Botanical Studies,* 11: 119–136.

Debarbieux, B. (1993) Du haut en général et du Mont Blanc en particulier, *L'Espace Géographie,* 1: 5–13.

Debarbieux, B. (1999) Is 'Mountain' a relevant object and/or a good idea?' in M. Price (ed.), *Global Change in the Mountains,* Carnforth: Parthenon, 7–9.

Delormé, R. (1997) The foundational bearing of complexity, in A. Amin and J. Hausner (eds), *Beyond Market and Hierarchy: Interactive governance and social complexity,* Cheltenham: Elgar, 32–56.

Denniston, D. (1995) *High Priorities: Conserving mountain ecosystems and cultures,* Washington, D.C.: Worldwatch Institute.

Dhar, U. and Kachroo, P. (1983) *Alpine Flora of the Kashmir Himalaya,* Jodhpur: Scientific Publishers.

Dick, E. (1964) *Vanguards of the Frontier: A social history of the North Plains and Rocky Mountains,* Lincoln, Nebraska: University of Nebraska Press.

Dietrich, W. (1992) *The Final Forest,* New York: Simon & Schuster.

Dollfus, O. (1982) Development of land-use patterns in the Central Andes, *Mountain Research and Development,* 2: 39–48.

Dore, A., Sobik, M. and Migala, K. (1999) The role of orographic cap clouds in pollutant deposition in the Western Sudety Mountains, in M. Price (ed.), *Global Change in the Mountains,* Carnforth: Parthenon, 89–91.

Dougherty, W. (1994) Linkages between energy, environment and society in the High Atlas Mountains of Morocco, *Mountain Research and Development,* 14: 119–135.

Douglas, M. (1985) *Risk Acceptability According to the Social Sciences,* London: Routledge & Kegan Paul.

Dovers, S. R. and Mobbs, C. D. (1997) An alluring prospect? Ecology, and the requirements of adaptive management, in N. J. Klomp and I. D. Lunt (eds), *Frontiers in Ecology: Building the Links,* Oxford: Elsevier Science, 100–137.

Dozier, E. (1970) *The Pueblo Indians of North America,* New York: Holt, Rinehart & Winston.

Dresch, J. (1941) *Recherches sur l'evolution du relief dans le Massif Central du Grand Atlas, le Haouz et le Sous*, Paris: Armand Colin.

Dressler, J. (1982) The organisation of erosion control in Morocco, *Quarterly Journal of International Agriculture*, 21: 62–79.

Driver, T. S. and Chapman, G. P. (1996) *Timescales and Environmental Change*, Routledge: London.

Dunaway, W. A. (1996) The incorporation of mountain ecosystems into the capitalist world-system, *Review*, 19: 355–381.

Eckholm, E. (1976) *Losing Ground*, Worldwatch Institute, New York: W. W. Norton & Co. Inc.

Edwards, D. (1996) The trade in non-timber forest products from Nepal, *Mountain Research and Development*, 16, 4: 383–394.

Ehlers, E. and Kreutzmann, H. (2000) *High Mountain Pastoralism in Northern Pakistan*, Stuttgart: Franz Steiner Verlag.

Elsasser, H., Seiler, C. and Scheurer, T. (1995) The regional economic impacts of the Swiss Mountain Park, *Mountain Research and Development*, 15: 77–80.

Emberger, L. (1938) *Les arbres du Maroc et comment les reconnaître*, Paris: Larose.

England, P. (1996) UNCED and the implementation of forest policy in Thailand, in P. Hirsch (ed.), *Seeing the Forests for Trees: Environment and environmentalism in Thailand*, Chiang Mai: Silkworm Books.

English, P. W. (1968) The origin and spread of qanats in the Old World, *Proceedings of the American Philosophical Society*, 112: 170–181.

Environment and Natural Resources Information Network (ERIN) (1998) State of the Environment Report, Tajikistan: UNEP (http://www.grida.no/prog/cee/erin/htmls/tadjik/index.htm)

Erickson, C. L. (1992) Prehistoric landscape management in the Andean highlands: raised field agriculture and its environmental impact, *Population and Environment*, 13: 285–300.

Erickson, C. L. and Candler, K.L. (1989) Raised fields and sustainable agriculture in the Lake Titicaca basin of Peru, in J. O. Browder (ed.), *Fragile Lands of Latin America*, Boulder, CO.: Westview Press.

Escobar, A. (1995) *Encountering Development: The making and unmaking of the Third World*, Princeton, N.J.: University of Princeton Press.

Escobar, A. (1996) Constructing nature: elements for a post-structural political ecology, in R. Peet and M. Watts (eds), *Liberation Ecologies*, London: Routledge, 46–68.

European Parliament: Committee on Agriculture and Rural Development (1998) *Report on a New Strategy for Mountain Regions*, Doc. EN\RR\363\363705.

Fadloullah, A. (1987) Evolution recente de la population dans le Haut Rif, *Etudes Méditerranéenes*, 11: 463–482.

Fadloullah, A. (1990) Evolution recente do a population et du peuplement au Maroc, in A. Bencherifa and H. Popp (eds), *Le Maroc: espace et société*, Passau: Passavia Universitatsverlag, 75–84.

Fairbridge, R. W. (1968) Mountain and hilly terrain: mountain systems, mountain types, in R. W. Fairbridge (ed.), *Encyclopedia of Geomorphology*, New York: Reinhold, 745–761.

FAO (1995) *Managing Fragile Ecosystems: Sustainable development*, Report of the Secretary-General to the Commission on Sustainable Development.

FAO (1997) *Managing Fragile Ecosystems: Sustainable development,* Report of the Secretary-General to the Commission on Sustainable Development.

Feierman, S. (1993) Defending the promise of subsistence: population growth and agriculture in the West Usambara Mountains, 1920–1980, in B. L. Turner II, G. Hyden and R. Kates (eds), *Population Growth and Agricultural Change in Africa,* Gainsville: University of Florida Press, 114–144.

Fioravanti-Molanié, A. (1982) Multi-levelled Andean society and market exchange: the case of Yucay (Peru), in D. Lehmann (ed.), *Ecology and Exchange in the Andes,* Cambridge: Cambridge University Press, 211–230.

Fioretti, G. (1996) *A Concept of Complexity for the Social Sciences,* Laxenburg, Austria: International Institute for Applied Systems Analysis.

Fischlin, A. and Gyalistras, D. (1997) Assessing impacts of climate change on forests in the Alps, *Global Ecology and Biogeography Letters,* 6: 19–37.

Fisher, R. J. (1989) *Indigenous Systems of Common Property Forest Management in Nepal,* Honolulu: East–West Environment and Policy Institute.

Flohn, H. (1968) Contributions to a meteorology of the Tibetan highlands, *Atmospheric Science Paper,* 130.

Fogg, W. (1935) Villages and suqs in the High Atlas mountains of Morocco, *Scottish Geographical Magazine,* 51: 144–151.

Fookes, P. G., Sweeney, H., Manby, C. N. D. and Martin, R. O. (1985) Geological and geotechnical engineering aspects of low cost roads in mountainous terrain, *Engineering Geology,* 21: 1–152.

Ford, R. (1993) Marginal coping in extreme land pressures: Ruhengeri, Rwanda, in B. L. Turner II, G. Hyden and R. Kates (eds), *Population Growth and Agricultural Change In Africa,* Gainsville: University of Florida Press, 145–186.

Forman, S. (1988) The future of the verticality concept: implications and possible applications in the Andes, in N. R. Allan, G. Knapp and C. Stadel (eds), *Human Impact on Mountains,* Towota, N.J.: Rowman & Littlefield, 133–153.

Forsyth, T. (1995) The Mu'and and the mountain: perception of environmental degradation in Upland Thailand, *South East Asia Research,* 3: 65–72.

Forsyth, T. (1998) Mountain myths revisited: integrating natural and social environmental science, *Mountain Research and Development,* 18: 107–116.

Fox, D. J. (1997) Mining in mountains, in B. Messerli and J. D. Ives (eds), *Mountains of the World: A global priority,* Carnforth: Parthenon, 171–198.

Fox, J. L., Nurbu, C., Bhatt, S. and Chandola, A. (1994) Wildlife conservation and land use changes in the Trans-Himalayan region of Ladakh, India, *Mountain Research and Development,* 14: 39–60.

Fox, J. M. (1983) *Managing Public Lands in a Subsistence Economy: The Perspective from a Nepali Village,* Department of Agricultural Economy, Madison: University of Wisconsin.

Foxall, L. (1996) Feeling the earth move: cultivation techniques on steep slopes in classical antiquity, in G. Shipley and J. Salmon (eds), *Human Landscapes in Classical Antiquity,* London: Routledge, 44–67.

Franz, H. (1979) *Ökologie der Hochgebirge,* Stuttgart: Verlag Eugen Ulmer.

Freytag, C. (1987) Results from the MERKUR experiment: mass budget and vertical motions in a large valley during mountain valley wind, *Meteorol. Atmos. Phys.,* 37: 129–140.

Fricke, T. (1994) *Himalayan Households: Tamang demography and domestic process*, New York: Colombia Press.

Friedl, J. (1974) *Kippel: A changing village in the Alps*, New York: Holt Rinehart & Winston.

Frisancho, A. R. (1993) *Human Adaptation and Accommodation*, Ann Arbor: University of Michigan Press.

Frutiger, H. (1980) History and actual state of legislation of avalanche zoning in Switzerland, *Journal of Glaciology*, 26: 313–324.

Frye, N. (1964) *The Educated Imagination*, Bloomington: Indiana University Press.

Frye, N. (1990) *Words with Power, being a Second Study of 'The Bible as Literature'*, San Diego: Harcourt Brace Jovanovich.

Funnell, D. C. and Parish, R. (1999) Complexity, cultural theory and strategies for intervention in the High Atlas of Morocco, *Geografisker Annaler*, 81B: 131–144.

Fürer-Haimendorf, C. v. (1964) *The Sherpas of Nepal: Buddhist highlanders*, London: John Murray.

Fürer-Haimendorf, C. v. (1975) *Himalayan Traders: Life in Highland Nepal*, London: John Murray.

Ganjanapan, A. (1996) The politics of environment in Northern Thailand: ethnicity and highland development programmes, in P. Hirsch (ed.), *Seeing Forests for Trees: Environment and environmentalism in Thailand*, Chiang Mai: Silkworm Books, 202–222.

García-Ruiz, J. M. and Lasanta-Martinez, T. (1990) Land-use changes in the Spanish Pyrenees, *Mountain Research and Development*, 10: 267–279.

García-Ruiz, J. M. and Lasanta-Martinez, T. (1993) Land-use conflicts as a result of land-use change in the central Spanish Pyrenees, *Mountain Research and Development*, 13: 295–304.

García-Ruiz, J. M., Lasanta-Martinez, T., Ortigosa, L., Ruiz-Flaño, P., Martí, C. and González, C. (1995) Sediment yield under different land uses in the Spanish Pyrenees, *Mountain Research and Development*, 15: 229–240.

Garnett, A. (1935) Insolation, topography and settlement in the Alps, *Geographical Review*, 25: 601–617.

Garr, C. E. and Fitzharris, B. B. (1994) Sensitivity of mountain runoff and hydro-electricity to changing climate, in M. Beniston (ed.), *Mountain Environments in Changing Climates*, London: Routledge, 366–381.

Gautam, K. H. (1999) Conclusions from three case studies in Nepal. Contribution to Mountain Forum Electronic Conference on Mountain People, Forests and Trees, April. (http://www2.mtnforum.org.mtnforum/archives.document/discuss99/mpft.mpft.htm)

Gentry, A. H. (1995) Patterns of diversity and floristic composition in neotropical montane forests, in S. P. Churchill, H. Balslev, E. Forero and J. L. Luteyn (eds), *Biodiversity and Conservation of Neotropical Montane Forests*, New York: New York Botanical Garden, 103–126.

Gerrard, A. J. (1990) *Mountain Environments: An examination of the physical geography of mountains*, London: Belhaven.

Gigon, A. (1983) Typology and principles of ecological stability and instability, *Mountain Research and Development*, 3: 95–102.

Gilles, J. L., Hammoudi, A. and Mahdi, M. (1986) Oukaimedene, Morocco: a high mountain *agdal*, in *Proceedings of a Conference on Common Property Resource Management*, Washington, D.C.: National Academy of Science/Board of Trade and Technology, 281–304.

Gilmour, D. A. (1988) Not seeing the trees for the forest: a re-appraisal of the deforestation crisis in two hill districts of Nepal, *Mountain Research and Development*, 8: 343–350.

Gilmour, D. A. (1995) Rearranging trees in the landscape in the Middle Hills of Nepal, in J. E. M. Arnold and P. A. Dewees (eds), *Tree Management in Farmer Strategies: Responses to agricultural intensification*, Oxford: Oxford University Press.

Gilmour, D. A. and Nurse, M. C. (1991) Farmer initiatives in increasing tree cover in central Nepal, *Mountain Research and Development*, 11: 329–337.

Gilmour, D. A., Bonell, M. and Cassells, D. S. (1987) The effects of forestation on soil hydraulic properties in the Middle Hills of Nepal: a preliminary assessment, *Mountain Research and Development*, 7: 239–249.

Giono, J. (1999) *Second Harvest*, London: Harvill Press.

Glaser, G. (1983) Unstable and vulnerable ecosystems. A comment based on MAB research in island ecosystems, *Mountain Research and Development* 3: 121–123.

Glazyrin, G. E. (1970) Fazovoe sostoyanie osadkov v gorakh v savisimosti ot prizemnoy temperaturiy vozdukha, *Met. I Gidrol.*: 30–4.

Godde, P. (1999) *Community-Based Mountain Tourism: Practices for linking conservation with enterprise*, Synthesis of an electronic conference of the Mountain Forum, Franklin, USA: The Mountain Institute.

González, C., Ortigosa, L., Martí, C. and García-Ruiz, J. M. (1995) The study of the spatial organization of geomorphic processes in mountain areas using GIS, *Mountain Research and Development*, 15: 241–249.

Goody, J. (1966) *The Development Cycle in Domestic Groups*, Cambridge: Cambridge University Press.

Goudie, A. (1995) *The Changing Earth: Rates of geomorphological processes*, Oxford: Blackwell.

Graf, W. L. (1985) *The Colorado River: Instability and basin management*, Washington, D.C.: Association of American Geographers.

Griffiths, G. A. (1981) Some suspended sediment yields from South Island catchments, New Zealand, *Water Resources Bulletin*, 17: 662–671.

Grillo, R. D. and Stirrat, R. L. (1997) *Discourses of Development: Anthropological perspective*, Oxford: Berg.

Groser, R. F. (1974) Man living at high altitudes, in J. D. Ives and R. G. Barry (eds), *Arctic and Alpine Environments*, London: Methuen, 813–830.

Grossjean, G. (1984) Visual and aesthetic changes in landscape, in E. A. Brugger, G. Furrer, B. Messerli and P. Messerli (eds), *The Transformation of the Swiss Mountain Ranges*, Berne: Verlag Paul-Haupt.

Grossjean, M., Hofer, T., Liechti, R., Messerli, B., Weingartner, R. and Zumstein, S. (1995) Sediments and soils in the floodplain of Bangladesh: looking up the Himalayas?, in H. Schreier, P. B. Shah and S. Brown (eds), *Challenges in Mountain Resource Management in Nepal: Processes, trends and dynamics in middle mountain watersheds*, Proceedings of a workshop, Kathmandu, Nepal, 10–12 April, 1995, Kathmandu: ICIMOD, 25–32.

Grötzbach, E. F. (1984) Mobility of labour in high mountains and the socio-economic integration of peripheral areas, *Mountain Research and Development*, 4, 3: 229–235.

Grötzbach, E. F. (1988) High mountains as human habitat, in N. J. R. Allan, G. W. Knapp and C. Stadel (eds), *Human Impact on Mountains*, Towota, N.J.: Rowman & Littlefield, 24–35.

Grove, J. (1988) *The Century Time Scale*, London: Methuen.

Grove, J. (1996) The century time scale, in T. S. Driver and G. P. Chapman (eds), *Time Scales and Environmental Change*, London: Routledge.

Grunow, J. and Tollner, H. (1969) Nebelneiderschlag im Hochgebirge, *Arch. Met. Geophys. Biokl. B*, 17: 201–228.

Guha, R. (1989) *The Unquiet Woods: Ecological changes and peasant resistance in the Himalayas*, New Delhi: Oxford University Press.

Guichonnet, P. (1948) L'émigration alpine vers les pays de langue allemagne, *Revue de Géographie Alpine*, 36: 533–576.

Guillet, D. (1983) Towards a cultural ecology of mountains: the Central Andes and the Himalayas compared, *Current Anthropology*, 24: 561–574.

Guillet, D. (1987) Terracing and irrigation in the Peruvian Highlands, *Current Anthropology*, 28: 409–430.

Guillet, D. (1991) *Covering Ground: Communal water management and the state in Highland Peru*, Michigan: University of Michigan Press.

Gurung, B. (1992) Towards sustainable development: a case in the European Himalyas, *Futures*, 24, 9: 907–916.

Haigh, M. J., Rawat, J. S., Rawat, M. S., Bartarya, S. K. and Rai, S. P. (1995) Interactions between forest and landslide activity along new highways in the Kumaun Himalaya, *Forest Ecology and Management*, 78: 173–189.

Haila, Y. and Levins, R. (1992) *Humanity and Nature: ecology, science and society*, London: Pluto Press.

Halladay, P. and Gilmour, D. A. (eds) (1995) *Conserving Biodiversity Outside Protected Areas: The role of traditional agro-ecosystems*, Gland, Switzerland: IUCN.

Halpin, P. N. (1994) GIS analysis of the potential impacts of climate change on mountain ecosystems and protected areas, in M. F. Price and D. I. Heywood (eds), *Mountain environments and GIS*, London: Taylor & Francis, 281–301.

Hamilton, L. S. (1987) What are the impacts of Himalayan deforestation on the Ganges–Brahmaputra lowlands and delta? Assumptions and facts, *Mountain Research and Development*, 7: 256–263.

Hamilton, L. S. (1995) Mountain cloud forest conservation and research: a synopsis, *Mountain Research and Development*, 15: 259–266.

Hamilton, L. S., Gilmour, D. A. and Cassells, D. S. (1997) Montane forests and forestry, in B. Messerli and J. D. Ives (eds), *Mountains of the World: A global priority*, Carnforth: Parthenon, 281–311.

Hanna, S. R. and Strimaitis, D. G. (1990) Rugged terrain effects on diffusion, in W. Blumen (ed.), *Atmospheric Processes over Complex Terrain, Meteorological Monograph*, Vol. 23 (45), Boston: American Meteorological Society, 109–143.

Harris, R. B. (1991) Conservation prospects for musk deer and other wildlife in southern Qinghai, China, *Mountain Research and Development*, 11: 353–358.

Harris, R. B. and Shilai, M. (1997) Initiating a hunting ethic in Lisu villages, Western Yunnan, China, *Mountain Research and Development*, 17: 171–176.

Hart, D. M. (1981) *Dadda 'Atta and his forty grandsons: The socio-political organisation of the Ait Atta of Southern Morocco*, Cambridge: Menas Press.

Haslett, J. R. (1997) Mountain ecology: organism responses to environmental change, an introduction, *Global Ecology and Biogeography Letters*, 6: 3–6.

Hein, W. H. (ed.) (1986) *Alexander von Humboldt: Leben und Werk*, Frankfurt: Weisbecker.

Heinen, J. T. and Yonzon, P. B. (1994) A review of conservation issues and programs in Nepal: from a single species focus toward biodiversity protection, *Mountain Research and Development*, 14: 61–76.

Hewitt, F. (1989) Woman's work, woman's place: the gendered life-world of a high mountain community in Northern Pakistan, *Mountain Research and Development*, 9: 335–352.

Hewitt, K. (1968) Geomorphology of mountain regions of the Upper Indus Basin, unpublished Ph.D. thesis, University of London.

Hewitt, K. (1988) The study of mountain lands and peoples: a critical overview, in N. R. J. Allan, G. Knapp and C. Stadel (eds), *Human Impact on Mountains*, Towota, N.J.: Rowman & Littlefield, 6–23.

Hewitt, K. (1997) Risk and disasters in mountain lands, in B. Messerli and J. D. Ives (eds), *Mountains of the World: A global priority*, Carnforth: Parthenon, 371–408.

Higgins, G., Kassam, A. H., Naiken, L., Fischer, G. and Shah, M. M. (1982) *Potential Population-supporting Carrying Capacity of Lands in the Developing World*, Rome: FAO/UNFPA/IIASA.

Hirst, P. Q. and Thompson, G. (1996) *Globalization in Question: The international economy and problems of governance*, Cambridge: Polity Press.

Hock, R. J. (1970) The physiology of high altitude, *Scientific American*, 222: 52–62.

Hofer, T. (1993) Himalayan deforestation, changing river discharge, and increasing floods: myth or reality?, *Mountain Research and Development*, 13: 213–233.

Hofer, T. (1997) Meghalaya, not Himalaya, *HIMAL South Asia*, Sept/Oct.: 52–56.

Hofer, T. (1998) Do land use changes in the Himalayas affect downstream flooding? Traditional understanding and new evidences, *Memoir of the Geological Society of India* 41: 119–141.

Hoinkes, H. (1954) Beiträge zur Kenntnis des Gletscherwindes, *Arch. Met. Geophys. Boikl. B*, 6: 36–53.

Holling, C. S. (1985) Perceiving and managing the complexity of ecological systems, in *The Science and Praxis of Complexity*, Tokyo: UNU.

Holling, C. S. (1994) Simplifying the complex, *Futures*, 26, 598–609.

Holling, C. S. (ed.) (1978) *Adaptive Environmental Management and Assessment*, Chichester: Wiley.

Holtmeier, F.-K. (1994) Ecological aspects of climatically caused timberline fluctuations: a review and outlook, in M. Beniston (ed.), *Mountain Environments in Changing Climates*, London: Routledge, 220–233.

Honey, M. (1999) *Ecotourism and Sustainable Development: Who owns paradise?*, Washington, D.C.: Island Press.

Houston, C. S. (1982) Return to Everest – a sentimental journey, *Summit*, 28: 14–17.

Huang, W. and Yao, B. (1992) The development and protection of natural resources in Hengduan Mountains, *Oecologia Montana*, 1: 37–40.

Huggett, R. J. (1995) *Geoecology: An evolutionary approach*, London: Routledge.

Hughey, K. (1997) *Big Business and the Mountain Environment: Focus on mining*, Lincoln University, New Zealand (http://www/mtnforum.org)

Hulme, M., Conway, D., Kelly, P. M., Subak, S. and Downing, T. E. (1995) *The Impact of Climate Change in Africa*, Stockholm: Stockholm Environment Institute.

Humboldt, A. v. and Bonpland, A. (1807) *Essai sur la géographie des plantes. Accompagne d'un tableau physique des regions equinoxales*, Mexico City: Editorial cultura.

Hunter, C. (1997) Sustainable bioprospecting: using private contracts and international legal principles and policies to conserve raw materials, *Boston College Environmental Affairs Law Review*, 25: 129–174.

Hurni, H. (1983) Soil erosion and soil formation in agricultural ecosystems. Ethiopia and Northern Thailand, *Mountain Research and Development*, 3: 131–142.

Hurni, H. (1988) Degradation and conservation of the resources in the Ethiopian highlands, *Mountain Research and Development*, 8: 123–130.

Ibn Khaldun (1852) *Histoire des Berberes* (trans M. De Slane), Paris.

International Centre for Alpine Environments (ICALPE) (1997) *Vers une Politique Européenne des Montagnes. European Parliament*, Série Agriculture, forêts et développement rural, AGRI III/A Luxembourg.

International Centre for Integrated Mountain Development (1998) *Renewable Energy Technologies for Mountain Communities*, Newsletter No. 30.

IUCN (1990) *UN list of National Parks and Protected Areas*, Gland, Switzerland: IUCN.

IUCN (1996) *Maintaining Biodiversity in Pakistan with Rural Community Development. Annual Report*, Karachi: IUCN.

IUCN (1997) *Maintaining Biodiversity in Pakistan with Rural Community Development. Annual Report*, Karachi: IUCN.

IUCN/World Bank (1997) *Large Dams: Learning from the past, looking at the future*, Gland, Switzerland: IUCN.

Ives, J. D. (1970) Himalayan highway, *Canadian Geographical Journal*, 80: 26–31.

Ives, J. D. (1988) Development in the face of uncertainty, in J. D. Ives and D. C. Pitt (eds), *Deforestation: Social dynamics in watersheds and mountain ecosystems*, London: Routledge, 54–74.

Ives, J. D. and Messerli, B. (1989) *The Himalayan Dilemma: Reconciling development and conservation*, London: Routledge.

Ives, J. D. and Messerli, B. (1990) Progress in theoretical and applied mountain research 1973–1989, and major future needs, *Mountain Research and Development*, 10: 101–127.

Ives, J. D. and Stites, A. (1975) Project 6. Impact of human activity on mountain and tundra ecosystems, in *Proceedings of the Boulder Workshop* (Vol. Special Publication Colorado), Institute of Arctic and Alpine Research.

Ives, J. D., Messerli, B. and Rhoades, R. (1997) Agenda for sustainable mountain development, in B. Messerli and J. D. Ives (eds), *Mountains of the World: A global priority*, Carnforth: Parthenon, 455–466.

Jackson, R. T. (1971) Periodic markets in southern Ethiopia, *Transactions IBG*, 53: 31–42.

Janiga, M. (1999) The lead cycle in the Alpine environment of the Tatra mountains: vertebrates as bioindicators, in M. Price (ed.), *Global Change in the Mountains*, Carnforth: Parthenon, 97–99.

Jansky, L. (1999) Report to Mountain Forum Electronic Conference on Mountain People, Forests and Trees, April (http://www.mtnforum.org)

Jeník, J. (1961) *Alpine vegetation of the High Sudetes: Theory of anemo-orographic systems*, Praha: Czech National Academy of Sciences.

Jeník, J. (1997) The diversity of mountain life, in B. Messerli and J. D. Ives (eds), *Mountains of the World: A global priority*. Carnforth: Parthenon, 199–235.

Jina, P. S. (1995) *High Pasturelands of Ladakh Himalaya*, New Delhi: Indus Publishing.

Jodha, N. S. (1990) Mountain agriculture: the search for sustainability, *Journal of Farming Systems Research Extension*, 1: 55–75.

Jodha, N. S. (1997) Mountain agriculture, in B. Messerli and J. D. Ives (eds), *Mountains of the World: A global priority*, Carnforth: Parthenon, 313–335.

Joffe, G. (1992) Irrigation and water supply systems in North Africa, *Moroccan Studies*, 2: 47–55.

Johnston, B. (ed.) (1990) Breaking out of the tourist trap: parts one and two, *Cultural Survival Quarterly*, 14, 1–2.

Johnstone, A. (1997) A flash flooding event in the High Atlas Mountains of Morocco, *Geography*, 82: 85–90.

Jolly, C. (1994) Four theories of population change and the environment, *Population and Development*, 16, 1: 61–90.

Juvík, J. O. and Juvík, S. P. (1984) Mauna Kea and the myth of multiple use. Endangered species and mountain management in Hawaii, *Mountain Research and Development*, 4: 191–202.

Kalin Arroyo, M. T., Squeo, F. A., Armesto, J. J. and Villigran, C. (1988) Effects of aridity on plant diversity in the northern Chilean Andes: results of a natural experiment, *Annals of the Missouri Botanical Garden*, 75: 55–78.

Kalter, J. (1991) *The Arts and Crafts of the Swat Valley: Living traditions in the Hindu Kush*, London: Thames & Hudson.

Kampe, K. (1992) Northern Highlands development, bureaucracy and life on the margins: the Akha case, *Pacific Viewpoint*, 33: Special Issue: 'Marginalisation in Thailand: Disparities, democracy and development intervention'.

Kariel, H. G. (1993) Tourism and society in four Austrian Alpine communities, *GeoJournal*, 31: 449–456.

Kauffman, S. (1993) *At Home in the Universe: The search for laws of complexity*, Oxford: Oxford University Press.

Kessler, J. J. (1995) 'Mahjur' areas: traditional rangeland reserves in the Dhamar Montane Plains (Yemen Arab Republic), *Journal of Arid Environments*, 29: 395–401.

Kharel, F. R. (1997) Agricultural crop and livestock depredation by wildlife in Langtang National Park, Nepal, *Mountain Research and Development*, 17: 127–134.

Kienholz, H., Hafner, H., Schneider, G. and Tamrakar, R. (1983) Mountain hazards mapping in Nepal's Middle mountains. Maps of land use and geomorphic damages (Kathmandu–Kakani area), *Mountain Research and Development*, 3: 195–220.

Kienholz, H., Schneider, G., Bichsel, M., Grunder, M. and Mool, P. (1984) Mapping of mountain hazards and slope stability, *Mountain Research and Development*, 4: 247–266.

Kirkby, J., O'Keefe, P. and Timberlake, L. (1995) *The Earthscan Reader in Sustainable Development*, London: Earthscan Publications.

Kitching, G. (1980) *Development and Underdevelopment in Historical Perspective*, London: Methuen.

Klötzli, F. (1997) Biodiversity and vegetation belts in tropical and subtropical mountains, in B. Messerli and J. D. Ives (eds), *Mountains of the World: A global priority*, Carnforth: Parthenon, 232–235.

Knight, J. (1994) Town making in rural Japan: an example from Wakayama, *Journal of Rural Studies*, 10: 249–261.

Konèek, M., Samaj, F., Smolen, F., Otruba, J., Murínová, G. and Peterka, V. (1973) Climatic conditions in the High Tatra Mountains, *Zbornik TANAP*, 15: 239–324.

Kraus, H. (1967) Da Klima von Nepal, *Khumbu Himal*, 1: 301–321.

Kreutzmann, H. (1988) Oases of the Karakorum: evolution of irrigation and social organization in Hunza, Northern Pakistan, in N. J. R. Allan, G. W. Knapp and C. Stadel (eds), *Human Impact on Mountains*, Towota, N.J.: Rowman & Littlefield, 243–254.

Kreutzmann, H. (1991) The Karakorum Highway: the impact of road construction on mountain societies, *Modern Asian Studies*, 25: 711–736.

Kreutzmann, H. (1993) Challenge and response in the Karakorum: socioeconomic transformation in Hunza, Northern Areas, Pakistan, *Mountain Research and Development*, 13: 19–39.

Kreutzmann, H. (1994) Habitat conditions and settlement processes in the Hindu-Kush–Karakorum, *Petermanns Geographische Mitteilungen*, 138: 337–356.

Kreutzmann, H. (1998) From watertowers of mankind to livelihood strategies of mountain dwellers: approaches and perspectives for high mountains research, *Erdkunde*, 52, 3: 185–200.

Krippendorf, J. (1984) The capital of tourism in danger, in E. A. Brugger, G. Furrer, B. Messerli and P. Messerli (eds) *The Transformation of the Swiss Mountain Regions*, Berne: Verlag Paul-Haupt, 427–450.

Krippendorf, J. (1986) *Alpsegen, Alptraum. Fur eine Tourismus-Entwicklung im Einklang mit Mensch und Natur*, Berne: Kummerly & Frey.

Krorick, C. (1999) International politics of climate change, *Ecologist*, 29: 104–107.

Kupfer, J. A. and Cairns, D. M. (1996) The suitability of montane ecotones as indicators of global climatic change, *Progress in Physical Geography*, 20: 253–272.

Kutsch, H. (1982) Principal features of a form of water-concentrating culture on small-holdings with special reference to the Anti Atlas, *Trierer Geographische Studien*, 5.

Kydd, J. and Thayer, S. (1993) Agricultural policy reform in Morocco 1984–1991, in I. Goldin (ed.), *Economic Reform, Trade and Agricultural Adjustment*, London: St Martin's Press, 135–164.

Laban, P. (1979) *Landslide Occurrence in Nepal*, Integrated Watershed Management Project, Phewa Tal Project Report, Kathmandu.

Lahiri, S. (1974) Physiological response and adaptation to high altitude, in D. Robertshaw (ed.), *Environmental Physiology*, Vol. 7, London: Butterworth, 271–311.

Lamb, H. F., Damblon, F. and Maxted, R. W. (1991) Human impact on the vegetation of the Middle Atlas, Morocco, during the last 5000 years, *Journal of Biogeography,* 18: 519–532.

Lamb, H. F., Eicher, U. and Switsur, V. R. (1989) An 18,000 year record of vegetation, lake-level and climatic change from Tigalmamine, Middle Atlas, Morocco, *Journal of Biogeography,* 16: 65–74.

Lamb, H. H. (1982) *Climate History and the Modern World,* London: Methuen.

Lamb, H. H. (1988) *Weather, Climate and Human Affairs,* London: Routledge.

Lampietti, J. A. and Dixon, J. A. (1995) *To See the Forest for the Trees: A guide to non-timber forest benefits,* Environment Department Paper No. 013, Washington, D.C.: The World Bank.

Laouina, A. (1995) Demographie et degradation des sols dans le Rif, in G. Denoni (ed.), *Environnement Humain de l'Erosion,* Reseau Erosion Bulletin 15: 69–77.

Lauer, W. (1973) Klimatische Grundzüge der Höhenstufung tropischer Gebirge, in *Tagungsbericht und wissenschaftliche Abhandlungen, 40 Deutscher Geographentag, Innsbruck,* Innsbruck: F. Steiner, 76–90.

Lauer, W. (1981) Eco-climatological conditions of the Paramo Belt in tropical high mountains, *Mountain Research and Development,* 1: 209–221.

Lauer, W. (1993) Human development in the Andes: a geoecological overview, *Mountain Research and Development,* 13: 157–166.

Lauscher, F. (1976a) Methoden zur Weltklimatologie der Hydrometeore. Der Anteil des festen Niederschlags am Gesamtniederschlag, *Arch. Met. Geophys. Biokl. B,* 24: 129–176.

Lauscher, F. (1976b) Weltweite Typen der Höhenarbhängigkeit des Niederschlags bei verschiedenen Witterungsladen im Sonnblick Gebiet, *Arbeiten, Zentralanst. für Met. Geodynam.* (Vienna), 28: 80–90.

Lauscher, F. (1980) Die Schwankungen der Temperatur auf dem Sonnblick seit 1887 im Vergleich zu globalen Temperaturschwankungen, in *16 International Tagung für Alpine Meteorologie Soc.,* Météorol. de France, Boulogne-Billancourt, Aix-les-Bains, 315–319.

Lawrence, A. (1999) Tree cultivation by upland farmers in the Philippines, Contribution to Mountain Forum Electronic Conference on Mountain People, Forests and Trees, April (http://www2.mtnforum.org/mtnforum/archives/document/discuss99/mpft/mpft.htm)

Lazaar, M. (1997) La crise du système montagnard du nord et l'immigration en Espagne, in M. Khettani (ed.), *L'aménagement du territoire et le développement de l'économie de montagne en Méditerranée,* Rabat: Editions le Fennec.

Le Roy Ladurie, E. (1972) *Times of Feast, Times of Famine: A history of climate since the year 1000,* London: Allen & Unwin.

Leach, M., Mearns, R. and Scoones, I. (1997) *Environmental Entitlements: A framework for understanding the institutional dynamics of environmental change,* Brighton: Institute of Development Studies.

Lee, K. N. (1993) *Compass and Gyroscope: Integrating science and politics for the environment,* Washington, D.C.: Island Press.

Leemans, R. and Cramer, W. P. (1990) *The IIASA Database for Mean Monthly Values of Temperature, Precipitation and Cloudiness on a Global Terrestrial Grid. WP-90–41,* Laxenburg: Austria International Institute for Applied Systems Analysis.

Lehmann, D. (ed.) (1982) *Ecology and Exchange in the Andes,* Cambridge: Cambridge University Press.

Levy Hynes, A., Brown, A. D., Grau, H. R. and Grau, A. (1997) Local knowledge and the use of plants in rural communities in the montane forests of northwestern Argentina, *Mountain Research and Development,* 17: 263–271.

Libiszewski, S. and Bächler, G. (1997) Conflicts in mountain areas – a predicament for sustainable development, in B. Messerli and J. D. Ives (eds), *Mountains of the World: A global priority,* Carnforth: Parthenon, 103–130.

Lichtenberger, E. (1988) The succession of an agricultural society to a leisure society: the high mountains of Europe, in N. J. R. Allan, G. W. Knapp and C. Stadel (eds), *Human Impact on Mountain Environments,* Towota, N.J.: Rowman & Littlefield, 218–227.

Linder, W. (1994) *Swiss Democracy,* London: St Martin's Press.

Lo Presti, A. (1996) A critical analysis from the perspective of social science, *Futures,* 28: 891–902.

Loevinsohn, M. E., Magarura, J. and Nkusi, A. (1992) Group intervention in utilizing land and water resources in Rwandan valleys, in 'News from the field: a collection of short papers', *ODI Irrigation Management Network Paper* No. 13, London: Overseas Development Institute, 3–14.

Long, A. (1995) The importance of tropical montane cloud forests for endemic and threatened birds, in L. S. Hamilton, J. O. Juvík and F. N. Scatena (eds), *Tropical Montane Cloud Forests,* New York: Springer Verlag, 79–106.

Long, N. (1975) *Intermediaries and Brokers in Highland Peru,* London: EIU Agricultural Marketing Working Group.

Lowe, P. (1992) Preface: Mountain areas; a challenge and opportunity, *Revue de Géographie Alpine,* 80: 8–19.

Lucas, P. H. C., Hardie, N. D. and Hodder, R. A. C. (1974) *Report of the New Zealand Mission on Sagarmatha (Mt Everest) National Park, Nepal,* Wellington: Ministry of Foreign Affairs.

Lynch, O. J. and Maggio, G. F. (2000) *Mountain Laws and Peoples: Moving towards sustainable development and recognition of community based property rights,* Synthesis of an electronic conference for the Mountain Forum, Harrisonburg, Va.: The Mountain Institute.

MacArthur, R. H. (1972) *Geographical Ecology,* New York: Harper.

Macfarlane, A. (1976) *Resources and Population: A study of the Gurungs of Nepal,* Cambridge: Cambridge University Press.

McCrae, S. D. (1982) Human ecological modelling for the Central Andes, *Mountain Research and Development,* 21: 97–110.

McKinnon, J. (1989) Structural assimilation and the consensus: clearing grounds on which to rearrange our thoughts, in J. McKinnon and B. Vienne (eds), *Hill Tribes Today: Problems in Development,* Bangkok: White Lotus.

McKinnon, J. (1997) The forests of Thailand: strike up the ban?, in D. McCaskill and K. Kampe (eds), *Development or Domestication? Indigenous peoples of Southeast Asia,* Chiang Mai: Silkworm Books, 117–131.

McNeill, J. R. (1992) *The Mountains of the Mediterranean World: An environmental history,* Cambridge: Cambridge University Press.

Mahdi, M. (1986) Private rights and collective management of water in a High Atlas Berber tribe, in *Proceedings of a Conference on Common Property Resource*

Management, Washington, D.C.: National Academy of Science/Board of Trade and Technology, 181–197.

Malthus, T. R. ([1803] 1986) An essay on the principle of population, in E. A. Wrigley and D. Souden (eds), *The Works of Thomas Malthus* (2nd edn, 1986), London: Pickering.

Marcoux, A. (1999) *Population and Environmental Change: From linkages to policy issues,* Special paper (http://www.fao.org/sd/wpdirect/wpre0089.htm)

Margreth, S. and Funk, M. (1999) Hazard mapping for ice and combined ice/snow avalanches – two case studies from the Swiss and Italian Alps, *Cold Regions Science and Technology* 30, 1–3: 157–173.

Martin, E. D. and Yoder, R. (1987) *Institutions for Irrigation Management in Farmer-managed Systems: Examples from the hills of Nepal,* Colombo, Sri Lanka: International Irrigation Management Institute.

Martinez-Alier, J. (1990) *Poverty as a Cause of Environmental Degradation,* Washington, D.C.: The World Bank.

Maselli, D. A. (1996) Constraintes d'une utilisation durable des ressources naturelles du Haut-Atlas: le cas du bassin intramontagnard de Tagoundaft, *Revue de Géographie Alpine,* 84: 109–119.

Maurer, G. (1992) Agriculture in the Rif and Tell mountains of North Africa, *Mountain Research and Development,* 12: 337–347.

Mawdsley, E. (1997) Nonsecessionist regionalism in India: the Uttarakhand separate state movement, *Environment and Planning A,* 29: 2217–2235.

Mawdsley, E. (1998) After Chipko: from environment to region in Uttaranchal, *Journal of Peasant Studies,* 25, 4: 36–54.

Mawdsley, E. (1999) A new Himalayan state in India: popular perceptions of regionalism, politics and development, *Mountain Research and Development,* 19: 101–112.

Meadows, D. H., Randers, J. and Behrens III, W. W. W. (1972) *The Limits to Growth,* New York: Potomac Books.

Meiggs, R. (1982) *Trees and Timber in the Ancient Mediterranean World,* Oxford: Clarendon Press.

Mendels, F. F. (1972) Proto-industrialisation: the first phase of the industrialisation process, *Journal of Economic History,* 32: 241–261.

Merlo, V. and Zaccherini, R. (1992) *Montagna 2000,* Milan.

Mesoscale Alpine Programme (http://www.map.ethz.ch/form.w4h/map.html)

Messerli, B. (1983) Stability and instability of mountain ecosystems: introduction to a workshop sponsored by the United Nations University, *Mountain Research and Development,* 3: 81–94.

Messerli, B. and Hofer, T. (1995) Assessing the impact of anthropogenic land use change in the Himalayas, in G. P. Chapman and M. Thompson (eds), *Water and the Quest for Development in the Ganges Valley,* New York: Mansell Publishing, 64–89.

Messerli, B. and Ives, J. D. (eds) (1997) *Mountains of the World: A global priority,* Carnforth: Parthenon.

Messerli, B. and Winiger, M. (1992) Climate, environmental change and resources of the African mountains from the Mediterranean to the Equator, *Mountain Research and Development,* 12: 315–336.

Messerli, B., Messerli, P., Pfister, C. and Zumbuhl, H. J. (1978) Fluctuations in climate and glaciers in the Bernese Oberland, Switzerland, and their geo-ecological significance, 1600–1975, *Arctic and Alpine Research,* 10: 247–260.

Messerli, P. (1989) *Mensch und Natur im alpinen Leibenraum: Risiken, Chancen, Perspectiven,* Berne: Verlag Paul-Haupt.

Messerschmidt, D. A. (1987) Conservation and society in Nepal: customary forest management and innovative development, in P. D. Little and M. M. Horowitz (eds), *Land at Risk in the Third World,* Boulder, Colo.: Westview Press, 373–397.

Meyer, P. (1984) Forestry, in E. A. Brugger, G. Furrer, B. Messerli and P. Messerli (eds), *The Transformation of the Swiss Mountain Ranges,* Berne: Verlag Paul-Haupt, 643–682.

Mezario, R. (1989) *Il capitalismo nelle montagne,* Bologna.

M'Hirit, O. (1999) Mediterranean forests: ecological space and economic and community wealth, *Unasylva,* 197, 50: 3–15.

Miehe, G. (1990) *Langtang Himal: Flora und Vegetation als Klimazeiger und-zeugen im Himalaya,* J. Cramer, Dissertationes botanicae 158, Berlin/Stuttgart.

Miehe, S., Cramer, T., Jacobsen, J-P. and Winiger, M. (1996) Humidity conditions in the Western Karakorum as indicated by climatic data and corresponding distri-bution patterns of the montane and alpine vegetation, *Erdkunde,* 50: 190–204.

Miller, D. (1996) Rangelands and range management, *ICIMOD Newsletter* No. 27, Kathmandu: International Centre For Integrated Mountain Development.

Miller, J. A. (1984) *Imlil: A Moroccan mountain community in change,* Boulder, Colo.: Westview Press.

Milton, K. (1996) *Environmentalism and Cultural Theory,* London: Routledge.

Minghi, J. V. (1963) Boundary states and national prejudices: the case of South Tyrol, *Professional Geographer,* 15: 4–8.

Mitchell, B. (1997) *Resource and Environmental Management,* Harlow: Longman, 134–154.

Moench, M. (1988) 'Turf' and forest management in a Gahrwal hill village, in L. Fortmann and J. W. Bruce (eds), *Whose Trees? Proprietary dimensions of forestry,* Boulder, Colo.: Westview Press, 127–136.

Moffat, I. (1996) *Sustainable Development: Principles, analysis and policies,* New York: Parthenon.

Mohan, R. (1994) *Understanding the Developed Metropolis: Lessons from the city study of Bogata and Cali, Colombia,* New York: World Bank.

Molnar, A. (1981) *Nepal – the Dynamics of Traditional Systems of Forest Management: Implications for the community development project,* Nepal: The World Bank.

Montagne, R. (1930) *Les Berbères et le Makhzan dans le sud du Maroc: Essai sur la transformation politique des Berbères sédentaires (groupe chleuh),* Paris: Librairie Felix Alcan.

Montagne, R. (1931) *La vie social et la vie politique des Berbères,* London: Cass.

Montero, G. and Canellas, G. (1999) Sustainable management of Mediterranean forests in Spain, *Unasylva* 197, 50: 29–34.

Moore, H., Fox, H. R., Harrouni, C. and El Alami, A. (1999) Kif in the Rif: bio-diversity under threat in northern Morocco, in M. Price (ed.), *Global Change in the Mountains,* Carnforth: Parthenon, 136–138.

Morriset, P. and Payette, S. (1983) *Tree-line Ecology,* Collection Nordicana 47: Québec.

Moser, P. and Moser, W. (1986) Reflections on the MAB 6 Obergurgl project and tourism in alpine environments, *Mountain Research and Development,* 6: 101–118.

Moss, L. A. G. (1993) *Notes on Amenity Migration in the Chiang Mai Bioregion,* Bangkok: Asian Institute of Technology.

Mosse, D. (1995) Social analysis in participatory rural development, *PLA Notes,* 24.

Mountain Agenda (1992) *An Appeal for Mountains,* Berne: Mountain Agenda.

Mountain Agenda (1997) *Mountains of the World: Challenges for the twenty-first century,* Berne: Mountain Agenda.

Mujica, E. and Rueda, J. L. (1995) *Intergovernmental Consultation Concerning the Sustainable Development of Mountains in Latin America* (trans. M. Price and C. Correa), Lima, Peru: Centro Internacionale de la Papa.

Müller, H. (1985) On the radiation budget of the Alps, *Journal of Climatology,* 5: 445–462.

Mundy, M. (1989) Irrigation and society in a Yemeni valley: on the life and death of a bountiful resource, *Peuples méditerranéens,* 46: 97–128.

Murra, J. V. (1972) El control verticale de un maximo pisos en la economia de las sociedades Andinas, in I. O. de Zuniga (ed.), *Vista de la Provinca de Leon de Huanuco en 1562,* Vol. 2, Universidade Nacional Hermilio Valdazan.

Murra, J. V. (1995) The limits and limitations of the 'vertical archipelago', in S. Masuda, I. Shimada and C. Morris (eds), *Andean Ecology and Civilisation,* Tokyo: University of Tokyo, 15–20.

Myers, N. (1983) A priority ranking strategy for threatened species?, *Environmentalist,* 3: 97–120.

Myska, M. (1996) Proto-industrialisation in Bohemia, Moravia and Silesia, in S. C. Ogilvie and M. Cerman (eds), *European Proto-industrialisation,* Cambridge: Cambridge University Press.

Naveh, Z. (1982) Mediterranean landscape erosion and degradation as multivariate biofunctions: theoretical and practical implications, *Landscape Planning,* 9: 125–146.

Negi, G. C. S. (1994) High yielding vs. traditional crop varieties: a socio-agronomic study in a Himalayan village in India, *Mountain Research and Development,* 14: 251–254.

Nepal, S. K. (1997) *Tourism Induced Environmental Changes in the Everest Region: Some recent evidence,* Centre for Development and the Environment, University of Berne.

Netting, R. M. (1972) Of men and meadows: strategies of Alpine landuse, *Anthropological Quarterly,* 45: 132–144.

Netting, R. M. (1974) The system nobody knows: village irrigation in the Swiss Alps, in T. E. Downing and M. Gibson (eds), *Irrigation's Impact on Society,* Tucson: University of Arizona Press, 67–75.

Netting, R. M. (1981) *Balancing on an Alp: Ecological change and continuity in a Swiss mountain community,* Cambridge: Cambridge University Press.

Netting, R. M., Wilk, R. R. and Arnould, E. J. (eds) (1984) *Households: Comparative and historical studies of the domestic group,* Berkeley: University of California Press.

Nicolis, G. and Prigogine, I. (1989) *Exploring Complexity,* New York: Freeman.

O'Connor, K. F. (1984) Stability and instability of ecological systems in New Zealand mountains, *Mountain Research and Development,* 4, 15–29.

O'Connor, M. (1995) *Emergent Complexity and Procedural Rationality: Post normal science for sustainability,* University of Auckland, Department of Economics.

O'Neill, C. A. and Sustein, C. R. (1992) Economics and environment: trading debt and technology for nature, *Colombia Journal of International Law,* 17: 93–151.

Ohler, F. (1999) Leasehold Forestry in Nepal. Contribution to Mountain Forum Electronic Conference on Mountain People, Forests and Trees, April (http://www.mtnforum.org/mtnforum/library/discuss99/mpft/mpft.htm)

Okahashi, H. (1996) Development of mountain village studies in post-war Japan: depopulation, peripheralisation and village renaissance, *Geographical Review of Japan,* 691: 60–69.

Oppitz, M. (1973) Myths and facts: reconsidering some data concerning the clan history of the Sherpa, in C. von Fürer-Haimendorf (ed.), *Contributions to the Anthropology of Nepal,* Warminster: Aris & Phillips, 232–243.

Ostrom, E. (1990) *Governing the Commons: The evolution of institutions for collective action,* Cambridge: Cambridge University Press.

Ott, E. (1984) Forest potential, in E. A. Brugger, G. Furrer, B. Messerli and P. Messerli (eds), *The Transformation of the Swiss Mountain Ranges,* Berne: Verlag Paul-Haupt, 157–166.

Oxfam (1998) Briefing report on the earthquakes in Afghanistan (http://www.oxfam.org.uk/atwork/emerg/afghan0698.htm)

Pacey, A. and Cullis, A. (1986) *Rainwater Harvesting: The collection of rainfall and runoff in rural areas,* London: Intermediate Technology Publications.

Parish, R. (1999) The unseen, unknown and misunderstood: complexities of development in Hunza, Pakistan, *International Journal of Sustainable Development and World Ecology,* 6: 1–16.

Parish, R. and Funnell, D. C. (1999) Climate change in mountain regions: some possible consequences in the Moroccan High Atlas, *Global Environmental Change,* 9: 45–58.

Parker, T. A. and Carr, J. L. (eds) (1992) *Status of Forest Remnants in the Cordillera de la Costa and Adjacent Areas of Southwestern Ecuador (Rapid Assessment Programme),* Washington, D.C.: Conservation International.

Pascon, P. (1977) *Le Haouz de Marrakech,* Rabat: Centre Universitaire de la Recherche Scientifique.

Patzelt, G. (1983) Die Berg-und Gletscherstürze von Huascaran, Cordillera Blanca, Peru, *Innsbruck, Hochgebirgsforschung,* 6: 110.

Pawson, I. G. (1976) Growth and development of high altitude populations: a review of Ethiopian, Peruvian and Nepalese studies, *Proceedings of the Royal Society of London, Series B,* 194: 83–98.

Pawson, I. G., Stanford, D. D., Adams, V. A. and Nurbu, N. (1984) Growth of tourism in Nepal's Everest region: impact on physical environments and the structure of human settlement, *Mountain Research and Development,* 4, 4: 237–246.

Pearce, D. (1993) *Economic Values and the Natural World,* London: Earthscan.

Peattie, R. (1936) *Mountain Geography: A critique and field study,* Cambridge, Mass.: Harvard University Press.

Peet, R. and Watts, M. (1996) *Liberation Ecologies,* London: Routledge.

Perez-Trejo, F., Clark, N. G. and Allen, P. M. (1993) An exploration of dynamical systems modelling as a decision tool for environmental policy, *Journal of Environmental Management*, 39: 305–319.

Peters, R. L. and Darling, J. D. (1985) The greenhouse effect and nature reserves, *Bioscience*, 35: 707–717.

Pézelet, L. (1996) Gîtes étape chez l'habitant dans le Haut-Atlas central: Logiques touristique et le sens de l'espace domestique, *Revue de Géographie Alpine*, 84: 133–148.

Pfister, C. (1978) Climate and economy in eighteenth-century Switzerland, *Journal of Interdisciplinary History*, 10: 719–723.

Pfister, C. (1983) Changes in stability and carrying capacity of highland agro-ecosystems in Switzerland in the historical past, *Mountain Research and Development*, 3: 291–297.

Pfister, U. (1996) Proto-industrialisation in Switzerland, in S. C. Ogilvie and M. Cerman (eds), *European Proto-industrialisation*, Cambridge: Cambridge University Press, 137–154.

Pietri, C. (1993) Rénovation de la carte de localisation probable des avalanches, *Revue de Géographie Alpine*, 81,1: 85–97.

Pils, M., Glauser, P. and Siegrist, D. (1996) *Green Paper on the Alps,* Vienna: Friends of Nature International.

Ploeg, J. D. v. d. (1993) Potatoes and knowledge, in M. Hobart (ed.), *An Anthropological Critique of Development: The growth of ignorance,* London: Routledge, 209–227.

Powell, J. W. (1876) *Report on the Geology of the Eastern Part of the Uinta Mountains,* Washington, D.C.: Government Printing Office.

Pratt, D. J. and Preston, L. (1998) The economics of mountain resource flows, *Unasylva*, 149: 31–38.

Prescott, J. R. V. (1987) *Political Frontiers and Boundaries,* London: Unwin Hyman.

Preston, L. (ed.) (1997) *Investing in Mountains: Innovative mechanisms and promising examples for financing conservation and sustainable development. Synthesis of a Mountain Forum electronic conference,* Franklin, USA: The Mountain Institute.

Preston, P. W. (1994) *Discourses of Development: State, market and polity in the analysis of complex change,* Aldershot: Avebury.

Price, L. W. (1981) *Mountains and Man: A study of process and environment,* Berkeley: University of California Press.

Price, M. F. (1987) Tourism and forestry in the Swiss Alps: parasitism or symbiosis?, *Mountain Research and Development*, 9, 3: 83–89.

Price, M. F. (1990) Temperate mountain forests: common-pool resources with changing, multiple outputs for changing communities, *Natural Resources Journal*, 30: 685–707.

Price, M. F. (1991) An assessment of patterns of use and management of mountain forests in Colorado, USA: implications for future policies, *Mountain Research and Development*, 11: 57–64.

Price, M. F. (1995) *Mountain Research in Europe: An overview of MAB research from the Pyrenees to Siberia,* Carnforth: Parthenon.

Price, M. F. (1996) People in Biosphere Reserves: an evolving concept, *Society and Natural Resources*, 9: 645–654.

Price, M. F. (1997) *People and Tourism in Fragile Environments,* Chichester: Wiley.

Price, M. F. (1998) Mountains: globally important ecosystems, *Unasylva,* 149: 3–12.

Price, M. F. (1999a) *Chapter 13 in Action 1992–97. A task manager's report,* Forestry Department, Rome: FAO.

Price, M. F. (1999b) Towards co-operation across mountain frontiers: the Alpine Convention, *European Environment,* 9: 83–89.

Price, M. F. and Butt, N. (eds) (2000) *Forest and Sustainable Mountain Development: A state of knowledge report for 2000,* Wallingford: CAB International (forthcoming).

Price, M. F. and Kim, E.-G. (1999) Priorities for sustainable mountain development in Europe. *International Journal of Sustainable Development and World Ecology,* 6: 203–219.

Price, M. F. and Thompson, M. (1997) The complex life: human uses in mountain ecosystems, *Global Ecology and Biogeographical Letters,* 6, 1: 77–90.

Price, M. F., Moss, L. A. G. and Williams, P. W. (1997) Tourism and amenity migration, in B. Messerli and J. D. Ives (eds), *Mountains of the World: A global priority,* Carnforth: Parthenon, 249–280.

Project Migrations et Développement (1994) *Electrification – Taroudant,* Morocco.

Puigdefabregas, J. and Fillat, F. (1986) Ecological adaptation of traditional land uses in the Spanish Pyrenees, *Mountain Research and Development,* 6: 63–72.

Quézel, P., Médial, F., Loisel, R. and Barbero, M. (1999) Biodiversity and conservation of forest species in the Mediterranean Basin, *Unasylva,* 197, 50: 21–28.

Ramble, C. (1990) The headman as a force for cultural conservation: the case of the Tepas of Nepal, in N. K. Rustomji and C. Ramble (eds), *Himalayan Environment and Culture,* Simla: Indian Institute of Advanced Study, 119–130.

Rangan, H. (1996) From Chipko to Uttaranchal: development, environment, and social protest in the Garhwal Himalayas, India, in R. Peet and M. Watts (eds), *Liberation Ecologies,* London: Routledge, 205–226.

Rangan, H. (1997) Property vs control: the State and forest management in the Indian Himalaya, *Development and Change,* 28: 71–94.

Rangel-Ch, J. O. (1995) La diversidade floristica en el espacio andino de Colombia, in S. P. Churchill, H. Balslev, E. Forero and J. L. Luteyn (eds), *Biodiversity and Conservation of Neotropical Montane Forests.* New York: New York Botanical Garden, 187–205.

Ratzel, F. (1882) *Anthropo-Geographie, oder Grunzuge der Anwendung der Erdkunde auf die Gesichte,* Stuttgart: Engelhorn.

Redclift, M. (1987) *Sustainable Development: Exploring the contradictions,* London/New York: Methuen.

Refass, M. A. (1993) Traditions migratoires dans le Rif Oriental avant le Travail en Europe, in A. Bencherifa (ed.), *African Mountains and Highlands: Resource use and conservation,* Vol. 2, Rabat: Université Mohammed V, 89–97.

Reij, C. (1988) Soil and water conservation in sub-Saharan Africa – a bottom up approach, *Appropriate Technology,* 14, 4: 14–16.

Rhoades, R. and Bebbington, A. (1995) Farmers who experiment: an untapped resource for agricultual research and development, in D. M. Warren, L. J. Slikkerveer and D. Brokensha (eds), *The Cultural Dimension of Development: Indigenous knowledge systems,* London: Intermediate Technology Publications, 296–307.

Richards, P. (1985) *Indigenous Agricultural Revolutions: Ecology and food production in West Africa,* London: Hutchinson.

Richter, M., Pfeifer, H. and Fickhert, T. (1999) Differences in exposure and altitudinal limits as climatic indicators in a profile from Western Himalaya to Tien Shan, *Erdkunde,* 53: 89–107.

Riebsame, W. E., Gosnell, H. and Theobold, D. M. (1996) Land use and landscape change in the Colorado mountains, 1: Theory, scale and pattern, *Mountain Research and Development,* 16, 4: 395–405.

Rieder, P. and Wyder, J. (1997) Economic and political framework for sustainability in mountain areas, in B. Messerli and J. D. Ives (eds), *Mountains of the World: A global priority,* Carnforth: Parthenon, 85–102.

Rinshede, G. (1988) Transhumance in European and American mountains, in N. R. J. Allan, G. W. Knapp and C. Stadel (eds), *Human Impact on Mountain Environments,* Towota, N.J.: Rowman & Littlefield, 96–108.

Robinson, D. A. and Williams, R. B. G. (1992) Sandstone weathering in the High Atlas, Morocco, *Zeitschrift für Geomorphologie N.F.,* 36: 413–429.

Robinson, N. A. (1987) Marshalling environmental law to solve the Himalaya–Ganges problem, *Mountain Research and Development,* 7: 305–315.

Roche, P. (1965) L'Irrigation et la Statut Juridique des Eaux au Maroc, *Revue Juridique et Politique d'Outre Mer,* 55–120, 537–561.

Romano, B. (1995) National Parks policy and mountain depopulation: a case study in the Abruzzo region of the central Apennines, *Mountain Research and Development,* 15: 121–132.

Rosman, A. and Rubel, P. G. (1995) *The Tapestry of Culture: An introduction to cultural anthropology,* New York: McGraw-Hill.

Rostom, R. S. and Hastenrath, S. (1994) Variations in Mount Kenya's glaciers 1987–1993, *Erdkunde,* 48: 174–180.

Rougerie, G. (1990) *Les Montagnes dans la Biosphere,* Paris: Armand Colin.

Rougier, H., Sanguin, A. L. and Schwabe, E. (1984) Frontiers et minorites, *Les Alpes,* 25.

Rundel, P. W., Smith, A. P. and Meinzer, F. C. (eds) (1994) *Tropical Alpine Environments: Plant form and function,* New York: Cambridge University Press.

Rusek, J. (1997) The impact of air pollution on soil fauna in the Tatras, in B. Messerli and J. D. Ives (eds), *Mountains of the World: A global priority,* Carnforth: Parthenon, 215–216.

Rusten, E. P. (1989) *An Investigation of an Indigenous Knowledge System and Management Practices of Tree Fodder resources in the Middle Hills of Central Nepal,* East Lansing: Michigan State University.

Rusten, E. P. and Gold, M. A. (1995) Indigenous knowledge systems and agroforestry projects in the central hills of Nepal, in M. D. Warren, L. J. Slikkerveer and D. Brokensha (eds), *The Cultural Dimension of Development: Indigenous knowledge systems,* London: Intermediate Technology Publications, 88–111.

Sacherer, J. (1987) The Sherpas of Rolwaling: a hundred years of economic change, in D. Seddon (ed.), *Nepal: A state of poverty,* New Delhi: Vikas Publishing House.

Sadki, A. (1990) La montagne marocaine et le pouvoir central: Un conflit séculaire mal élucide (ac), *Hesperis-Tamuda,* 28: 15–28.

396

Sarmiento, L., Monasterio, M. and Montilla, M. (1993) Ecological bases, sustain-ability, and current trends in traditional agriculture in the Venezuelan High Andes, *Mountain Research and Development*, 13: 167–176.

Schelling, D. (1991) Flooding and road destruction in Eastern Nepal, *Mountain Research and Development*, 11: 78–79.

Schofield, R. (1976) The relationship between demographic structures and environ-ment in pre-industrial Western Europe, in W. Conze (ed.), *Sozialgesichte der Familie in der Neuzeit Europas*, Stuttgart: Kletl-Cotta, 147–160.

Schumm, S. A. (1979) Geomorphic thresholds: the concept and its applications, *Transactions of the Institute of British Geographers, NS*, 4: 485–515.

Schwabe, E. (1984) Development of settlement structures in Swiss mountain areas, in E. A. Brugger, G. Furrer, B. Messerli and P. Messerli (eds), *The Transformation of Swiss Mountain Regions*, Berne: Verlag Paul-Haupt, 125–144.

Schwarzl, S. (1990) Causes and effects of flood catastrophes in the Alps – examples from summer 1987, *Energy and Buildings*, 15–16, 1085–1103.

Schweinfurth, U. (1954) Cie horizontale und vertikale Verbreitung der Vegetation in Himalaya, *Bonner Geographische Abhandlung*, H20.

Schweinfurth, U. (1984) The Himalaya: complexity of a mountain system mani-fested by its vegetation, *Mountain Research and Development*, 4: 339–344.

Schweizer, G. (1985) Social and economic change in the rural distribution system: weekly markets in the Yemen Arab Republic, in B. R. Pridham (ed.), *Economy, Society and Culture in Contemporary Yemen*, London: Croom Helm, 107–121.

Schweizer, G. (1984) Traditional distribution systems under the influence of recent development processes: periodic markets in the Yemen Arab Republic as an example, *Applied Geography and Development*, 24: 24–37.

Scoones, I. (1998) *Sustainable Rural Livelihoods: A framework for analysis*, IDS, University of Sussex.

Scott, J. (1998) *EC Environmental Law*, London: Longman.

Scott, P. (1997) *People–Forest Interactions on Mount Elgon, Uganda: Moving upwards towards a collaborative approach to management*, Nairobi: IUCN.

Selby, M. J. (1985) *Earth's Changing Surface*, Oxford: Oxford University Press.

Semple, E. C. (1923) *Influences of Geographic Environment in the Basis of Ratzel's System of Anthropo-geography*, London: Holt.

Shackley, S., Wynne, B. and Waterton, C. (1996) Imagine complexity: the past, present and future potential of complexity thinking, *Futures*, 28, 3: 201–225.

Shakya, T., Rabgyas, T. and Crook, J. (1994) Monastic economics in Zangskar 1980, in J. Crook and H. Osmaston (eds), *Himalayan Buddhist Villages*, Bristol: University of Bristol Press, 601–630.

Shara, T. D. (1999) *Environment and Tourism in Ladakh*. (Lead article and comments by S. Walker and M. van Beek. (http://www.mtnforum.org/library/discuss.htm)

Sharma, C. K. (1983) *Water and energy resources of the Himalayan block: Pakistan*, Kathmandu: Sangeeta Sharma.

Sharma, N. P., Rowe, N. P., Openshaw, K., Jacobsen, M. (1992) World forests in perspective, in N. P. Sharma (ed.), *Managing the World's forests: Looking for the balance between conservation and development*, Dubuque: Kendall/Hunt (for the World Bank), 17–31.

Sharma, P. (1998) Sustainable tourism in the Hindukush–Himalaya: issues and approaches, in P. East, K. Luger and K. Inmann (eds), *Sustainability in Mountain Tourism: Perspectives for the Himalayan countries,* Delhi: Book Faith, and Innsbruck: Studienverlag, 47–69.

Sher, M. S. (1993) Can lawyers save the rainforest? – exploring the second generation of debt for nature swaps, *Harvard International Law Review,* 17: 151–224.

Sherpa, L. (1999) Contribution to Mountain Forum Electronic Conference on Mountain People, Forests and Trees, April. (http://mtnforum.org/resources/library/mpft_01.html)

Shiva, V. (1989) *Staying Alive: Women, ecology and development,* London: Zed Books.

Shiva, V. and Bandyopadhyay, J. (1986) The evolution, structure and impact of the Chipko movement, *Mountain Research and Development,* 6, 2: 133–142.

Shrestha, N. (1990) *Landlessness and Migration in Nepal,* Boulder, Colo.: Westview Press.

Sidky, M. H. (1993) Subsistence, ecology and social organization among the Hunzakut: a high-mountain people in the Karakorams, *The Eastern Anthropologist,* 46: 145–170.

Sierra Nevada Ecosystem Project (1996) *Final Report to Congress: Assessment Summaries and Management Strategies,* Centre for Water and Wildlife Resources, Davis: University of California.

Sillitoe, P. (1998) It's all in the mound: fertility management under stationary shifting cultivation in the Papua New Guinea Highlands, *Mountain Research and Development,* 18, 2: 123–134.

Singh, S. (1997*) Taming the Waters: The political economy of large dams in India,* Calcutta: Oxford University Press.

Singh, T. V. and Kaur, J. (1986) The paradox of mountain tourism: case references from the Himalayas, *UNEP Industry and Development,* 9: 21–26.

Skeldon, R. (1985) Population pressure, mobility and socio-economic change in mountainous environments: regions of refuge in comparative perspective, *Mountain Research and Development,* 5, 3: 233–250.

Slaymaker, O. (1990) Climate change and erosion processes in mountain regions of Western Canada, *Mountain Research and Development,* 10: 171–182.

Slayter, R. O., Cochrane, P. M. and Galloway, R. W. (1984) Duration and extent of snow cover in the Snowy Mountains and a comparison with Switzerland, *Search,* 15: 327–331.

Slocombe, D. S. (1992) The Kluane/Wrangel–St Elias National Parks, Yukon and Alaska: seeking sustainability through Biosphere Reserves, *Mountain Research and Development,* 12: 87–96.

Smadja, J. (1992) Studies of climatic and human impacts on a mountain slope above Salme in the Himalayan Middle Mountains, Nepal, *Mountain Research and Development,* 12: 1–28.

Smith, K. (1992) *Environmental Hazards,* London: Routledge.

Sreedhar, R. (1995) *Mountain Tourism in Himachal Pradesh and the Hill Districts of Uttar Pradesh,* Kathmandu: ICIMOD.

Stadel, C. (1992) Altitudinal belts in the tropical Andes: their ecological and human utilisation, in T. L. Martinson (ed.), Bench Mark 1990: *Conference of Latin American Geographers,* Vol. 17/18: 45–60.

Starkel, L. (1972) The role of catastrophic rainfall in shaping the relief of the lower Himalaya (Darjeeling Hills), *Geographica Polonica*, 21: 103–147.

Stevens, S. F. (1993) *Claiming the High Ground: Sherpas, subsistence and environmental change in the highest Himalaya*, Berkeley, Calif.: University of California Press.

Stone, P. B. (1992) *The State of the World's Mountains – a global report*, London: Zed Books.

Streufert, S. (1997) Complexity: an integration of theories, *Journal of Applied Social Psychology*, 27: 2068–2095.

Suryanata, K. (1994) Fruit trees under contract: tenure and landuse in upland Java, Indonesia, *World Development*, 22: 1567–1578.

Swagman, C. F. (1988) *Development and Change in Highland Yemen*, Salt Lake City: University of Utah Press.

Swearingen, W. D. (1988) *Moroccan Mirages: Agrarian dreams and deceptions 1912–1986*, London: Taurus.

Symanski, R. and Webber, M. J. (1974) Complex periodic market cycles, *Annals of the Association of American Geographers*, 64: 203–213.

Tapp, N. (1989) *Sovereignty and Rebellion: The White Hmong of Northern Thailand*, Singapore: Oxford University Press.

Taylor, P. J. (1993) *Political Geography: World economy, nation state and locality*, Harlow: Longman,

Templeton, S. R. and Scherr, S. J. (1997) *Population Pressure and the Microeconomy of Land Management in Hills and Mountains of Developing Countries*, EPTD Discussion Paper No. 26, Washington, D.C.: International Food Policy and Research Institute.

Templeton, S. R. and Scherr, S. J. (1999) Effects of demographic and related micro-economic change on land quality in hills and mountains of developing countries, *World Development*, 27: 6, 903–918.

Thacker, P. (1991) Migration: a strategy for survival in the mountains, *Appropriate Technology*, 17: 26–28.

Thirgood, J. V. (1981) *Man and the Mediterranean Forest*, London: Academic Press.

Thompson, M. (1997) Security and solidarity; an anti-reductionist framework for thinking about the relationship between us and the rest of nature, *Geographical Journal*, 163: 141–149.

Thompson, M. and Rayner, S. (1998) Risk and governance. Part 1: The discourses of climate change, *Government and Opposition*, 33: 139–166.

Thompson, M. and Warburton, M. (1985a) Uncertainty on a Himalayan scale, *Mountain Research and Development*, 5: 115–135.

Thompson, M. and Warburton, M. (1985b) Knowing where to hit it: a conceptual framework for the sustainable development of the Himalaya, *Mountain Research and Development*, 5: 203–220.

Thompson, M., Ellis, R. and Wildavsky, A. (1990) *Cultural Theory*, Boulder, Colo.: Westview Press.

Thompson, M., Rayner, S. and Ney, S. (1998) Risk and governance. Part 2: Policy in a complex and plurally perceived world, *Government and Opposition*, 33: 330–354.

Thompson, M., Warburton, M. and Hatley, T. (1986) *Uncertainty on a Himalayan Scale*, London: Ethnographica.

Thorsell, J. W. (1997) Protection of nature in mountains, in B. Messerli and J. D. Ives (eds), *Mountains of the World: A global priority,* Carnforth: Parthenon, 237–248.

Thorsell, J. W. and Harrison, J. (1992) National parks and nature reserves in mountain environments and development, *GeoJournal,* 27: 113–126.

Ticehurst, D. and Cameron, C. (1998) *Review of the Current Status of Impact Monitoring Systems for Rural Development and Rural Livelihood Programmes,* Chatham: NRI & World Bank.

Tiffen, M., Mortimore, M. J. and Guchuki, F. (1994) *More People, Less erosion: Environmental recovery in Kenya,* Chichester: Wiley.

Tinau Watershed Management Project (1980) Tinau Watershed Management Plan. Swiss Association for Technical Assistance, Internal Report (cited in Ives and Messerli, 1989), Kathmandu.

Tobias, M. (1986) Dialectical dreaming, in M. Tobias (ed.), *Mountain People,* Norman and London: University of Oklahoma Press,183–202.

Trakarnsuphakorn, P. (1997) The wisdom of the Karen in natural resource conservation, in D. McCaskill and K. Kampe (eds), *Development or domestication? Indigenous peoples of Southeast Asia,* Chiang Mai: Silkworm Books, 205–218.

Treacy, J. M. (1989) Agricultural terraces in Peru's Colca valley: promises and problems of an ancient technology, in J. O. Browder (ed.), *Fragile Lands of Latin America,* Boulder, Colo.: Westview Press, 209–229.

Troll, C. (1959) Die tropischen Gebirge: Ihre dreidimensionale klimatische und planzengeographische Zonierung, *Bonner Geographischer Abhandlunge,* 25: 169–204.

Troll, C. (ed.) (1968) *Geo-ecology of the Mountainous Regions of the Tropical Americas,* Mexico City: Ferdinand Dummlers Verlag.

Troll, C. (1971) Geoecology and biogeography, *Geoforum,* 8: 43–46.

Troll, C. (ed.) (1972) *Geoecology of the High Mountain Regions of Eurasia,* Weisbaden: Franz Steiner Verlag.

Troll, C. (1973a) The upper timberlines in different climatic zones, *Arctic and Alpine Research,* 5: 3–18.

Troll, C. (1973b) High mountain belts between the polar caps and the equator: their definition and lower limit, *Arctic and Alpine Research,* 5: 19–28.

Troll, C. (1988) Comparative geography of the high mountains of the world in the view of landscape ecology: a development of three and a half decades of research and organization, in N. J. R. Allan, G. W. Knapp and C. Stadel (eds), *Human Impact on Mountains,* Towota, N.J.: Rowman & Littlefield, 36–45.

Tuchy, E. (1982) Forestry and ecology in mountainous areas, in R. Heinrich (ed.), *Logging of Mountainous Forests,* Forestry Paper No. 33, Rome: FAO.

Turner, H. (1980) Types of microclimate at high elevations, in U. Benecke and M. R. Davis (eds), 'Mountain environments and subalpine tree growth' Wellington Technical Paper No. 70, Forest Research Institute, New Zealand Forest Service, pp. 21–26.

Turner, R. K., Pearce, D. and Bateman, I. (1994) *Environmental Economics: An elementary analysis,* New York: Harvester.

Uhlig, H. (1973) Wanderhirten in Weslichen Himalaya: Chopan, Gujars, Bakerwals, Gaddi, in C. Rathjens, C. Troll and H. Uhlig (eds), *Vergleichende Kulturgeographie der Hochgeberg de sudlichen Asiene,* Tübingen: Steiner, 157–167.

Uhlig, H. (1978) Geoecological controls on high altitude rice cultivation in the Himalayas and mountain regions of Southeast Asia, *Mountain Research and Development*, 10: 515–529.

Uhlig, H. (1995) [H. Kreutzmann] Persistence and change in high mountain systems, *Mountain Research and Development*, 15, 3: 199–212.

van Beek, M. (1999) Hill councils, development and democracy: assumptions and experience from Ladakh, *Alternatives – Social Transformation and Humane Governance*, 24, 4: 435–459.

van Beek, M. and Bertelsen, M. (1995) Ladakh: independence is not enough, *Himal*, 8: 7–15.

Vander Velde, E. J. (1992) Farmer-managed irrigation systems in the mountains of Pakistan, in N. S. Jodha, M. Banskota and T. Partap (eds), *Sustainable Mountain Agriculture*, Vol. 2, London: Intermediate Technology Publications, 569–587.

Varela, M. C. (1999) Cork and the cork oak system, *Unasylva*, 197, 50: 42–44.

Varisco, D. M. (1983) 'Sayl' and 'Ghayl': the ecology of water allocation in Yemen, *Human Ecology*, 11: 365–383.

Veyret, P. and Veyret, G. (1962) Essai de definition de la montagne, *Revue de Géographie Alpine*, 50: 5–35.

Veyret-Verner, G. (1949) Le probleme de l'equilibre demographique en montagne, *Revue de Géographie Alpine*, 37: 331–342.

Viazzo, P. (1989) *Upland Communities: Environment, population and social structure in the Alps since the sixteenth century*, Cambridge: Cambridge University Press.

Vincent, L. (1995) *Hill Irrigation: Water and development in mountain agriculture*, London: Intermediate Technology Publications.

Vita-Finzi, C. (1969) *The Mediterranean Valleys*, Cambridge: Cambridge University Press.

Vivier, D. (1992) Avant-propos, *Revue de Géographie Alpine*, 80: 8–25.

Vogel, H. (1987) Terrace farming in Yemen, *Journal of Soil and Water Conservation*, January–February: 18–21.

Vogel, H. (1988) Impoundment-type bench terracing with underground conduits in Jibal Haraz, Yemen Arab Republic, *Transactions of the Institute of British Geographers NS*, 13: 29–38.

Vogt, E. Z. (1990) *The Zinacantecos of Mexico. A modern Maya way of life*, Fort Worth: Harcourt Brace Jovanovich College Publishers,

Walker, A. R. (1992) North Thailand as Geo-Ethnic Mosaic – an introductory essay, in A. R. Walker (ed.), *The Highland Heritage: Collected essays on upland north Thailand*, Singapore: Suvarnabhumi Books, 1–93.

Wardle, P. (1973) New Zealand timberlines, *Arctic and Alpine Research*, 5: 127–135.

Warsinsky, S. (1996) Cheese production in the Beaufort Valley, in L. Preston (ed.), *Investing in Mountains: Innovative mechanisms and promising examples for financial conservation and sustainable development*, Franklin, USA: Mountain Forum, 25.

Warsinsky, S. (1997) *Notes on the Alpine Convention*, Franklin, USA: The Mountain Institute.

Welford, R. (1995) *Environmental Strategy and Sustainable Development: The corporate challenge for the twenty-first century*, London: Routledge.

West, J. B. (1998) *High Life: A history of high altitude physiology and medicine*, Oxford: Oxford University Press.

White, R. (1991) *A New History of the American West,* Norman: University of Oklahoma Press.

Whiteman, P. T. S. (1988) Mountain agronomy in Ethiopia, Nepal and Pakistan, in N. J. R. Allan, G. W. Knapp and C. Stadel (eds), *Human Impact on Mountains,* Towota, N.J.: Rowman & Littlefield, 57–82.

Wiegandt, E. (1977) Inheritance and demography in the Swiss Alps, *Ethnohistory,* 24: 133–148.

Williams, A. S. and Jobes, P. C. (1990) Economic and quality of life considerations in urban–rural migration, *Journal of Rural Studies,* 6: 187–194.

Williams, P. W. and Todd, S. E. (1997) Towards an environmental management system for ski areas, *Mountain Research and Development,* 17: 75–90.

Winiger, M. (1983) Stability and instability of mountain ecosystems. Definitions for evaluation of human systems, *Mountain Research and Development,* 3: 103–111.

Winkler, D. (1999) Tibetan Plateau Forest Ecosystems. Contribution to Mountain Forum Electronic Conference on Mountain People, Forests and Trees, April (http://www.mtnforum.org/library/discuss99/mpft/mpft.htm)

Witmer, U., Filliger, P., Kunz, S. and Küng, P. (1986) Erfassung, Bearbeitung und Kartierung von Schneedaten in der Schweiz, *Geographica Bernensia,* G25 (Berne).

Wolf, E. R. (1970) The inheritance of land among Bavarian and Tyrolese peasants, *Anthropologica, NS,* 12: 99–114.

World Bank (1979) *Nepal: Development performance and prospects. A World Bank country study,* Washington, D.C./South East Asia Regional Office: World Bank.

World Bank (1996) *World Bank Lending for Large Dams: A preliminary review of impacts,* Operations Evaluation Department, Precis No. 125, Washington, D.C.: World Bank.

World Commission on Environment and Development (1987) *Our Common Future,* Oxford: Oxford University Press.

World Conservation and Monitoring Centre (WCMC) (2000) *Mountains of the World – 2000,* Cambridge, UK: WCMC.

Wyckoff, W. and Dilsaver, L. M. (eds) (1995) *The Mountainous West: Explorations in historical geography,* Lincoln: University of Nebraska Press.

Yaffee, S. L. (1994) *The Wisdom of the Spotted Owl: Political lessons for a new century* Washington, D.C.: Island Press.

Yeh, D.-Z. (1982) Some aspects of the thermal influences of Quinghai-Tibetan plateau on the atmospheric circulation, *Arch. Met. Geophys. Biocl.,* A31: 205–220.

Zimmerer, K. S. (1992) The loss and maintenance of native crops in mountain agriculture, *GeoJournal,* 27: 61–72.

Zimmerer, K. S. (1994) Human geography and the 'New Ecology': the prospect and promise of integration, *Annals of the Association of American Geographers,* 84: 108–125.

Zingari, P. C. (1998) French forest communes and sustainable development in mountain areas, *Unasylva,* 195, 46: 55–57.

Zingari, P. C. (1999) Balancing values: Europe. Contribution to Mountain Forum Electronic Conference on Mountain People, Forests and Trees, April (http://www.mtnforum.org/library/discuss99/mpft/mpft.htm)

Zorilla, C. (1999) The struggle to save Intag's forests and communities from Mitsubishi. Toisin Range, Ecuador (http://www.mtnforum.org/library/zorrc 99a.htm).

INDEX

Note: Page numbers in **bold** type refer to **figures**
Page numbers in *italic* type refer to *tables*
Page numbers followed by 'P' refer to plates